3D

CELL-BASED BIOSENSORS IN DRUG DISCOVERY PROGRAMS

Microtissue Engineering for High Throughput Screening

3D

CELL-BASED BIOSENSORS IN DRUG DISCOVERY PROGRAMS

Microtissue Engineering for
High Throughput Screening

WILLIAM S. KISAALITA

CRC Press
Taylor & Francis Group
Boca Raton London New York

CRC Press is an imprint of the
Taylor & Francis Group, an **informa** business

CRC Press
Taylor & Francis Group
6000 Broken Sound Parkway NW, Suite 300
Boca Raton, FL 33487-2742

First issued in paperback 2017

ISBN 13: 978-1-138-11202-5 (pbk)
ISBN 13: 978-1-4200-7349-2 (hbk)

Library of Congress Cataloging-in-Publication Data

Kisaalita, William Ssempa, 1953-
    3D cell-based biosensors in drug discovery programs : microtissue engineering for high throughput screening / William S. Kisaalita.
        p. ; cm.
    Includes bibliographical references and index.
    Summary: "This book is based upon cutting-edge research conducted in the authors lab (Cellular Bioengineering), which over the past decade has developed a number of sophisticated techniques to facilitate use of 3D cell based assays or biosensors. This book uses data from peer-reviewed publications to conclusively justify use of 3D cell cultures in cell-based biosensors (assays) for (HTS). The majority of assays performed in accelerated drug discovery processes are biochemical in nature, but there is a growing demand for live cell-based assays. Unlike biochemical ones, cellular assays are functional approximations of in vivo biological conditions and can provide more biologically relevant information"--Provided by publisher.
    ISBN 978-1-4200-7349-2 (hardcover : alk. paper)
    1. Pharmaceutical biotechnology. 2. Biosensors. 3. High throughput screening (Drug development) I. Title.
    [DNLM: 1. Biosensing Techniques--methods. 2. Cells, Cultured. 3. Drug Discovery. QT 36 K61z 2010]

RS380.K537 2010
615'.19--dc22   2010018119

Visit the Taylor & Francis Web site at
http://www.taylorandfrancis.com

and the CRC Press Web site at
http://www.crcpress.com

*To Rose, my wife, and Christine and
Christopher, my parents, without whom
this book would not be possible*

Contents

PART I *Introduction*

PART II   3D versus 2D Cultures

PART III   Emerging Design Principles

PART IV Technology Deployment Challenges and Opportunities

Preface

The idea for this project was planted during the 2007 Annual Society of Biomaterials Meeting in Chicago by an individual who at that time was with an HTS lab for a major pharmaceutical company. This individual, after visiting three posters from my research group on 3D cell-based biosensors, mused that, "it would be very helpful if a book was available that would integrate the knowledge bases critical to serious consideration of 3D cell-based systems in a discovery program and especially if this book would provide evidence that such systems have the potential to lower the discovery cost." I thought deeply about what this individual said and the next day I ran the project idea by Michael Slaughter (an executive editor with Taylor & Francis) who was also at the meeting at the exhibitor stand. Michael suggested that I send him a proposal, which I did. To test the project idea, I offered a one-semester hour seminar at my institution that fall, which attracted nine graduate students with research interests in bioengineering, drug discovery, and bionanotechnology. At the end of the semester, the students were unanimous about developing a full 3-hour course on the subject, which I am currently teaching every fall of odd years.

This book is intended to serve as a catalyst for the widespread adoption of 3D cell-based systems. In addition to pharmaceutical and biopharmaceutical industry bioengineers and bioscientists involved in HTS, the book should be of value to those outside the industry with interest in tissue engineering and/or cell-based biosensors. It has been written to provide the latest—from theory to practice—on the challenges and opportunities for incorporating 3D cell-based biosensors or assays in drug discovery programs. Furthermore, the book provides evidence in support of embracing 3D cell-based systems. It goes to the root of the issue, first by comparing 2D and 3D culture from genomic to functional levels, establishing the 3D cell-based biosensor physiological relevance. Second, the bioengineering principles behind successful 3D cell-based biosensor systems are assembled in one place. Third, the challenges and opportunities for incorporating 3D cell-based biosensors or cultures in current discovery and preclinical development programs are addressed. The book will also be useful as a reference in graduate courses on biosensors and biotechnology. To maintain a broad appeal for the book, the advanced mathematical treatment I am using in my class in several chapters has been left out or kept to a minimum.

As in any book written for an audience from several disciplines, the coverage may be mundane in some areas while adequate in others, depending on the reader's background. For this I apologize in advance. I also apologize to authors whose works may not have been cited. It has been difficult to exhaustively cite all important works and at the same time meet the publisher's page limit for the project.

A number of individuals deserve acknowledging for their invaluable help. First, I express my gratitude to my college at the University of Georgia for unwavering support. Second, I thank the graduate students who enthusiastically participated in the class, offering to test the material contained herein. I am especially grateful to the students in my research group: Yinzhe Lai, Ke Cheng, Lina Wang, Amish Asthena, and Angela Zachman who have contributed to this effort in more ways than I can thank them. Special thanks go to Yinzhi Lai, who in addition prepared/assembled the majority of the figures. Finally, I am grateful to my family for their patience and support of my seeming obsession with this project.

William S. Kisaalita

Author

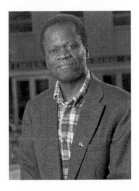

William S. Kisaalita, PhD, is professor and former coordinator of graduate engineering programs at the University of Georgia, where he also directs the Cellular Bioengineering Laboratory. The main research focus of his laboratory is cell-surface interactions with applications in cell-based biosensing in drug discovery. He has published more than 80 peer reviewed and trade press papers and made more than 100 poster and podium presentations. He has received numerous instructional awards including membership in the University of Georgia Teaching Academy. He is a member of ACS, AAAS, ASEE, and SBS. Dr. Kisaalita serves on the editorial boards of *The Open Biotechnology Journal* and *The Journal of Community Engagement and Scholarship.*

Part I

Introduction

1 Biosensors and Bioassays

1.1 CONVENTIONAL BIOSENSORS

A conventional biosensor is composed of closely coupled biological sensing and signal processing elements. The biological sensing element converts a change in an immediate environment to a signal conducive to processing by the signal processing component. The characteristics of an ideal biosensor are presented in Figure 1.1 and, as shown, include specificity—especially in complex matrices; fast measurements—in millisecond ranges that are consistent with biological signaling timescales, such as the duration of a nerve cell action potential; small size—to provide portability as well as *in vivo* implantation capabilities; continuous measurement or reversibility—to provide multiple use with little sensor regeneration or renewal; electronic processing—or integration with other devices or into larger systems; and sensitivity—in picomolar ranges that are consistent with biological detection ranges, such as concentrations of signaling molecules (e.g., cyclic AMP).

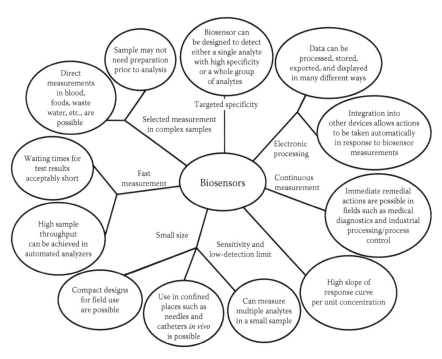

FIGURE 1.1 Characteristics of "ideal" biosensors. (Adapted from Griffiths, D. and Hall, G. 1993. *Trend Biotechnol.* 11: 122–130.)

It is a challenge to come up with a biosensor that exhibits all the characteristics outlined in Figure 1.1. As such, practical biosensors embody only a fraction of these characteristics. For example, most of the home-use blood glucose biosensors (meters and strips) inspired by the first biosensor that was described by Clark and Lyons (1962a, 1962b), termed *enzyme electrode*, meet the specificity but not the reversibility requirements. In the Clark and Lyons enzyme electrode, an oxidoreductase enzyme was held next to a platinum electrode in a membrane sandwich (Figure 1.2); the electrode and the enzyme layer allowed the passage of H_2O_2 but prevented the passage of ascorbate or other intereferents. The platinum anode polarized at +0.6 V responded to the peroxide produced by the enzyme reaction with the substrate. The first commercial glucose analyzer for the measurement of glucose in whole blood from Yellow Springs Instrument Company of Ohio was successfully launched in 1974. Key defining events in the history of glucose biosensor development, 50 years after the invention of the oxygen electrode, are presented in Table 1.1. The demand for specific, low cost, rapid, sensitive, and easy detection of biomolecules is huge. Glucose meters as used by diabetics to monitor blood glucose have successfully met these requirements, which probably explains why glucose monitoring is the largest biosensor market with

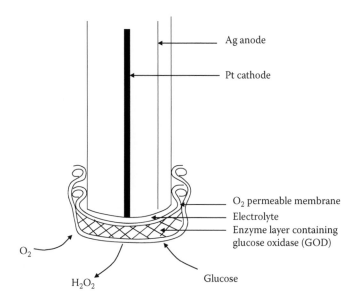

The glucose oxidation reaction, catalyzed by GOD is:

$$\text{Glucose} + O_2 + H_2O \xrightarrow{\text{GOD}} \text{Gluconic acid} + H_2O_2$$

At the electrode:

$$H_2O_2 \longrightarrow O_2 + 2e^- + 2H^+$$

FIGURE 1.2 The Clark and Lyons enzyme electrode and glucose oxidation reaction catalyzed by GOD. (From Hall, E.A.H. 1991. *Biosensors*. Englewood Cliffs, New Jersey: Prentice Hall. With permission.)

TABLE 1.1
Key Defining Events in the History of Glucose Biosensor Development

Date	Event
1916	First report on the immobilization of protein—adsorption of invertase on activated charcoal
1922	First pH electrode
1953	First use of mammalian cells in diagnosis—utilization of HeLa (human epithelial cells) for diagnosis of poliomyelitis
1956	Invention of the oxygen electrode
1962	First description of a biosensor—an amperometric enzyme electrode for glucose
1969	First potentiometric biosensor—urease immobilized on an ammonia electrode to detect urea
1970	Invention of the ion-selective field-effect transistor (ISFET)
1972–1975	First commercial biosensor—Yellow Springs Instrument glucose biosensor
1975	First microbe-based biosensor
	First immunosensor—ovalbumin on a platinum wire
	Invention of the pO_2/pCO_2 optode
1976	First bed-side artificial pancreas (Miles)
1980	First fiberoptic pH sensor for *in vivo* gases
1982	First fiberoptic-based biosensor for glucose
1983	First surface plasmon resonance (SPR) immunosensor
1984	First mediated amperometric biosensor—ferrocene used with GOD for the detection of glucose
1987	Launch of the MediSense ExacTec blood glucose system
1990	Launch of the Pharmacia BIACore SPR-based biosensor
1992	Launch of the i-STAT handheld blood analyzer
1996	Launch of Glucocard
	Abbott acquires MediSense for $867 million
1998	Launch of LifeScan FastTake blood glucose biosensor
	Merger of Roche and Boehringer Mannheim to form Roche Diagnostics
2001	LifeScan acquires Inverness Medical's glucose testing business for $1.3 billion
2003	Abbot acquires i-STAT for $392 million
2004	Abbott acquires TheraSense for $1.2 billion
2006	DexCom received approval for its STS continuous monitor
	MinMed Paradigm® received FDA approval for sale of its Paradigm Real-Time Revel which integrates ×22 series of paradigm pumps with a continuous glucose monitor

Source: Modified from Newman, J.D. and Setford, S. 2006. *Mol. Biotechnol.* 32(3): 252. With permission.

2004 worldwide sales of $5.9 billion. Roche Diagnostics lead with 2004 sales of over $2.0 billion followed by Life Scan/Johnson & Johnson with a little less than $2.0 billion. As shown in Table 1.1, DexCom (San Diego, CA) is offering a short-term sensor that wirelessly transmits glucose readings to a handheld receiver. Medtronic Diabetes (Northridge, CA) is providing a real-time continuous biosensor (MinMed Paradigm®), the only FDA-approved product for integration with an insulin pump.

Biological sensing elements come in three basic forms: molecular, cellular, and tissue. Molecular biosensors utilize biomolecules such as nucleic acids (see, e.g.,

Rusling et al., 2007), antibodies (see, e.g., Goodchild et al., 2006), enzymes (see, e.g., Newman and Setford, 2006), and ion channels and/or receptors (see, e.g., Hirano and Sugawara, 2006). Nucleic acid detection relies on the hybridization of molecular DNA probes or sequences with complementary strands in a sample, whereas antibody detection relies on specific interaction with antigenic regions of the analyte. Enzyme detection utilizes the natural specific interaction with substrates and the subsequent catalyzation of the biochemical reaction involving the substrate. Ion channel detection relies on the ability of transmitters, toxins, and potential pharmaceutical agents to bind to ion channel receptors and modulate their function, for example, opening and closing.

Cellular and tissue-based biosensors incorporate isolated cells or tissue from a wide range of sources. Figure 1.3 shows the cell types used in implementing cell-based biosensors (Racek, 1995). A microbial sensor was first commercialized by professor Isao Karube of Tokyo University in 1977 for the biochemical oxygen demand (BOD) of waste water. A comprehensive review of microbial sensors and their applications has recently been published by Nakamura et al. (2008). The trend in bacterial sensors is to genetically engineer the organism to respond to the presence of a chemical or physiological stressor by synthesizing a reporter protein, such as luciferase, β-galactosidase, or green fluorescence protein (GFP) (Yagi, 2007). Cells express and sustain an array of potential molecular sensors (receptors, channels, and enzymes) that are maintained in a physiologically relevant manner by the cellular machinery.

Biosensors incorporating mammalian cells have a distinct advantage of responding in a manner that offers insight into the physiological effect of an analyte. Groups of cells and surrounding substances that function together to perform one or more specialized activities are called tissues. There are four primary types of tissue in the human body: epithelial (e.g., skin), connective (e.g., blood, ligaments, and tendons), muscle (e.g., cardiac, skeletal, and smooth), and nervous (e.g., neurons and glial cells). Typical functions of epithelial tissue include absorption (lining of the small intestine), secretion (glands), transport (kidney tubules), excretion (sweat glands), protection (skin), and sensory reception (taste buds). Connective tissue is the most abundant, widely distributed, and can be loose (loosely woven fibers found around

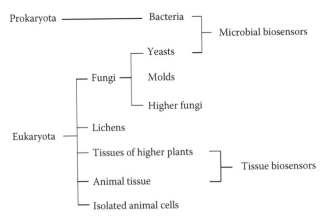

FIGURE 1.3 Cell types used for the preparation of cell-based biosensors.

and between organs), irregular (protective capsules around organs), regular dense (ligaments and tendons), or specialized (blood, bone, cartilage, and adipose). The function of muscle tissue is to provide movement through its ability to contract upon stimulation and then return to the relaxed state. Muscle types include skeletal (attached to bones), smooth (found in the wall of blood vessels), and cardiac (found in the heart). The function of neurons is to generate and conduct electrical impulses that carry information, and glial cells protect, support, and nourish the neurons.

The use of mammalian cells as sensing elements, especially in drug discovery, is an integral component of the larger discipline of tissue engineering. Cellular or tissue-based drug screening is projected to become a more accurate assessment of drug action, and three-dimensional (3D) human tissue models will provide a better understanding of human diseases early in discovery and preclinical development and thus minimize the use of animals in research. Tissue science and engineering is defined as the use of physical, chemical, biological, and engineering processes to control and direct the aggregate behavior of cells. The history of tissue engineering has been eloquently narrated by Vacanti (2006). Since the first recorded use of the term "tissue engineering" as it is applied today (Vacanti and Vacanti, 1991), the focus has been and is still mainly on the regeneration of new tissue for therapeutic applications.

The interaction between the environment (or analyte) and the biological sensing element results in one of four measurable changes (electrochemical, electromagnetic, mass, and thermal, outlined in more detail in Figure 1.4). The principles of signal processing elements (or transducers) are based on the type of resulting change. Potentiometric transducers measure potential at zero current and the potential is usually proportional to the logarithm of the concentration of the substance being detected. Changing the potential across the sensing element results in a sharp rise in the current and the peak current is directly proportional to the concentration of the electroactive

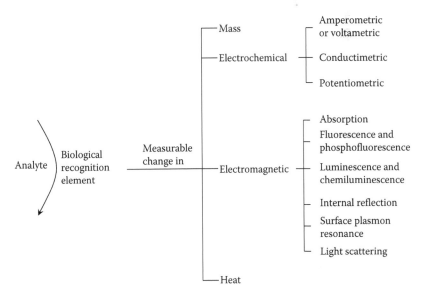

FIGURE 1.4 Transduction or readout principles.

material. In amperometric transducers, the potential is stepped up to the known oxidation/reduction potential values while observing the current. Solutions containing ions conduct electricity and, based on the reaction, the relationship between conductance and concentration can provide reliable sensing modality. The range of optical detection techniques is wide, as suggested by the electromagnetic spectrum with important radiation such as ultraviolet, visible, and infrared. The basic principles of optical detection techniques (readouts) will not be repeated here as there are many texts that cover the subject at varying levels of difficulty. See Chapter 9 for the more commonly used configuration in high-throughput screening (HTS) and for an excellent introductory reference (see Eggins, 1996). Piezoelectric crystals have the natural resonant frequency of oscillations, which can be modulated by the environment. The usual frequency is in the radio range of 10 MHz. The actual frequency is dependent on the mass of the crystal together with any other material coated on it. The change in resonant frequency resulting from the adsorption of an analyte on the surface can be measured with high sensitivity (500–2500 Hz/g), resulting in a pictogram detection limit. There are various ways in which this mass detection modality has been implemented. One of these is the acoustic wave mode, where waves are not generated in the bulk but on the surface. The various acoustic wave modes include surface acoustic waves, the plate mode wave, the evanescent wave, the lamb mode device, and the thickness shear mode device. Nearly all biochemical reactions are exothermic, that is, enzymatic conversion of a substance is accompanied by heat production. Detection normally employs a thermistor as a temperature transducer. Thermistors are resistors with a very high negative temperature coefficient of resistance.

Coupling of the sensing and signal processing elements is a field of investigation with a broader appeal. For biosensor applications, the approach adopted is largely dictated by the chemistry of the two elements to be conjugated. Conjugation can be achieved by simple adsorption, microencapsulation, entrapment, covalent attachment, and cross-linking. In a classic review, Scouten (1987) outlined covalent attachment approaches for conjugating enzyme or protein to a variety of inorganic surfaces. Generally, the approaches involve a three-stage process, starting with activating the inorganic surface with a linker molecule that reacts with the protein or enzyme of interest without affecting its biological activity. Recent developments in microarray technology as well as nanobiotechnology—a field of research at the interface between physics, biotechnology, and materials science and engineering—are yielding innovative conjugation approaches that are overcoming some of the problems encountered in classical methods. For example, Hazarika et al. (2004) described a protocol for fabricating gold nanoparticles functionalized with thio-modified single-stranded oligonucleotides. Nanomaterials functionalized with proteins are used as tools in biosciences and biosensors (Wang, 2008).

1.2 CONVENTIONAL BIOSENSOR APPLICATIONS

Biosensors are being developed and commercialized for applications in a wide range of markets, including bioprocess monitoring and control, food quality control, environmental monitoring, military, clinical diagnostics, and pharmaceutical markets. Over 90% of the market is blood glucose monitoring (Alocilja and Radke, 2003).

1.2.1 Bioprocess Monitoring and Control

A variety of sensing systems have been developed for bioprocess monitoring and control applications (Becker et al., 2007). For example, Katrlík et al. (2007) recently described a novel microbial biosensor based on cells of *Gluconobacter oxydans* for the selective determination of 1,3-propanediol (1,3-PP) in the presence of glycerol. The sensor performed exceedingly well when integrated into a flow loop to monitor the concentration of 1,3-PD during a real bioprocess. In recent years, there have been many exciting developments relating to bioprocess monitoring and control; however, most of these new biosensing systems and related control strategies have not been commercialized, mainly because of poor stability, sterilization issues, and inaccuracy (Becker et al., 2007).

1.2.2 Food Quality Control

A well-known food quality biosensor is the fish freshness enzyme sensor that was developed by professor Isao Karube's group (Karube et al., 1984). The target market was raw fish (sashimi), a popular dish in Japan and other Asian countries. Freshness is indicated by a color change of thiazole blue due to the redox reaction accompanying the oxidation of hypoxanthine by xanthine oxidase. The decomposition of ATP in fish meat starts after death, following the reaction below:

$$ATP \rightarrow ADP \rightarrow AMP \rightarrow IMP \rightarrow inosine \rightarrow hypoxanthine \rightarrow uric\ acid \quad (1.1)$$

The total amount of ATP-related product is constant; the reaction moves to the right with product degradation and as such freshness can be accurately determined by the ratio of fish meat concentrations of (inosine + hypoxanthine) to (IMP + inosine + hypoxanthine). The inherent specificity, selectivity, and adaptability of biosensors have made them attractive for investigating their use throughout the food industry. The majority of applications investigated have been enzyme-based, coupled to amperometric detectors. Industry is calling for adapting the screen-printed meter-strip blood glucose format for use in the food industry (see Table 1.2 for a list of potential applications). Detecting the presence of contaminating microorganisms on or in food is an area of active investigation. Examples of target organisms include *Escherichia coli*, staphylococcal enterotoxins A and B, *Salmonella typhimurium*, and *Salmonella* groups B, D, and E. Other potential opportunities include improved quality control assurance of food-derived raw materials, testing the absence/presence of genetically modified constituents, and incorporation in "smart" packaging (Terry et al., 2005) or added functionality.

1.2.3 Environmental Monitoring

Public concern and legislation are fueling the application of biosensors toward environmental monitoring. Rodriguez-Mozaz et al. (2004) have listed the proposed representative biosensors. Despite the high number of biosensors in development, few practical systems have been commercialized (Table 1.3). This has been attributed to limitations of sensitivity, response time, and life time. Many researchers are

TABLE 1.2

Analytes Monitored in the Fresh Produce Food Matrices Suitable for the Meter-Strip Biosensing Format Well Developed for Blood Glucose

Analyte	Food Matrix	Enzyme	Detection Limit
Fructose	Citrus fruits	Fructose dehydrogenase	10 µM
Amines	Apricots and cherries	Diamine oxidase and polyamine oxidase	2×10^{-6} mol/L
L-Ascorbic acid	Fruit juices	Ascorbate oxidase	5.0×10^{-5} M
Sucrose	Fruit juices	Sucrose phosphorylase, phosphoglutaminase and glucose-6-phosphate-1-dehydrogenase	1.25 g/L in pineapple juice
Malic acid	Apples, potatoes, and tomatoes	Malate dehydrogenase	0.028 mM
Polyphenols	Vegetables	Horseradish peroxidase	1 µmol/L
β-D-Glucose, total D-glucose, sucrose, and L-ascorbic acid	Tropical fruits (mango, pineapple, and papaw)	Invertase, mutarotase, GOD, and ascorbate oxidase	
Cysteine sulfoxides	Alliums (e.g., onion and garlic)	Allinase	5.9×10^{-6} M
Pyruvic acid	Onion	Pyruvate oxidase	2 µmol/g
Essential fatty acids	Fats and oils	Lipoxygenase, lipase, and esterase	0.04 M in a flow injection analysis system
Lysine	Range of foods	Lysine oxidase	1×10^{-5} mol/L
Glucose and maltose	Beer	GOD and amyloglucosidase	40 mM (upper limit)
Glucose and glutamate	Beverages	GOD and glutamate oxidase	3 µM for glutamate
Rancification indicators	Olive oils	Tyrosinaser	0.2–2.0 µM in different oils
Lactate	Wine and yoghurt	Lactate oxidase	1.4×10^{-6} mol/L
D- and L-amino acids		Amino acid oxidase	0.1 or 0.2 mM for L- and D-amino acids
Choline	Dairy products	Choline oxidase	5 µmol/L
Organophosphate pesticides	Range of foods	Acetylcholinesterase	0.2–1.8 ppm
Insecticide residues	Infant food	Acetylcholinesterase	5 µg/kg
Laminarin	Seaweed	1,3-Glucanase and GOD	50 µg/mL
Alcohol	Beer and wine	Alcohol oxidases and horseradish peroxidases	5.3×10^{-6} mol/L

Source: From Terry, L.A., White, S.F., and Tigwell, L. 2005. *J. Agric. Food Chem.* 53: 1309–1316. With permission.

TABLE 1.3
Commercially Available Biosensor Systems for Environmental Applications

Analyte	Biorecognition	Detection	System (Company)
Sulfonamides and pathogens	Biomolecular interaction	SPR	BIACORE (Biocore AB, Uppsala, Sweden)
Same as above	Same as above	Same as above	IBIS (Windsor Scientific, Ltd., Berks, UK)
Same as above	Whole cells or macromolecules	Same as above	SPR-CELLIA (Nippon Laser and Electronics Lab, Nagoya, Japan)
Same as above	Biomolecular interaction	Same as above	Spreeta (Texas Instruments, Inc. Dallas, Texas)
Same as above	Biomolecular interaction	Same as above	BIOS-1 (Artificial Sensing Instruments, Zurich, Switzerland)
Same as above	Antibody–antigen	Same as above	Pathogens (Amersham International, Buckinghamshire, UK)
Same as above	Biomolecular interaction	Same as above	Not Listed (XanTech Bioanalytical GmbH Münster, Germany)
Same as above	Same as above	Same as above	Kinomics Plasmon™ (BioTul AG, Munich, Germany)
Same as above	Same as above	Same as above (evanescent wave)	IASys plus™ (NeoSensors Limited, Sedgefield, UK)
Toxicity	Whole cell	Bioluminescence	REMEDIOS (Remedios, Aberdeen, Scotland)
Toxicity	E. coli	Amperometric	Cellsense (Euroclon, Ltd., Yorkshire, UK)
Toxicity	Biomolecular interaction	Electrochemical	ToxSen™ (Abtech Scientific, Inc., Yardley, Pennsylvania)
Pathogens	Antibody–antigen	Piezoelectric	PZ 106 (Universal Sensors, Kinsale, Ireland)
BOD	Not reported	Not reported	ARAS BOD (Bruno Lange GmbH, Dusseldorf, Germany)
Nitrate	Nitrate reductase	Amperometric	Nitrate Elimination Co., Inc. (Michigan)

Source: Modified from Rodriguez-Mozaz, S., et al. 2004. *Pure Appl. Chem.* 76(4): 739. With permission.

actively pursuing whole-cell bacterial sensors not only for biotechnological applications but also for environmental monitoring. The most common approach is to use reporter genes fused with a DNA response element (RE) for an analyte or molecules induced by environmental stress. Commonly used reporter genes include *lux* (bacterial luciferase), *luc* (firefly luciferase), *gfp* (GFP—jelly fish), and *lacZ* (β-gal—*E. coli*). As illustrated in Figure 1.5, the bacteria receive extracellular signals through a receptor-dependent or independent pathway, and the signal-mediated activation of the RE induces the transcription of the reporter gene. The reporter protein exhibits specific luminescence, fluorescence, or color development as the detectable signal. A key limitation is that none of the bacterial sensing elements developed to date detects below 100 nM. Efforts are needed to increase the detection limits. Also, the

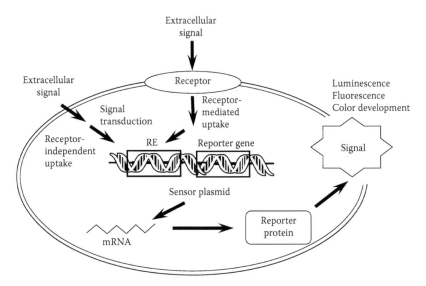

FIGURE 1.5 Whole-cell bacterial biosensors made using recombinant DNA technology. (From Yagi, K. 2007. *Appl. Microbiol. Biotechnol.* 73: 1252. With permission.)

time required for reporter genes to be transcribed and translated makes them less desirable in situations requiring real-time monitoring.

1.2.4 MILITARY BIODEFENSE APPLICATIONS

There are many challenges faced by various military and civilian defense forces in rapid, accurate, and sensitive detection and identification of infectious or harmful agents in the field. Biosensors and related technology have been proposed and developed as potential solutions. The summary of approaches for biowarfare agents' molecular recognition by Iqbal et al. (2000) underscores the emphasis placed on the detection of agents that conform to expectations with respect to biochemical, immunological, or genetic traits. What is needed more are biosensor systems that can function in much the same manner as canaries once served as sensors for coal miners and troops in trenches in World War I. The system will respond to a dangerous biological agent due to its toxic effect rather than an inherent biochemical, immunological, or genetic signature. In this way, biological threats engineered to evade detection will be sensed, because to be a threat, they must retain their capacity to induce toxic effects. Cell-based biosensors, with their capacity to sense the microenvironment and integrate signals into a complex response such as changes in viability, proliferation, differentiation, or migration, are being investigated for this role; however, field implementation is likely to be a challenge. A biosensor consisting of three parts, a freeze-dried biosensing strain within a vial, a small light-proof test chamber, and an optical-fiber connected between the sample chamber and an illuminometer, has been reported (Shin et al., 2005) as a way of overcoming portability. It is easy to envision the use of micro- and nanofabrication technologies to further miniaturize the sensor to a field-deployable device.

TABLE 1.4

Enzyme-Based Biosensors Used in Systems for Critical Care Analysis

Analyte	Detection	System	Manufacturer
Glucose, lactate	Electrochemical, thick film	RapidLab 800	Bayer
Glucose	Electrochemical	RapidPoint 400	Bayer
Glucose, lactate, urea	Electrochemical	Critical Care	Nova Biomedical
Creatinine	Electrochemical	Xpress, Nove 16	Nova Biomedical
Glucose, lactate	Electrochemical	PHOx	Nova Biomedical
Glucose, lactate	Electrochemical	ABL 725	Radiometer
Glucose, lactate	Electrochemical	Synthesis 1745	Instrumentation Laboratory
Glucose, lactate	Electrochemical, planer	GEM Premier 3000	Instrumentation Laboratory
Glucose, lactate, urea, creatinine	Electrochemical, single use, thin film	I-STAT1 and PCA	i-STAT
Glucose, lactate, urea	Electrochemical, single use, thick film, optical test strip (glucose option)	IRMA	Diametrics
Glucose, lactate, urea	Electrochemical, thick film	OMNI 9	Roche
Glucose, urea	Optical, single use	OPTI CCA	Roche

Source: Modified from D'Orazio, P. 2003. *Clin. Chim. Acta* 334: 41–69. With permission.

1.2.5 CLINICAL DIAGNOSTICS

Clinical analyses are no longer carried out exclusively in the clinical chemistry laboratory. Measurements of analytes in biological fluids are routinely performed in various locations, including hospital point-of-care settings, by caregivers in nonhospital settings and by patients at home. These types of measurements are highly suited to being accomplished with biosensors. Biosensors for the measurement of blood metabolites such as glucose, lactate, urea, and creatine, using both electrochemical and optical modes of transduction, are commercially available and routinely used in the laboratory. Table 1.4 lists a few popular systems. Active research in synthetic recognition elements, for example, aptamers (Balamurugan et al., 2007), to allow broader application to different classes of analytes and modes of transduction will likely yield new biosensors. In general, there is a growing tendency toward miniaturization that is permitting biosensors to be constructed as arrays and incorporated into lab-on-a-chip devices.

1.3 CELL-BASED BIOSENSORS VERSUS CELL-BASED ASSAYS (BIOASSAYS)

In the 1990s, the explosion in bioscience knowledge yielding many potential points for therapeutic intervention (targets), coupled to pharmaceutical and biotechnological industries widely adopting combinatorial chemistry, gave birth to a new, target-driven, drug discovery paradigm, discussed in more detail in Chapter 2. The resultant explosion in the number of novel chemical compounds fueled the expansion of HTS technologies that could rapidly "assay" a large number of compounds for receptor (biological sensing element)–drug (analyte) interaction with high sensitivity. The target–compound interactions are either biochemical or cell-based. Biochemical approaches provide limited information on the potency and specificity of compounds because the target's activity is assessed using isolated targets. Unlike biochemical interactions, cellular interactions are functional approximations of the *in vivo* biological content and can, therefore, provide more biologically relevant information. Traditionally, the HTS industry uses the word "assay" in reference to biochemical or cell-based interaction experimentation. The book title suggests a break from this tradition for the simple reason that in most cell-based instances, the cell at the heart of the assay is a biosensor. The following paragraphs expand on this argument.

The word assay comes from the French word "essai," which in turn comes from the Latin word "exagium," meaning "the act of weighing." In a more current biological context, Webster's New World Medical Dictionary (p. 30) offers the definition of assay as an analysis done to determine the following:

1. The presence of a substance and the amount of that substance. Thus an assay may be done, for example, to determine the level of thyroid hormones in the blood of a person suspected of being hypothyroid (or hyperthyroid).
2. The biological or pharmacological potency of a drug. For example, an assay may be done of a vaccine to determine its potency.

Based on the above definition, there is overlap between assays and biosensors, so what is the difference? Traditional biosensors have been associated with a reagent-less single-step detection, whereas assays have been associated with multiple steps. Where automation has successfully reduced the number of steps into a single step by the "operator" (e.g., bioanalytical instruments), some have found the use of biosensors to describe such systems appropriate, but others have not. For example, Patel (2002, p. 96) stated that, "Biosensors should be distinguished from assay or bioanalytical systems, which require additional processing steps, such as reagent addition and where the assay design is permanently fixed in the construct of the device." While this author is in agreement with the need to minimize the confusion, he is not totally agreeable to the definition that precludes instrumentation that has been cleverly built to provide biosensing in a manner similar to the nose, the most sophisticated biosensor, which in connection to the brain embodies all the characteristics outlined in Figure 1.1. For the nose, the chemicals to be detected pass through the olfactory membrane to the olfactory bulbs, which are the biological sensing elements. The response of the olfactory membrane is then converted by the olfactory nerve cell, which is equivalent to the transducer, into an electrical signal that passes along the nerve fiber to the brain for interpretation. The brain acts as a microprocessor, turning the signal into a sensation that we call smell. For prolonged exposure, where the ability of the nose to detect compounds can be compromised due to desensitization, a step may be required (e.g., replacing the air with compound-free air or air that contains a structurally different compound—a step that may be equivalent to exposing a high-affinity immune sensor to low pH to remove the ligand for the next detection). This author contends that at a minimum and for the purpose of this book, for a sensing instrument or platform to qualify as a biosensor, it should exhibit specificity, should possess, in principle, the capacity for reversibility, and requires one step (detection) after compound/ligand/analyte addition. With the advent of nanotechnology, miniaturization will probably in the near future deliver today's bioanalytical instruments in sensor pen sizes, fitting very well with the sizes associated with traditional biosensors.

To further illustrate, four examples are presented in Table 1.5. Two of the examples are related to second messenger cAMP activity and three of the four rely on engineered cells. Examples that rely on fluorescence or radioactive labels were deliberately chosen because current HTS detection or readout technologies rely predominantly on these modes (Chapter 9). In the first example (second column), the multistep procedure involves culturing cells in the presence of a radiolabeled substrate that is incorporated in cAMP following stimulation. Cells are lysed and the lysate is chromatographed to separate the radiolabeled cAMP followed by scintillation counting of the chromatographic fractions. This procedure represents a true assay based on the criteria outlined above. Especially since the cells are destroyed in the process, there is no potential for reversibility. While the remaining examples meet the criteria, some (e.g., column 5) are more qualified than others at different points on a sliding scale. The heterologous expression of a fluorescent protein in column 3 (EGFP) captures a commonly employed procedure that has revolutionalized imaging in cell biology. This author considers this and similar constructs on the borderline between assay and biosensor. While in the application presented the detection is not reversible, it is easy to imagine a situation where the aggregation of the fluorescent protein can change, in which case the system will respond with the change. In the biarsenic–tetracysteine (Cys–Cys–Pro–Gly–Cys) example presented in column 4, Adams et al. (2002) pointed out that the association rates are fast, and take place in seconds with dissociation constants <10 pM. Timescales for dissociation are weeks, but can be accelerated to minutes with millimolar dithiol concentrations. In principle, this is not any different from the desensitization example outlined for the nose above. The single-use configuration of cell-based biosensors similar to the meter-strip practice for blood glucose detection is almost the same here with the increasing use of frozen cells that are delivered just in time, as discussed further in Chapter 11. Obviously not all HTS applications will be amenable to the just-in-time frozen cell approach.

1.4   3D CULTURES

1.4.1   Two-Dimensional (2D) Culture Systems

When he invented the ubiquitous dish that bears his name in 1887, Julius Petri, a technician in Robert Koch's laboratory, fundamentally transformed our ability to culture, manipulate, and analyze cells. Tissue culture as a technique came into being 20 years after the Petri invention, with Ross Harrison (1907) of Rockefeller Institute culturing frog embryo tissue in plasma clots for one to four weeks in order to observe

TABLE 1.5
Biosensor versus Bioassay

Attribute	Radioactive-Labeled Analyte	Heterologous Expression of GFP	Recombinant Protein Bearing a Tetracysteine Target for Fluorogenic Biarsenic Compounds	Bioluminescence Resonant Energy Transfer
Reference(s)	Evans et al. (1984)	Apostol et al. (2003)	Roberti et al. (2007) and Adams et al. (2002)	Prinz et al. (2006)
Cell construct	1321N1 astrocytoma and NG108 neuroblastoma x glioma	PC12 rat pheochromocytoma expressing engineered human huntigtin (containing a range of polyglutamine repeats) fused in frame to the coding sequence of enhanced GFP (EGFP)	SH-SY5Y human dopaminergic neuroblastoma expressing engineered α-synuclein-C4 bearing a tetracysteine target for FlAsH or ReAsH	COS-7 African green monkey kidney cells engineered to expressing an isoform-specific protein kinase A whose subunits are targed with Renilla luciferase or GFP
Potential drug discovery or toxicity sensing objective	cAMP accumulation reducing activity	Disruption of polyglutamine-containing aggregates and inclusions—hallmarks of pathogenesis in Huntington's disease	Disruption or activation of α-synuclein, a major component of intraneuronal protein aggregates constituting a distinctive feature of Parkinson's disease	Deregulation of protein kinase A (PKA), the main effector of cAMP—demonstrated to contribute to diabetes, cancer, and cardiovascular diseases

Exposure and detection	Cells were cultured in [3H]adenine containing media and were exposed to agonists (e.g., PGE1, 1 µM) and antagonists (e.g., Arecoline); cells were lysed and chromatographically separated followed by scintillation counting	Compounds (e.g., cystamine bitartrate, 1–50 µM) were added directly to the culture medium; cell images were viewed by fluorescence microscopy	Cells were stressed by incubating in 5–10 mM $FeCl_2$ and were stained with FlAsH and detected by fluorescence microscopy	Cells were exposed to increasing amounts of the PKA against (8-Br-cAMP) or antagonist (Rp-8-Br;cAMP); bioluminescence resonant energy transfer (BRET) was measured in a plate reader
Readout	dpm [3H] cAMP/(dpm [3H]ATP + dpm[3H]cAMP	Aggregation was expressed as the percentage of cells with aggregate versus total EGFP-positive cells	From positively transfected cells, the fraction displaying fluorescence aggregation was calculated	BRET values were calculated as (em515—n.t cell em515) divided by (em410—n.t cell em410)
Steps after treatment	6	1	1	1
Reversibility	Not possible	Remotely possible (only if the aggregation can be reversed)	Possible (timescales of dissociation are weeks in the absence of 1,2-dithiols, yet can be accelerated to minutes at mM dithiol concentrations)	Yes (PKA dissociation is reversible)
Suitability for HTS	Not possible	Yes if all cells are EGFP-positive	Yes if all cells are positively transfected	Yes

the development of nerve fibers. Alexis Carrel (1912), also at the Rockefeller Institute, expanded the possibilities of cell culture by keeping fragments of chick embryo heart alive and beating into the third month of culture and growing chick embryo connective tissue for three months. For a detailed account of the history of tissue/cell culture, see Landecker (2007), who has narrated and organized the various phases from an anthropologist's point of view. Since the Petri invention, the dish has become a staple in laboratory work on both prokaryotic and eukaryotic cells. The 96-, 384-, and 1536-well plastic plate vessels of choice for cell-based studies in HTS format are derivatives of the initial invention, with the exception that they have high density. The cultures in these vessels are referred to as 2D, meaning that cells can only proliferate in the x–y (flat) directions, in contrast to specialized vessels (scaffolds) where cells can proliferate in x–y–z (3D) directions, similar to the *in vivo* situation.

1.4.2   3D Culture Systems

Elsdale and Bard (1972) pioneered 3D culture with a model system for fibroblastic cells in the body using collagen I matrices polymerized *in vitro* to form a 3D fibrous network. Collagen hydrogels have been used widely since then for studying cell motility and the roles of the physical 3D state. More recently, synthetic polymer scaffolds for 3D cell growth have included poly(ethylene oxide), poly(vinyl alcohol), poly(acrylic acid), and poly(propylene fumarate-co-ethylene glycol). For excellent scaffold reviews, see Lozinsky and Plieva (1998) and Drury and Mooney (2003). The main application of both natural and synthetic scaffold research has been either controlled drug delivery or engineered tissue replacement (Langer and Vacant, 1993). Also, see Appendix A for a list of related patents issued between 1996 and 2006 that attest to the engineered tissue and controlled drug delivery applications focus. *In situ*, cells in a living organism have 3D architecture, which contrast starkly with cells in a 2D environment in flat culture plates. The promise of 3D cell-based biosensors in HTS is rooted in their presumed *in vivo* relevance when compared to their 2D counterparts. This book provides evidence in support of the validity of this assumption.

1.4.3   Tissue Engineering versus Microtissue Engineering

As previously mentioned, tissue science and engineering is defined as the use of physical, chemical, biological, and engineering processes to control and direct the aggregate behavior of cells. The history of tissue engineering has been eloquently narrated by Vacanti (2006), mainly as it has been applied on the regeneration of new tissue for therapeutic applications. While engineering tissue for sensing, especially for HTS in drug discovery, presents similar challenges as traditional tissue engineering, there are several important differences. First, in HTS there are construct size limitations (microscale) imposed by the need to harmonize with high-density multiwell plate platforms. Second, there are new challenges of sensing modality incorporation and signal readout that are absent in traditional tissue engineering. Third, the scaffolding material does not have to be biodegradable and FDA approved or approvable as there are no *in vivo* implantation safety concerns. The term "microtissue engineering" has been used in the subtitle, and the word "micro-tissue" is used in the book to capture these differences and accurately define this subfield of tissue engineering.

1.5 CONCLUDING REMARKS

The intent of this first chapter was to familiarize the reader with the field of traditional biosensors and their applications. The second intent was to show that the cells used in most assays are biosensors. This helps in understanding the difference between cell-based biosensors and cell-based assays—a cell-based biosensor can be a sensing element in a cell-based assay. The establishment of the main problem the book addresses, of incorporating 3D cell-based biosensors in drug discovery programs, has been left to Chapter 2, which together with this chapter comprises the first part of the book (Part I, Introduction). A comparison of 2D and 3D cultures is presented in Part II and engineering of systems that facilitate the incorporation is presented in Part III. Challenges to incorporation and the commercial state of the technology are presented in Part IV. This is the first book to assemble information in one place that provides answers to questions that are important to drug discovery and nonclinical development professionals and should serve as a catalyst for the widespread adoption of 3D cell-based biosensors by the pharmaceutical, biopharmaceutical, and biotechnology industries.

REFERENCES

Adams, S.R., Campbell, R.E., Gross, L.A., Martin, B.R., Walkup, G.K., Yao, Y., Llopis, J., and Tsien, R.Y. 2002. New biarsenical ligands and tetracysteine motifs for protein labeling *in vitro* and *in vivo*: Synthesis and biological applications. *J. Am. Chem. Soc.* 124: 6063–6076.

Alocilja, E.C. and Radke, S.M. 2003. Market analysis of biosensors for food safety. *Biosens. Bioelectron.* 18: 841–846.

Apostol, B.L., Kazantsev, A., Raffioni, S., Illes, K., Pallos, J., Bodai, L., Slopko, N., et al. 2003. A cell-based assay for aggregation inhibition as therapeutics of polyglutamine-repeat disease and validation in *Drosophila. PNAS* 100(10): 5950–5955.

Balamurugan, S., Obubuafo, A., Soper, S.A., and Spivak, D.A. 2007. Surface immobilization for aptamer diagnostic applications. *Anal. Bioanal. Chem.* 390(4): 3436–3445.

Becker, T., Hitxmann, B., Muffler, K., Portner, R., Reardon, K.F., Stahl, F., and Ulber, R. 2007. Future aspects of bioprocess monitoring. *Adv. Biochem. Eng. Biotechnol.* 105: 249–293.

Carrel, A. 1912. On the permanent life tissues outside of the organism. *J. Exp. Med.* 15: 516–528.

Clark, L.C., Jr. and Lyons, C. 1962a. Monitoring and control of blood tissue O_2 tension. *Trans. Am. Soc. Artif. Intern. Organs* 2: 41–48.

Clark, L.C., Jr. and Lyons, C. 1962b. Electrode systems for continuous monitoring in cardio-vascular surgery. *Ann. N. Y. Acad. Sci.* 102: 29–33.

D'Orazio, P. 2003. Biosensors in clinical chemistry. *Clin. Chim. Acta* 334: 41–69.

Drury, J.L. and Mooney, D.J. 2003. Hydrogels for tissue engineering: Scaffold design variables and applications. *Biomaterials* 24: 4337–4351.

Eggins, B. 1996. *Biosensors: An Introduction.* New York: Wiley.

Elsdale, T. and Bard, J. 1970. Collagen substrates for the study of the cell behavior. *J. Cell Biol.* 41: 298–311.

Evans, T., Smith, M.H., Tanner, L.I., and Harden, T.K. 1984. Muscarinic cholinergic receptors of two cell lines that regulate cyclic AMP metabolism by different molecular mechanisms. *Mol. Pharmacol.* 26: 395–404.

Goodchild, S., Love, T., Hopkins, N., and Mayers, C. 2006. Engineering antibodies for biosensor technologies. *Adv. Appl. Microbiol.* 58: 185–226.

Griffiths, D. and Hall, G. 1993. Biosensors—what real progress is being made? *Trend Biotechnol.* 11: 122–130.

Hall, E.A.H. 1991. *Biosensors.* Englewood Cliffs, New Jersey: Prentice Hall.

Harrison, R. 1907. Observations on the living developing nerve fiber. *Proc. Soc. Exp. Biol. Med.* 4: 140–143.

Hazarika, P., Giorgi, T., Reibner, M., Ceyhan, B., and Niemeyer, C.M. 2004. Synthesis and characterization of deoxyribonucleic acid-conjugated gold nanoparticles. In *Bioconjugation Protocols: Strategies and Methods*, C.M. Niemeyer (ed.), pp. 295–304. Totowa: Humana Press.

Hirano, A. and Sugawara, M. 2006. Receptors and enzymes for medical sensing of L-glutamate. *Mini Rev. Med. Chem.* 6(10): 1091–1100.

Iqbal, S.S., Mayo, M.W., Bruno, J.G., Bronk, B.V., Batt, C.A., and Chambers, J.P. 2000. A review of molecular recognition technologies for detection of biological threats. *Biosens. Bioelectron.* 15: 549–578.

Karube, I., Matsuoka, H., Suzuki, S., Watanabe, E., and Toyama, K. 1984. Determination of fish freshness with an enzyme sensor system. *J. Agric. Food Chem.* 32: 314–319.

Katrlík, J., Vostiar I., Sefcovicová, J., Tkác, J., Mastihuba, V., Valach, M., Stefuca, V., and Gemeiner, P. 2007. A novel microbial biosensor based on cells of *Gluconobacter oxydans* for the selective determination of 1,3-propanediol in the presence of glycerol and its application to bioprocess monitoring. *Anal. Bioanal. Chem.* 388(1): 287–295.

Landecker, H. 2007. *Culturing Life.* Cambridge, Massachusetts: Harvard University Press.

Langer, R. and Vacanti, J.P. 1993. Tissue engineering. *Science* 260(5110): 920–926.

Lozinsky, V.I. and Plieva, F.M. 1998. Poly(vinyl alcohol) cryogels employed as matrices for cell immobilization. 3. Overview of recent research and development. *Enzyme Microb. Technol.* 23: 227–242.

Nakamura, H., Shimomura-Shimizu, M., and Karube, I. 2008. Development of microbial sensors and their applications. *Adv. Biochem. Eng. Biotechnol.* 109: 351–394.

Newman, J.D. and Setford, S. 2006. Enzyme biosensors. *Mol. Biotechnol.* 32(3): 249–269.

Patel, P.D. 2002. (Bio)sensors for measurement of analytes implicated in food safety: A review. *Trends Anal. Chem.* 21(2): 96–115.

Prinz, A., Diskar, M., Erlbruch, A., and Herberg, F.W. 2006. Novel, isotype-specific sensors for protein kinase A subunit interaction based on bioluminescence energy transfer (BRET). *Cell. Signal.* 18: 1616–1625.

Racek, J. 1995. *Cell-Based Biosensors*, 5pp. Lancaster, PA: Technomic Publishing Co., Inc.

Roberti, M.J., Bertoncini, C.W., Klement, R., Jares-Erijman, E., and Jovin, T.M. 2007. Fluorescence imaging of amyloid formation in living cells by a functional, tetracysteine-tagged α-synuclein. *Nat. Methods* 4(4): 345–351.

Rodriguez-Mozaz, S., Marco, M.-P., Lopez de Alda, M.J., and Barcelo, D. 2004. Biosensors for environmental applications: Future development trends. *Pure Appl. Chem.* 76(4): 723–752.

Rusling, J.F., Hvastkovs, E.G., and Schenkman, J.B. 2007. Toxicity screening using biosensors that measure DNA damage. *Curr. Opin. Drug Discov. Dev.* 10(1): 67–73.

Scouten, W.H. 1987. A survey of enzyme coupling techniques. In *Methods in Enzymology*, K. Mosback (ed.), Vol. 135, pp. 19–45. New York: Academic Press.

Shin, H.J., Park, H.H., and Lin, W.K. 2005. Freeze-dried recombinant bacteria for on-site detection of phenolic compounds by color change. *J. Biotechnol.* 119: 36–43.

Terry, L., White, S.F., and Tigwell, L.J. 2005. The application of biosensors for fresh produce and the wider food industry. *J. Agric. Food Chem.* 53: 1309–1316.

Vacanti, C.A. 2006. The history of tissue engineering. *J. Cell. Mol. Med.* 10(3): 569–576.

Vacanti, C.A. and Vacanti, J.P. 1991. Functional organ replacement: The new technology of tissue engineering. In *Surgical Technology International*, M.H. Braverman and R.L. Tawes (eds), pp. 43–49. London: Century Press.

Wang, J. 2008. Amplified transduction of biomolecular interactions based on the use of nanomaterials. *Adv. Biochem. Eng. Biotechnol.* 109: 239–254.

Yagi, K. 2007. Application of whole-cell bacterial sensors in biotechnology and environmental science. *Appl. Microbiol. Biotechnol.* 73: 1251–1258.

2 Target-Driven Drug Discovery

2.1 DRUG DISCOVERY AND DEVELOPMENT

The literature is full of excellent writings on the drug discovery and development process. More recent examples include Ng (2009, 2004), Evens (2007), and Chorghade (2006). The present author believes that Chapter 4 ("Discovery and Nonclinical Development" by Stephen Carroll) and Chapter 5 ("Types of Clinical Studies" by Lewis J. Smith) in Evens (2007) are excellent, especially for a reader new to the industry or the field. The main stages are summarized below. But, before summarizing (Section 2.4), two discovery study cases for Taxol® (paclitaxel) and Gleevec® (imatinib mesylate), removed in time, are presented in a time-line format. These cases highlight the difference between the conventional and new target-driven drug discovery paradigms and provide a preface for the narrative that follows on the below-expectation outcomes of the new paradigm as well as what the future holds. The take-home message from this book is that 3D cell-based biosensors are an essential component of this future. To better appreciate the cases, six terms frequently encountered in discovery and nonclinical development (Carroll, 2007) are defined and discussed below.

2.1.1 TARGET

A target is a protein, enzyme, receptor, ion channels, signaling, or other molecule that may play a role in a particular disease. The discovery and therapeutic strategy is typically focused on the target molecule or process. In Appendix B, the nature of currently known targets that play roles in diseases and examples of the corresponding approved drugs are presented. Based on the level of target definition presented, Imming et al. (2006) has estimated 218 targets for approved drugs. In a more recent paper, Landry and Gies (2008) have found that there are about 330 targets for approved drugs; 270 of them are encoded by the human genome and 60 belong to pathogenic organisms. In contrast, an analysis of the human genome carried out in 2002 led to the estimation that there are 6000–8000 targets of pharmacological interest (cited in Landry and Gies, 2008).

2.1.2 HIT

A compound (protein, peptide, or chemical compound—usually a small molecule) that produces a significant response in a screen designed to reveal promising compounds—compounds that appear to act on the target. Depending on the target and the screen (biological or chemical), many hits may be further evaluated in a secondary screen for the most promising compounds for further testing.

2.1.3 LEAD

Hits or their "descendants" that produce promising activity in the whole cell, an intact organ system, or the whole animal, confirming the relevance of the activity found in the initial screen. Lead compounds are further examined in greater detail.

2.1.4 CANDIDATE

A compound that is a candidate for development; usually, one of a few compound that graduate from lead status by displaying most or all of the properties of the desired therapeutic in more than one relevant animal species suffering from the model disease. To become a serious contender for clinical evaluation, the candidate must meet other nonactivity requirements, such as how easy it is to manufacture, how safe it is, and does it meet the medical need. To become a clinical candidate, for establishing safety and then efficacy, the compound must be judged worthy of the major cost involved in evaluation of it in both healthy and diseased humans.

2.1.5 INVESTIGATIONAL NEW DRUG (IND) APPLICATION

An IND application is filed for the initial testing of any new drug in humans. This document is filed with the Food and Drug Administration (FDA) to request approval to begin clinical testing in humans. The application provides a summary of the compound to be tested, especially all the animal pharmacology and toxicology results, the rationale for testing in a particular indication (basis for treatment based on knowledge of cause) in humans, a detailed description of the clinical protocol, and the methods used to manufacture and test the compound.

2.1.6 DRUG OR PRODUCT

A marketed therapeutic that has successfully passed through all the above stages and is in human use. But to obtain marketing approval from the FDA or an equivalent agency outside the United States, the following four clinical studies must be performed (Smith, 2007): Phase I Studies are normally conducted with 20 to 100 healthy volunteers. The objective of the trial is to assess the safety of the new drug in addition to pharmacokinetics (absorption, distribution, metabolism, and excretion—ADME). The overall duration is typically 1.5 years. Phase II Studies are typically conducted with 100 to 300 patient volunteers. The objective of the trial is to determine the initial effectiveness of the drug in patients with the disease as well as the short-time side effects and risks associated with the drug. Another major focus is to establish the dose(s) to use in the larger subsequent Phase III study. The overall duration is approximately 2 years. Phase III Studies are typically conducted with 1000 to 3000 patient volunteers. The objective is to gather more information to establish efficacy and safety in a large number of patients. Because of the high cost involved, the decision to go ahead with Phase III is done carefully. The overall duration is between 2.5 and 5 years. Phase IV Studies are clinical trials, sometimes referred to as postmarket surveillance trials, conducted after marketing approval to monitor the efficacy and side effects of the drug in an uncontrollable real-life situation. The effectiveness of the drug when compared to alternative treatments, side effects of the drug, the patient's quality of life, and cost effectiveness are evaluated. If serious adverse effects are found, the drug is recalled. The drugs proposed for use in Chapter 13 were all recalled after approval.

2.2 THE TAXOL (PACLITAXEL) DISCOVERY CASE

The detailed discovery and development case or story is presented in Box 2.1 in a time-line format. As the story reveals, Taxol (see Figure 2.1 for the chemical structure) was originally isolated from the bark of the Pacific yew tree, *Taxus brevifolia*,

---

BOX 2.1   DISCOVERY AND DEVELOPMENT TIME LINE FOR TAXOL

1945: The Sloan Kettering Institute is founded and becomes the largest private cancer research institute in the United States. Its goal was to "concentrate on the organization of industrial techniques for cancer research."

1945–1970: Chemotherapy becomes more popular than radiation and surgery to treat or palliate cancer.

1947: The Sloan Kettering Institute screens for potential anticancer compounds with tumor cell lines.

1948: Cancer funding increases to turn cancer into America's best-funded disease.

1950: President Truman decrees a search for plants native or easily imported to the United States that could be used to produce cortisone.

1952: A group from Western Ontario begins work on the leaves of the Madagascar periwinkle (*Catharanthus rosea*) for the treatment of diabetes without insulin.

1953: U.S. Congress directs the National Cancer Institute (NCI) to investigate the possibility of starting a program to research the effect of chemotherapy on acute leukemia. Congress grants $1 million for this research despite public disapproval.

1955: The Cancer Chemotherapy National Service Center (CCNSC) is formed within the NCI. At this point only six anticancer drugs (all of which are synthetic) had been approved for use in the United States.

1955: The Western Ontario group changes the focus of their research on the Madagascar periwinkle from diabetes to leukemia.

1956: The Roscoe B. Jackson Laboratories begins supplying 100,000 in-bred mice annually for tumor testing to the NCI.

continued

1955–1960: The CCNSC screens 115,000 voluntarily submitted, synthetic compounds with known chemical structures.

1957: Jonathan Hartwell, an organic chemist, visits a research branch that was trying to produce cortisone (under Truman's order from 1950) in order to test the rejected but saved plant samples for possible anticancer uses.

1957: Charles Beer, from the W. Ontario Group researching the Madagascar periwinkle, isolates an active fraction of *Catharanthus rosea* named vincaleukoblastine, later renamed vinblastine (Noble et al., 1958).

1958: Jonathan Hartwell becomes assistant chief for program analysis activities of the CCNSC and, the following year, he requests that the USDA send him large amounts of fresh samples of *Camptotheca acuminata*, a plant that he found from the discarded samples at the cortisone research center, for further testing after finding that the leaf samples were active.

1960: The CCNSC branches out their research to screening natural products with unknown structures, screening over 30,000 compounds per year.

1960: Hartwell sends a proposal to the Agricultural Research Service asking for the New Crops Research Branch to procure plants for anticancer screening.

1960: The National Cancer Institute and the United States Department of Agriculture (USDA) starts an interagency program to procure and screen plant products for potential anticancer agents.

1961: A group of researchers from the Eli Lilly laboratory, also studying the Madagascar periwinkle, identifies another active fraction of *Catharanthus rosea* named leurocristine, or vincristine (Neuss et al., 1967; Svoboda, 1961).

1961: The Wisconsin Alumni Research Foundation (WARF) begins a contract with NCI to prepare collected plant extracts.

1962: Starting in 1962, plant extracts are tested on KB culture, L1210 (a lymphoid leukemia), and two random transplantable tumors.

1962: In the plant screening program's sweep of Washington, *Taxus brevifolia*, the Pacific yew, is sampled by Arthur Barclay and three graduate students in the Gifford Pinchot National Forest.

1964: Vinblastine and vincristine become available for clinical use in the treatment of lymphomas and acute childhood leukemia, respectively (Neuss et al., 1964).

1964: The stem bark from *Taxus brevifolia* is found to be cytotoxic at the WARF laboratories.

1964: Barclay is sent to gather 30 lbs more *Taxus brevifolia* samples from the same area in Washington.

1965: Monroe Wall, head of the fractionation and isolation laboratory in the Research Triangle Institute in North Carolina, is able to increase the potency of the compound from *Taxus brevifolia* 1000 times.

1965: At the end of the year, Wall runs out of plant samples.

1966: S-180 (a sarcoma) and CA-755 (an adenocarcinoma), both previously used as implantable mice tumors for testing, are found to be sensitive to extensive plant constituents that produce many false positives and hence S-180 and CA-755 are replaced by Walker 256.

1966: Botanist Robert Perdue, the head of the plant screening program for the USDA, calls Richard Chase, a park ranger, who sends Bruce Smith to collect 45 lbs of stem bark, 135 lbs of twigs and leaves, and 55 lbs of stem wood for Wall's testing.

July 1966: The active compound from *Taxus brevifolia* is found to be in the stem bark.

September 1966: The active compound is isolated, with 0.5 g of the compound capable of being recovered from 12 kg of bark. This corresponds to a yield of only 0.004%.

1966: Research shows activity of the cytotoxic compound from the stem bark of *Taxus brevifolia* in a number of tumor systems, including P1534.

1966: Perdue begins testing the differences between *Taxus* species and finds that many of them are active, but not as active or as high yielding as *Taxus brevifolia* (even though *Taxus brevifolia* did not have a high yield).

1967–1970: Hartwell publishes his work on "cancer plants" from reports of folk remedies for cancer in 11 installments (Hartwell, 1969).

1967: Perdue outlines detailed procedures for the procurement of a random sweep of plants to be tested for anticancer properties.

1967: Perdue arranges for the collection of almost 2500 lbs of *Taxus brevifolia* from Idaho, California, Oregon, and Washington.

June 1967: The cytotoxic compound isolated from *Taxus brevifolia* is named taxol: "tax-" from *Taxus* and "-ol" from the alcohols known to be present in its structure.

1967: At the annual meeting of the American Chemical Society, Wall only mentioned the cytotoxicity and antitumor activity of *Taxus brevifolia*. No other *Taxus* species were mentioned.

1968: Perdue tests the effect of location of *Taxus brevifolia* on its activity.

1968: Wall turns over all information, data, and samples to Robert Engle of the CCNSC.

1969: Sensitivity to extensive plant properties is found in Walker 256 as well and hence Walker 256 is replaced by P388.

1969: Taxol is officially adopted by the NCI.

1969: Hartwell and Perdue publish an article on the supply problem of *Taxus brevifolia*, suggesting that it should not be the sole future source of taxol since all *Taxus* species contain taxol. As an alternative, they recommend that a faster growing, higher yielding type should be developed (Perdue and Hartwell, 1969).

continued

1970: The molecular formula of taxol ($C_{47}H_5NO_{14}$) is discovered along with the structure.

1970: Wall tells Engle at the CCNSC the total structure of taxol. The CCNSC does not commit to continue working on taxol.

1971: Wall and his coworkers publish their work in the *Journal of American Chemical Society* showing the antileukemic and antitumor activity of taxol (Wani et al., 1971).

1971: Wall asks the CCNSC what he can do to help taxol's drug development.

1972: Wall's question is finally answered with a request for him to produce 15 g of taxol.

1973: Almost a year later, Wall completes the CCNSC's order for 15 g of taxol.

1975: Hartwell retires from the plant screening project at NCI and is succeeded by John Douros.

1977: Susan Horowitz, a molecular pharmacologist at the Albert Einstein College of Medicine in New York, and David Fuchs and Randall Johnson of the Division of Cancer Treatment at the NCI begin working on taxol's mechanism of action.

1978: The NCI writes a letter to the USDA warning that the interagency agreement would not be renewed unless certain problems with USDA's operations are fixed. A month later, Robert Perdue is reassigned to the Plant Taxonomy Laboratory to be replaced by James Cook.

November 1978: Taxol is found to cause regression in MX-1, a mammary zenograft.

1978: David Fuchs and Randall Johnson at the NCI publish an article that reports that taxol does have antimitotic effects, but is less efficient at arresting mitosis than compounds such as those found in the Madagascar periwinkle (Fuchs and Johnson, 1978).

1979: Susan Horwitz and her team publish their finding that taxol has antimitotic effects by stabilizing microtubules and therefore preventing depolymerization. This method of arresting mitosis was different from any previously known antimitotic compound (Schiff et al., 1979).

June 1980: After taxol is found to have solubility problems that interfere with developing a method to administer the drug, a formulation of polyoxyethylated castor oil is found to be an acceptable solvent, allowing the continued development of taxol.

October 1980: Taxol begins toxicology studies.

December 15, 1980: The USDA asks for bids to supply Pacific yew bark in 500–10,000 lbs increments.

1981: The NCI-USDA plant program is abolished after having screened over 114,000 extracts from an estimated 15,000 plant species. At this time, no plant-based anticancer drug has been approved as a result of the program.

1981: Robert Warner, a lumber collector, fulfills the USDA contract with 3366 lbs of bark.

1981: Pierre Potier, the head of the Institut de Chimie des Substances Naturelless, publishes a semi-synthesis of taxol from 10-DAB (a compound easily isolated with a high yield from the needles of *Taxus baccata*) by attaching a side chain at the carbon 13 position (Chauvière et al., 1981).

February 1981: A study by Jerry Franklin on the old-growth forests of the Pacific Northwest is published showing that almost all the old-growth forest are on Federal lands and that if current logging practices continued, 95% of the old-growth forests would be destroyed in 50 years (Franklin et al., 1981).

1982: John Douros leaves the NCI to go to Bristol-Myers. Matthew Suffness is his replacement as Chief of the Natural Products Branch of the NCI.

January 1982: The 3366 lbs of bark is finally turned into 890 lbs of extract.

June 1982: 16,000 lbs of bark is sent for purification to Polysciences.

June 1982: Toxicology studies on taxol are completed.

October 1982: The 890 lbs of extract is purified to 264 g of taxol and delivered to the NCI.

April 1984: Phase I clinical trials for taxol begin.

December 27, 1984: a 54-year-old man with renal cell carcinoma, and no previous record of heart disease, dies from cardiorespiratory arrest on his second course of taxol. Phase 1 trials are stopped until modifications are made to the administration of taxol.

April 16, 1985: Taxol's toxic effects (including myelosuppression and leukopenia), proper dosing regimen (continuous infusion to prevent cardiorespiratory problems), and all other researched information are presented to the Decision Network Committee. Taxol passed the requirements to enter Phase 2 of testing.

July 10, 1985: Robert Warner is contracted to collect 12,000 lbs of *Taxus brevifolia* bark to the NCI by November 8. Due to an unusually hot and dry summer, Warner was unable to complete the contract order until the fall of 1986.

September 1986: Warner delivers the dried bark to Polysciences for purification into taxol.

December 1, 1986: Phase II clinical trials begin accepting patients, but due to taxol shortages, some trials are delayed. In response, the Natural Products Branch places a bid for 60,000 lbs of bark, the largest amount to be collected to date.

December 1986: Newspapers nationwide print stories about the environmental impact of supplying taxol in addition to information about its success in clinical trials on various types of cancer.

continued

1987: Green World and the Sierra Club Defense Fund petition the U.S. Fish and Wildlife Service to list the northern spotted owl, whose habitat is the old growth forest of the Pacific Northwest, on the endangered species list.

March 1987: The second annual report on taxol to the FDA states that partial responses had been seen in melanoma and refractory ovarian cancer patients. The response rate for refractory ovarian cancer patients at Johns Hopkins is just under 20%, the acceptable response rate to justify further, serious research.

May 6, 1987: Patrick Connolly wins the bid for collecting all 60,000 lbs of bark at a price of \$3.58 lbs^{-1}.

July/August 1987: Forest Services closes access to Umpqua National Forest due to fire risk. At this time, Connolly has collected a little more than 37,000 lbs.

September 1987: Connolly asks for an extension of his contract, which he receives until September 6, 1988.

November 1987: First delivery of bulk, purified taxol to the NCI.

December 1987: Hauser Chemical Research makes an agreement with the NCI to extract taxol from the bark.

March 1988: The response rate in refractory ovarian cancer at Johns Hopkins reaches 30%.

April 1988: Two French groups, headed by Pierre Potier and Andrew Greene, send an article to the *Journal of the American Chemical Society* on a semisynthetic method of producing taxol from *Taxus baccata*, a more common *Taxus* species. Their research was previously published in France in 1981 (Dennis et al., 1988).

April 18–21, 1988: Gordon Cragg of the NCI travels to the Pacific Northwest to meet Connolly, representatives from Weyerhaeuser (the largest timber company in the area), and members of a "Protect the Yew" group. He discusses the amount of bark that can be stripped from a tree, the numbers of trees present in the area, the possibility of propagating yew seedlings, and the environmental considerations of harvesting the Pacific yew.

June 1988: The remainder of the purified taxol from Connolly's collection is delivered to the NCI.

June 23, 1988: The first meeting of the Taxol Working Group decides to start only two clinical trials of taxol due to limited supply: a Phase I trial of cisplatin and taxol and a Phase II trial on advanced ovarian cancer.

July 1988: Weyerhaeuser considers branching out from lumber to forest by-products.

July 6, 1988: The NCI seeks help from private pharmaceutical companies for the funding of taxol drug development.

August 1988: Connolly finishes collecting the 60,000 lbs of bark.

September 1988: The Taxol Working Group decides to solicit a drug company to take over taxol and form a partnership with Weyerhaeuser for continued supply of taxol.

April 1989: The U.S. Fish and Wildlife Service finally list the northern spotted owl as threatened, but do not mention its habitat.

April 17, 1989: John Destito, the owner of Advanced Molecular Technologies, a supplier of biologically active products, receives a bid for collecting the next 60,000 lbs of bark at $2.74 lbs^{-1} despite resentment from Connolly. Destito contracts help from Weyerhaeuser to collect the timber.

August 1, 1989: A Cooperative Research and Development Agreement (CRADA) announcement for a pharmaceutical company partner with the NCI for taxol is placed in the *Federal Register* and is sent to 23 interested groups.

September 1989: Destito has only collected 8000 lbs of bark, forcing him to extend his contract.

September 15, 1989: Four pharmaceutical companies submitted proposals by this deadline for the CRADA: Bristol-Myers, Rhone-Poulenc, Unimed, and Xechem.

October 23, 1989: Bristol-Myers Squibb is informed that they are one of the finalists for the CRADA.

October 27, 1989: Rhone-Poulenc is informed that they are the other finalist for the CRADA.

December 1989: Bristol-Myers Squibb is chosen as the CRADA pharmaceutical partner.

1990: A study of the taxol content by tree part is published by the Frederick Cancer Research Facility and the Natural Products Branch of the NCI. The study concludes that there is little difference between taxol contents of *Taxus* needles and *Taxus brevifolia* bark (Witherup et al., 1990).

1990: A different study by Vidensek and his associates concludes that the bark has 10 times as much taxol as the needles (Vidensek et al., 1990).

April 4, 1990: Bristol-Myers Squibb and the NCI arrange with Hauser to supply timber, cutting the current NCI lumber contractor, Destito, out of the arrangement.

June 14, 1990: Destito's lawyers warn the NCI that they will appeal to members of Congress to investigate the NCI's activities regarding the contract with Hauser.

June 26, 1990: The NCI holds its first taxol workshop, inviting all interested parties concerning taxol to attend. At the meeting, Destito announces that he is dropping the legal action against the NCI, Bristol-Myers Squibb, and Hauser.

August 12, 1990: The Native Yew Conservation Group, formed from attendees at the taxol workshop who were upset that the sustainability

continued

of the yew was not mentioned at the workshop, meets to discuss how to preserve the yew as a species and for sustainable harvesting of taxol.

September 19, 1990: Michael Bean and Bruce Manheim, lawyers for the Environmental Defense Fund, petition to list *Taxus brevifolia* as threatened to the Secretary of the Department of the Interior, Manuel Lujan.

October 26, 1990: Hauser buys Connolly's lumber plant, ending Connolly's job as a subcontractor for Hauser.

November 7, 1990: The USDA invites the NCI, the Forest Service, the FDA, Zelenka Nurseries, Secrest Arboretum at Ohio State University, the University of Mississippi, and Bristol-Myers Squibb to a meeting to work on solving the taxol supply problem through agriculture.

December 1990: Advanced Molecular Technologies, Destito's business, falls apart after it is purchased by a Seattle-based investment firm.

December 20, 1990: Bristol-Myers Squibb applies for the trademark for the name "taxol."

1991: Hauser and Bristol-Myers Squibb plan to collect 750,000 lbs of bark (four times more than the total amount collected since 1962).

January 1991: Phyton Catalytic, a biotech company, enters a CRADA with the USDA to make taxol from plant cell and tissue cultures.

January 1991: The Phase II response rate of taxol in breast cancer patients rises to 48%.

January 9, 1991: The Fish and Wildlife Service decline the petition to list *Taxus brevifolia* as threatened because they claim that 130 million yews are still present on Federal lands.

January 19–23, 1991: The NCI and Bristol-Myers Squibb sign the CRADA.

February 1, 1991: Bristol-Myers Squibb applies for a trademark on "taxol" in Canada.

March 1991: The final report of the Phase II breast cancer trials shows a wonderful response rate of 56%. Oncologist Frankie Ann Holmes publishes her study's findings.

March 22, 1991: The first meeting of the Alliance for the Production of Taxol is held, at which nurseries report their *Taxus brevifolia* inventories. The result of these inventories showed that over 30 million Pacific yew trees are currently growing in nurseries.

May 1991: The Agricultural Research Service in New Orleans receives a patent for a method of making taxol from plant cell and tissue cultures.

May 6 1991: Judge Thomas Zilly orders the Fish and Wildlife Service to declare 11.6 million acres as critical habitat for the northern spotted owl.

June 1991: The USDA and the Department of the Interior sign a CRADA regarding taxol.

July 29, 1991: The first Congressional hearing on taxol is held over the agreements between Bristol-Myers Squibb and the NIH, and the USDA and the Department of the Interior.

August 1991: The publisher of *The Cancer Letter* (vol. 17, no. 32, August 1991 p. 2) accuses Robert Wittes, Chief of the Medicine Branch of the Clinical Oncology Program who had left the NCI to go to Bristol-Myers before returning to the NCI, of unjustly favoring Bristol-Myers Squibb for the CRADA.

September 1991: A project is started through the USDA to collect 100,000 lbs of fresh weight or 35,000 lbs of dry weight of needles, stems, and twigs from the cultivated species *Taxus x media*, developed at the University of Mississippi.

November 20, 1991: Gerry Studs, a Democratic Congressman from Massachusetts, introduces the Pacific Yew Act to provide for government management of the yew.

March 1992: A Congressional hearing is held on the Pacific Yew Act, which would provide for Federal management of *Taxus brevifolia* to ensure the continued supply of taxol.

March 4, 1992: The second Congressional hearing on taxol discusses the Pacific Yew Act of 1991.

May 26, 1992: Bristol-Myers Squibb receives a trademark on the name "Taxol®" with the generic name of "paclitaxel."

July 1992: After an investigation, Wittes is cleared of any interference with the CRADA.

July 22, 1992: Bristol-Myers Squibb files a New Drug Application for taxol to be used on refractory ovarian cancer.

August 7, 1992: The Pacific Yew Act becomes law.

September 1992: The NCI holds its second taxol workshop in Alexandria, Virginia.

December 12, 1992: An inventory by the Forest Services finally reports an educated approximation of 41 million yews left on Federal lands.

December 29, 1992: Taxol receives FDA approval, making it the only drug to receive FDA approval from the plant screening program. Bristol-Myers Squibb receives five years of exclusive marketing rights to the taxol compound without having a patent.

December 29, 1992: The Canadian Health Protection Board grants the trademark on Taxol® to Bristol-Myers Squibb.

January 25, 1993: The third Congressional hearing on taxol discusses the pricing of drugs developed by both Federal laboratories and private companies.

January 25, 1993: Zola Horovitz, vice-president of the Business Development and planning at Bristol-Myers Squibb, states that they will not take any Pacific yew bark from Federal lands in 1993. Instead, they will produce taxol semisynthetically by Potier's method.

continued

1994: Bristol-Myers Squibb sues a small pharmaceutical company from Quebec named Corporation Biolyse Pharmacopée Internationale for using the trademarked name Taxol® (Giles, 1993).

February 17, 1994: An article in *Nature* announces the first published total synthesis of taxol by K.C. Nicolaou and his associates at Scripps Research Institute in California. A week later, another article in *Nature* announces that Bob Holton and a group from Florida State University successfully created a different method of synthetically producing taxol (Holton et al., 1994a, 1994b).

August 1994: Paul Wender and his team make taxol synthetically by a third path: from pinene, with funding from the National Science Foundation, the NCI, and Bristol-Myers Squibb (Wender et al., 1995).

February 2, 1995: An editorial in *Nature* criticizes Bristol-Myers Squibb for placing a trademark on the already commonly used name "taxol" (Anonymous in Nature, 1995—"Names for hi-jacking").

February 14, 1995: The Secretary of Health and Human Services ends the Pacific Yew Act under the "sunset clause" provided in the act. This clause allowed for the yew act to end when taxol demands could be satisfied by other sources than yews on Federal lands. As taxol is now being made semisynthetically, there is no need for the management of the Pacific yew.

March 16, 1995: Bristol-Myers Squib write an article in defense of the February 2, 1995 attack, saying that whether or not taxol had been a commonly used name for the anticancer drug had "no bearing on whether Taxol is a recognized trademark" (Chesnoff, 1995).

December 15, 1995: The Federal Court rules in favor of Bristol-Myers Squibb in the lawsuit against Biolyse Pharmacopée Internationale.

Source: Developed from Goodman, J. and Walsh, V. 2001. The Story of Taxol: *Nature and Politics in Pursuit of Anti-Cancer Drug.* New York: Cambridge University Press.

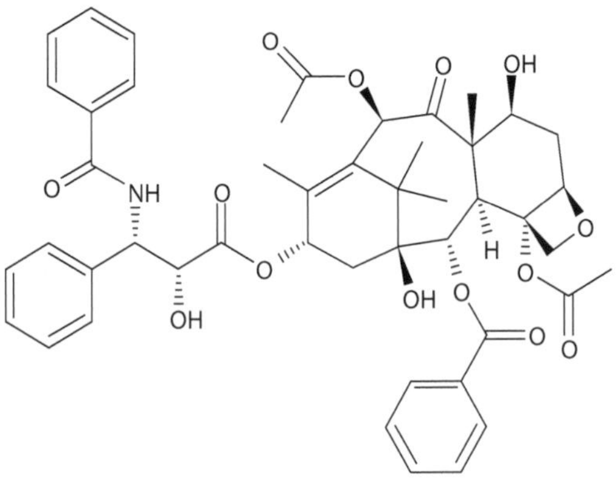

FIGURE 2.1 Taxol chemical structure.

TABLE 2.1
β-Tubulin Isotypes

Class	Human Isotype	Expression
I	M40	Major constituitively expressed β-tubulin isotype
II	hβ9	Major neuronal isotype, expressed mainly in the brain, but at low concentrations in various cell types
III	hβ4	Minor neuronal isotype, expressed only in neurons and the brain, at lower concentrations than class II isotype
IVA	hβ5	Neuro-specific
IVB	hβ2	Constitutive
VI	hβ1	Hemopoietic-specific

Source: From Berrieman, H.K., Lind, M.J., and Cawkwell, L. 2004. *Lancent Oncol.* 5: 158–164. With permission.

and was shown to have antileukemic and antitumor activity in 1971. More than 20 years later, it was initially approved by the FDA for the treatment of advanced ovarian cancer in 1992 and subsequently endorsed for the treatment of metastatic breast cancer in 1994. Due to the limited supplies from a natural source, research effort has gone into developing synthetic routes. To date, it is derived semisynthetically from the inactive taxane precursor, 10-deacetylbaccatin III, found in the needles of the European yew tree, *Taxus baccata* (Gueritte, 2001).

The cellular target for Taxol is a site on β-tubulin, called the taxane site, and there are seven isotypes of β-tubulin in humans (Table 2.1). α- and β-tubulin heterodimers are the building blocks of microtubules' hollow cylindrical cores. When microtubules are being formed, α- and β-tubulin heterodimers associate together in a head-to-tail fashion to form a microtubule nucleus and the nucleus elongates linearly into protofilaments, which then associate laterally to form microtubules (Jordan and Wilson, 2004). Microtubules have important functions in cellular activities such as maintenance of cell shape, cellular movement, cell signaling, division, and mitosis (Jordan and Wilson, 2004). The taxane-binding site on microtubules is only present in assembled tubulin (Abal and Andreu, 2003). Microtubule-targeting drugs, including Taxol, function by suppressing mitotic spindle microtubule dynamics (Figure 2.2), thus inhibiting the metaphase–anaphase transition, blocking mitosis, and inducing apoptosis. Essentially, Taxol stabilizes microtubules by binding to polymeric tubulin and preventing disassembly (Zhou and Giannakakou, 2005). For an in-depth review of targets for Taxol and other antimitotic drugs, see McGrogan et al. (2008). Sales figures for Taxol from Bristol-Myers Squibb annual reports are $747 (2005), $563 (2006), and $422 (2007) millions.

2.3   THE GLEEVEC (IMATINIB MESYLATE) DISCOVERY CASE

The Gleevec discovery and development case or story is presented in Box 2.2 in a time-line format as well. Gleevec, or compound STI-571 as it was known during

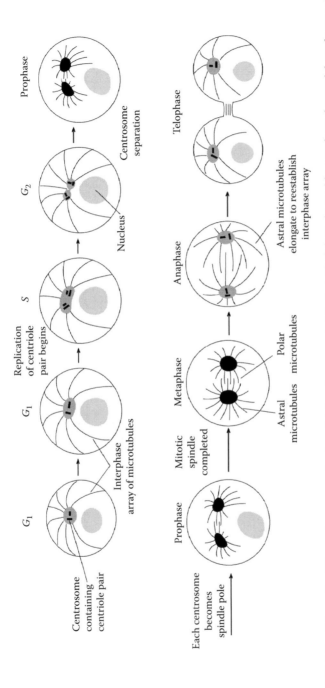

FIGURE 2.2 Schematic of the centrosome cycle. A centriole pair represented by two black bars, located in centrisomal material, nucleates microtubule outgrowth. During G_1 phase, the two centrioles separate and during the S phase, a daughter centriole begins to grow near the base of each mother centriole at right angles. The elongation of the daughter centriole is normally completed by G_2 phase at the end of G_2 phase (the beginning of M phase); each centriole pair becomes part of a separate microtubule-organizing center that nucleates a radial array of microtubules, known as aster (meaning star). The two asters initially lie side by side close to the nuclear envelope and then move apart. Preferential elongation of the microtubule bundles that interact between the astars occurs by late prophase, forming mitotic spindles very rapidly. (From Alberts, B., et al. 1989. *Molecular Biology of the Cell*, 2nd edition, New York: Garland Publishing. With permission.)

BOX 2.2   DEVELOPMENT TIME LINE FOR GLEEVEC

1960: Dr. Peter Nowell from the University of Pennsylvania School of Medicine and Dr. David Hungerford from the Institute of Cancer Research identify the "Philadelphia chromosome" (a shortened 22nd chromosome) as being present in the majority of chronic myeloid leukemia (CML) cells. This was the first time a cancer-related genetic abnormality had been identified (Nowell and Hungerford, 1960, 1961).

December 23, 1971: Richard Nixon signs the National Cancer Act to strengthen the National Cancer Institute to more effectively carry out the National effort against cancer.

1973: Dr. Janet Rowley of the University of Chicago noted extra DNA on chromosome 9 and discovered the swap between chromosomes 9 and 22 in CML cells. The "Abl" part of chromosome 9 attaches to the "Bcr" part of chromosome 22 forming the Bcr-Abl gene in 95% of CML patients (Rowley, 1980).

The early 1980s: One of the first oncogenes (genes that dispose normal cells to change to tumor cells) is identified: EGF—epidermal growth factor. In cancer growth factor receptors (GFRs), the internal segment is a tyrosine kinase that activates signaling proteins that signal for cell replication by attaching phosphate groups.

1983: Alex Matter heads the cancer research at Ciba-Geigy (the large pharmecuetical company in Basal, Switzerland).

1985: Nick Lydon joins Alex Matter on research of kinases, while Brian Druker, a postdoc medical oncologis, tests for abnormalities that are responsible for cancer in Boston.

1988: Nick Lydon travels to Dana-Farber Cancer Institute in Boston to meet Brian Druker about working together.

Late 1988: Professor Alexander Levitzki publishes an article in *Science* on the ability to inhibit only a specific EGF receptor. This article gives Druker hope for developing a molecule to inhibit only the specific kinase responsible for CML (Yaish et al., 1988).

1986 and 1987: Dr. David Baltimore of the Whitehead Institute for Biomedical Research in Cambridge, Massachusetts, and Dr. Owen N. Witte, Professor in developmental immunology at UCLA, publish two articles showing their discovery that CML is caused by an activated tyrosine kinase from the mutated Philadelphia chromosome (Daley et al., 1987).

1990–1993: Dana-Farber signs an exclusivity agreement with Sandoz ending Druker's relationship with Matter and Lydon.

August 2, 1990: Jürg Aimmerman is hired as a chemist for Ciba-Geigy to develop the specific inhibiting molecule that must block the ATP pocket of the kinase. He develops 10 compounds per week (probably

continued

with the aid of combinatorial chemistry techniques), which are sent to be tested by biologists to see if they are selective enough to inhibit only the Bcr-Abl kinase while leaving other vital kinases intact.

June 1992: Vasella moves to Sandoz Pharmecueticals.

August 26, 1992: Jürg has a series of compounds that were selectively active against Bcr-Abl.

Early 1993: Buchner's tests prove that the compound prevented phosphorylation, and therefore cell growth, on the Bcr-Abl oncogene.

Spring 1993: The compound reaches "drug candidate status." The search begins for doctors to run trials in humans.

February 1994: Druker shows the results that the compound inhibits 90% of CML cells *in vitro* to Swiss scientists. They decide to pursue further tests on the compound.

1995: Animal studies begin on STI571.

March 1996: Sandoz and Ciba-Geiga merge to form Novartis.

1996: Vasella becomes CEO of Novartis.

April 1996: Druker begins planning for patient trials.

May 1, 1996: The first paper on STI571 is published by Brian Druker in *Journal of Nature Medicine*. The paper shows that the studies on dogs reveal problems with blood clots at the catheter site, delaying the trials (Druker et al., 1996).

November 1996: Animal trials of rats and dogs reveal high liver toxicity.

1997: Pediatric hematologist Renaud Capdeville joins Novartis as International Clinic leader.

Early 1998: Novartis grants approval for patient trials to start in early summer.

June 1998: Druker lines up 12 patients for trials. Eventually 149 patients are enrolled in Phase I trials.

June 22, 1998: A minister for Bakersfield, California, is the first patient to receive the drug at a dose of 25 mg/day.

June 25, 1998: Bud Romine of Telmec, Oregon, begins trials at the same 25 mg dose. He is the second patient to receive Gleevec. A third patient also receives the 25 mg dose soon after Romine.

July 16, 1998: The minister's white cell count normalized, and then within a week rose back. The temporary stabilization was explained by typical "cycling." The other two patients showed no response.

July/August 1998: Three patients start at 50 mg/day, but with climbing white cell counts they are forced to stop.

August/September 1998: Four patients take 85 mg/day.

September 1998: The four 85 mg patients all show stabilized white blood cell counts.

October 1998: Three patients begin 140 mg /day trials. In all of the patients their white cell counts fall, with one patient reaching a normal white cell count.

December 1998: A patient with difficult CML is admitted into the trials.

1999: Novartis begins to consult with the FDA and European health authorities to get approval for Gleevec based only on the Phase I and II trials, which is only allowed in exceptional circumstances.

January 1999: One of the 140 mg patients, a doctor from Argentina, withdraws to be with her family, despite the fact that she showed a positive response to the drug.

Late January 1999: For the first time, a patient in the trials reaches a white cell count below 3000/mm^3.

February 1999: Planning for Phase II trials begins, for a speedy development.

Early February 1999: The patient with difficult CML reaches a normal white cell count and maintains this level.

March 1999: A patient taking 300 mg/day proves to be 50% Philadelphia chromosome negative, but only on a test of 10 cells.

April 1999: All 31 patients at the 300 mg/day dose show a complete hematologic response. One-third also has a complete cytogenetic response (the absence of Philadelphia chromosome).

April 1999: The results from the Phase I trials on CML show that all 31 patients at 300 mg doses had white blood cell counts that were reduced to normal levels.

April 1999: Patients in the final, blast imature white cells phase enter Phase I trials.

April 1999: Vasella beomes the chairman of Novartis.

April 26, 1999: The first patient to ever receive the STI571 compound (Gleevec) is allowed to re-enter the program at the effective dose of 300 mg/day. His white cell count becomes normal very quickly.

Spring 1999: The Phase I trials are expanded to include 59 blast phase CML patients. The first three to start soon have their white cell count controlled and then suddenly get worse and die.

June 1999: A report shows that one patient had a complete cytogenetic response after only 5 months on Gleevec treatment.

June 1999: Phase II trials begin just 1 year after Phase I trials began.

July 1999: Druker holds a meeting in Bordeaux, France, to tell other scientists about the positive results from blast phase patients.

July 28, 1999: Maeve Devlin, the head of the pharmaceutical production center called Ringaskiddy, is asked to take over the entire mass production of Gleevec.

September 1999: After her disease worsens, the doctor from Argentina who withdrew from the trial is accepted back into the study and is allowed to spend most of her time in Argentina with her family. She later shows a complete hematologic response.

continued

September 1999: The production at Ringaskiddy begins, bypassing the "prototype adequate" phase.

September 21, 1999: A determined CML patient named Suzan McNamara, aged 32, who was deteriorating but not yet at the accelerated or blast phase, starts an online petition for Novartis to produce more Gleevec to allow more people to be placed in the clinical trials.

September–October 1999: The Ringaskiddy plant takes all raw materials and focuses on 7 of the 12 manufacturing steps. The Stein plant then manufactures the drug in capsule form. The workers at Ringaskiddy use suits from NASA called "doverpacs" to protect them from the dangerous starting materials of Gleevec.

October 11, 1999: After 3 weeks, McNamara's petition has 3030 signatures.

October 12, 1999: McNamara sends the petition to Novartis.

October 1999: Phase II trials begin with 1000 patients enrolled.

November 7, 1999: McNamara receives a letter from James S. Shannan, the head of Clinical Research and Development at Novartis, telling her that Novartis is actively trying to produce large quantities of the drug.

December 1999: The results from the Gleevec trials are for the first time reported to the public at the American Society of Hematology in New Orleans, Louisianna. At this meeting, Druker gives a speech about the continued development of Gleevec. During the meeting and for 1 month after the meeting, Novartis's clinical trials phone line receives 2000 calls per day. After that the number drops to 600 per month, compared to the 15 calls per month before Druker's speech.

December 1999: The Gleevec group, consisting of patients on Gleevec, family, and friends and visits by doctors and nurses, appears on the television show 20/20 for 2 min.

December 1999: Phase II trials begin on CML patients not in the accelerated or blast phase, but in the chronic phase who could no longer continue interferon treatment. Susan McNamara is one of these patients in the trials.

June 2000: A deadline is set for 1 ton of Gleevec to be produced, a year earlier than generally planned for development of a drug.

June 2000: Novartis starts an Expanded Access Program to allow entry into Phase II trials to more than a set number of patients, in response to the clamoring desire of patients to join the trials. 7000 patients enroll in the Phase II trials, as compared to the typical number of 50–100 patients. The expanded program includes 32 countries.

July 2000: The FDA grants fast-track designation to Gleevec. This designation is only given to drugs to treat life-threatening conditions that fulfill unmet needs or improve therapy.

August 2000: On the deadline set for the Rigaskiddy team to produce 1400 kg of Gleevec, expectations are exceeded as 1536 kg of Gleevec are completed.

February 2001: A newsletter from a gastrointestinal stromal tumor (GIST) Gleevec support group, Life Raft Group, started by Norman Schezer, presents data on the side effects and response of 16 GIST patients taking Gleevec. 87.5% of the patients had a response (defined as tumor shrinkage).

February 27, 2001: The New Drug Application is filed for Gleevec for CML in all three phases (chronic, accelerated, and blast phase) after interferon failure.

Early Spring 2001: The FDA inspects Novartis sites around the world producing or testing Gleevec.

March 26, 2001: The FDA grants Gleevec priority review status.

Late April 2001: The FDA says Novartis must change the European spelling of "Glivec" because it may be easily confused with the current medications Glynase or Glyset. Novartis quickly changes the spelling to "Gleevec" to avoid the delay of a totally different new name.

May 1–4, 2001: The final audit of Novartis facilities involved with Gleevec is conducted by the FDA.

May 10, 2001: A special press conference with the Secretary of the Department of Health and Human Services, the FDA, Vasella, and Susan McNamara is held in Washington, DC, announcing the FDA's approval of Gleevec for patients who no longer respond to interferon therapy.

May 11, 2001: 5500 bottles of Gleevec are shipped and cleared by the FDA and sent to wholesalers, just 1 day after FDA approval (this generally takes about 14 days).

May 14/15: Gleevec is available at pharmacies at a price of $2200/month for patients taking 400 mg/day in the chronic phase, and $3500/month for patients taking 600 mg/day in the accelerated or blast phase. These prices are set worldwide, but a Patient Assistance plan offered by Novartis helps those who cannot afford the drug. (In comparison, interferon costs were $1700–$3300/month in the United States, $1250/month in Australia, and $4750/month in Japan.)

June 2001: Enrollment ends for Phase III trials of those who had never taken interferon. 1106 patients enroll in the trials. The results of these trials show that 83% and 68% of patients on Gleevec had a major or complete cytogenetic response, respectively, while only 20% and 7% had a major or complete cytogenetic response on a combination of interferon and Ara-C. 96% of the patients on Gleevec showed a complete hematologic response, while only 67% of the patients on interferon and Ara-C showed a complete response. (Acra-C is a chemotherapy drug typically given to leukemia patients.)

continued

June 22, 2001: *The Wall Street Journal* publishes an article titled "Gleevec Shows a Weakness in Fighting Cancer: New Study Details How Disease Resists Drug." In the article they point out that 80% of late-stage patients who initially show a response to Gleevec relapse within 6 months, often resulting in death.

February 28, 2002: *The New England Journal of Medicine* publishes the results of 454 patients in a study of late-chronic phase CML that have failed interferon therapy. The study showed that after 18 months on Gleevec, 89% of the patients' CML had not progressed to the accelerated or blast phase and 95% of the patients were still active. Also, when used as a first-line defense (before interferon has been administered), 68% of the patients showed a complete cytogenetic response (Kantarjian et al., 2002).

September 2002: Gleevec is available in over 80 countries for treatment of CML.

December 23, 2002: The FDA approves Gleevec as a first-line treatment of CML.

Source: Developed from Vasella, D. and Slater, R. 2003. *Magic Cancer Bullet: How a Tiny Orange Pill is Rewriting Medical History.* New York: HarperCollins.

preclinical and clinical development phases, is one of the first examples of rational drug design arising from studies on the human genome. See Figure 2.3 for the chemical structure of Gleevec. It is also the first signal transduction inhibitor (STI) drug. It works by interrupting the growth pathway of cancer cells by inhibiting three kinases: Bcr-Abl, which is coded by the Philadelphia chromosome (a shortened 22nd chromosome) and is responsible for the majority of CML (chronic myeloid leukemia) cells; PDGF-R, which is a platelet-derived growth factor and is responsible for a variety of cancers; and c-KIT, which is responsible for GISTs (gastrointestinal stromal tumors) and possibly small cell lung cancer. Protein kinases represent a large family of ATP-dependent phosphotransferases from as many as 518 putative kinase genes that make up the human kinome (Manning et al., 2002). Kinases phosphorylate

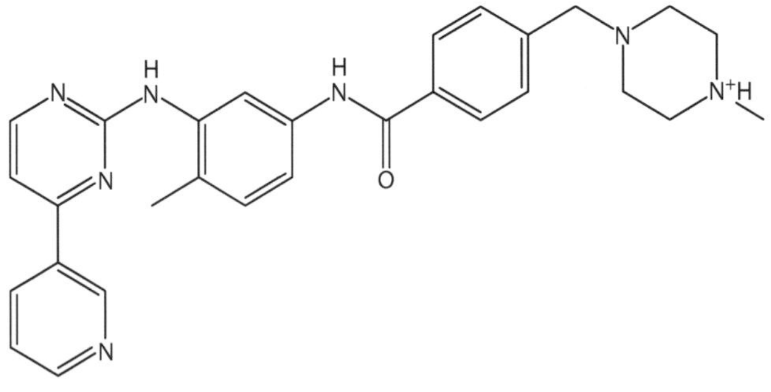

FIGURE 2.3   Gleevec chemical structure.

proteins and as such are involved in many signaling pathways. Protein kinase inhibitors are an emerging class of targeted therapeutic agents (Grant, 2009).

CML is characterized by an abnormally high white cell count, which is the result of a genetic mutation. The 22nd chromosome in CML cells is abnormally short, and is called the Philadelphia chromosome, with a large part of the chromosome translocated to chromosome 9, while part of chromosome 9 is translocated to chromosome 22. Cells with the shortened 22nd chromosome are called Philadelphia positive. This mutation results in the mutated Bcr-Abl region, which codes for a tyrosine kinase with unregulated replication outcomes in white cells.

CML is characterized by three phases: chronic phase, the first to develop, shows few if any symptoms but patients do have an elevated white cell count; the accelerated phase, in which the patient's white cell count continues to rise and blast cells (immature white cells) account for 15% or more of the blood; and the blast phase, the final phase in which blast cells account for 30% or more of the blood.

In 1995, before Gleevec approval, 70% of bone marrow transplant (the then only known cure for CML) patients were dying within one year of transplant, and 25% died each year after that. The other treatment options were chemotherapy or interferon, both of which had serious side effects and often were not effective. Hydroxyurea, a type of chemotherapy previously frequently used to treat CML, lowered the white cell count in patients but did nothing to reduce the number of Philadelphia chromosome-positive cells. Gleevec sales figures for 2007 were $3.05 billion.

2.4 TARGET-DRIVEN DRUG DISCOVERY PARADIGM

In comparing the two cases presented above, three observations can be made. First, the discovery and development of Taxol represent the conventional approach, where thousands of potential compounds from natural sources are screened for a hit via a specific assay that represents the target. Some have compared this approach to trying out many keys to find one that will lock or open. With this approach chances for failure are high; as described in the Taxol discovery and development time line (Box 2.1), only one product made it to the market from the expansive screening effort. However, it should be pointed out that the majority of drugs on the market today were discovered in this manner. On the contrary, the discovery of Gleevec represents the target-driven approaches, where the process starts with discovering the target and then designing a drug that interacts with the target to yield desirable therapeutic outcomes. Second, natural product sources can add complications to the discovery and development process owing to multiple stakeholders whose interests may be harmed or enhanced by the process. Third, the target-driven approach has potential to reduce the time from discovery to the market. For example, it took 20 years to get to IND application for Taxol compared with seven years for Gleevec. The important steps in the target-driven drug discovery and preclinical development are presented in Table 2.2. The new approach has been made possible by the advent of whole-genome sequencing, combinatorial chemistry, and HTS. Figure 2.4 presents a schematic of how the three technologies can be brought together in discovery and preclinical development processes. The role of

TABLE 2.2

Important Steps in Drug or Biologic Discovery and Nonclinical Development

Step	Comment
Discovery Steps	*Identification of Disease Targets and Potential Therapeutic Compounds*
Target identification	*In vitro* process by which potential targets are investigated, screened, and prioritized
Target validation	*In vitro* process by which the role a target plays in a disease is characterized and established
Lead identification	*In vitro* process by which potential therapeutics are screened and prioritized
Lead optimization/validation	*In vitro* process by which potential actions of compounds on disease are characterized and confirmed; if needed, improvements in compounds are made and the new compounds are evaluated
Pharmacokinetics and dynamics	Both *in vitro* and *in vivo* studies of the potential effect that the body has on the drug (e.g., ADME) and the potential effect the drug has on the body (e.g., lowering the heart rate)
Efficacy and safety studies	Both *in vitro* and *in vivo* establishment of how efficient a given dose is in producing a given therapeutic outcome
Preclinical development steps	Translate potential therapeutic compounds into therapeutic candidates
Process development	As needs for the drug move from grams to large quantities needed for toxicological studies, a commercial manufacturing process needs to be conceptualized to tens or hundreds of liter pilot capacities
Formulation and stability	Stability must be established for the active pharmaceutical ingredient to ensure the integrity throughout the proposed shelf life
Manufacturing	Changes in batch size often call for changes in processing and formulation
Analytical testing	To be able to ensure purity, potency, and consistency of the product
Toxicology	Many studies (~10) are prescribed in regulatory guidelines for IND submission, for example, acute toxicity, mutagenic potential, carcinogenic potential, and so on.
IND submission	Filed with the FDA (USA) prior to clinical trials or studies

Source: Adapted from Carroll, S.F. 2007. In *Drug and Biological Development*, R. Evens (ed.). New York: Springer.

each of these technologies in discovery and preclinical development is further explained below.

2.4.1 GENOMICS AND PROTEOMICS

A genome of an organism is the complete genetic makeup, the entire DNA complement of that organism. Genomics is a somewhat vague term describing the study of the entire genome including the organization, the detection of open reading flames, by genomic sequencing, and the subsequent characterization of the corresponding genes

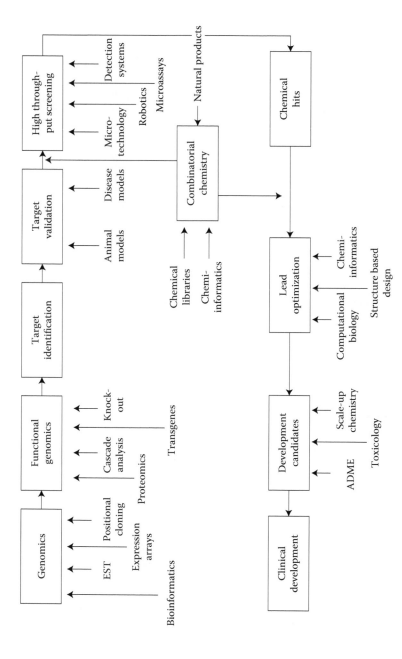

FIGURE 2.4 Target-driven drug discovery paradigm from genomics to clinical development. *Abbreviations:* ADME = absorption, distribution, metabolism, and excretion; EST = expressed sequence tags.

and gene products (Kahl, 2009). The proteome is the complete set of proteins of an organism. And proteomics is the systematic study of the proteome. The term "proteomics" refers to the study of RNA and proteins and to a set of techniques and methods that enable the study. From the discovery point of view, sequencing and analysis of the entire human genome enable us to perform a rapid search for targets and for biomarkers of disease and therapeutic activity. For example, comparative expressions of healthy and diseased subjects may reveal the absence or presence of disease-associated genes (Pienaar et al., 2008). Translating such information into targets is not straightforward, mainly because gene expression results can be quite different from protein expression results (Ideker et al., 2001). Jones and Warren (2006) have provided the following possible reasons for these discrepancies: (1) RNA may get degraded before translation by, for example, RNA interference (RNAi). (2) There are more proteins than genes present in any genome because alternative splicing might give rise to RNAs that are different from the original RNAs, and most importantly, proteins are frequently modified by co- or post-translational modifications. Post-translation modifications (such as phosphorylation and methylation), which occur after the transcript has been translated into a protein, cannot be inferred from transcription profiles. A protein may become a disease target due to the way in which it is modified post-translationally. (3) The disease-causing effect of a protein may be due to its location within the cell, which cannot be predicted from transcripts. (4) Transcripts do not provide any information on protein interactions with nucleotides, phospholipids, ligands, or other proteins that may be important in disease. As such, deciphering the progression information (DNA) to function (protein), referred to as functional genomics, is a critical step to target identification as shown in Figure 2.4.

2.4.2 COMBINATORIAL CHEMISTRY

As the Taxol case suggests, historically, the main source of biologically active compounds in drug discovery has been natural products, isolated from plant, animal, or fermentation sources. With many targets identified through the use of genomics and proteomics tools, another source of compounds to screen against the targets was needed, not only to get around natural product problems such as those in the Taxol case, but also to increase volume and diversity enough to improve chances of identifying compounds from hit to lead. Combinatorial chemistry was a natural solution, where a large collection of structurally distinct molecules that are chemically related, but designed to display a highly diverse combination of chemical reactivities, are synthesized. Such a collection is referred to as a combinatorial library. It should be pointed out that the history of combinatorial chemistry pre-dates the target-driven drug discovery. The paper by Merrifield (1963), which was a study on utilizing functionalized cross-linked polystyrene beads as solid support for the synthesis of peptides, is widely recognized as the pioneering work in this field. For readers interested in the historical development of combinatorial chemistry, Hughes (2006) has provided a table of key developments and references to key papers from 1963 to 1999. As shown in Figure 2.4, combinatorial chemistry comes after the target 3D structure is known, making possible the *de novo* design of a library of compounds. Usually, computational models of the interaction between candidate compounds and the target active site guide the library design. Even after leads are identified, their optimization may call for combinatorial chemistry with the goal of optimizing the structure to meet the criteria for progression to the next development stage. There are many combinatorial libraries that have been documented in the literature, for example, see Miller (2006) and Dolle et al. (2006, 2007).

2.4.3 HTS/uHTS

With genomics and proteomics providing new targets and combinatorial chemistry providing compounds that interact with the targets, a third technology, namely HTS, was crucial in completing the loop. In the initial stages, HTS was synonymous with the use of the now traditional 96-well plates in conjunction with automation. To increase throughput while reducing the reagent volumes needed, higher density plates, for example, 386-, 1536-, and even 3456-well plates, have gained widespread use. A newer term, ultra HTS (uHTS), has come to symbolize the use of high-density well plates, enabling screening in excess of 100,000 samples per day (Wunder et al., 2008).

2.5 THE NEW DISCOVERY PARADIGM PROMISE

Genomics and proteomics have improved the identification of targets or useful points of therapeutic intervention; combinatorial chemistry has generated massive numbers of molecules for testing against the targets; and screening in HTS or uHTS has automated the process of conducting a large number of cell-based biosensors or bioassays. In the early 1990s, when the target-driven drug discovery approach was first implemented, there was excitement in both the pharmaceutical and biotechnological industries with regard to the promise of many new "designer" drugs that were anticipated to enter the development pipeline. Almost two decades later, the promise has not been realized. For example, the effect of the interaction of these three technologies has not reduced the discovery-to-market time substantially from the conventional 10 to 14 years (DiMasi et al., 2003; Kola and Landis, 2004). Additionally, the past several years have seen the lowest number of drugs approved in the industry's history (Owens, 2006) as illustrated in Figure 2.5. From 1990 to 1999, the average number of drugs (new molecular entities, NMEs, and new biologicals, BLAs) was 30.6 versus 21.7 approvals from 2000 to 2008. Given that the implementation of the change in drug discovery and preclinical developments started in the early 1990s, with Gleevec being the first resulting drug to be approved, drug approvals in this decade should have been much higher than in the previous decade as a result of the target-driven discovery paradigm. The questions that need answering are: why there are fewer drug approvals under the new paradigm, and most importantly, where does the industry go from here? The drug discovery pyramid presented in Figure 2.6 is very well known. Since the number of hits and leads has been increased by the new discovery approach, it is reasonable to assume that the new approach is associated with a higher attrition rate. In fact, skeptics have considered the target-driven approach a failed paradigm (Landers, 2004). Three possible explanations have been cited in the literature from industry leaders. First, the period (approximately 20 years) during which the new drug discovery paradigm has been seriously implemented is

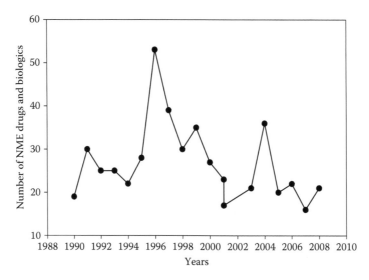

FIGURE 2.5 New molecular entities and biologics approved by the FDA. Biologics include recombinant proteins, monoclonal antibodies, vaccines, and blood products. Figures were obtained from Drugs@FDA, Drug Approval Reports (http://www.accessdata.fda.gov/scripts/cder/drugsatfda/index.cfm). Please note that not all biologics are included in Drugs@FDA Reports. Biologics or biopharmaceuticals include recombinant proteins, monoclonal antibodies, vaccines, blood products, and other biopharmaceuticals.

relatively short. Given the time it takes for a drug to move from discovery to the market, the benefits are ahead of us. The problem with this explanation is that what is in the drug industry pipeline is public knowledge—the drugs are just not there (Lawrence, 2007). Second, the credibility of the targets is low. In other words, target–disease association does not mean that the target is druggable or represents a

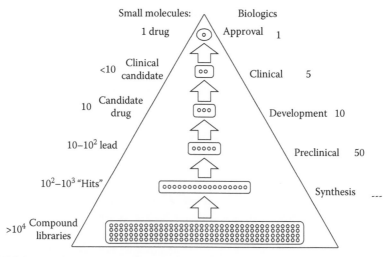

FIGURE 2.6 The drug discovery pyramid.

suitable point of therapeutic intervention (Macarron, 2006). Third, lack of under-standing of, and inattention to, how to culture cells specifically so that they pheno-typically represent their *in vivo* counterparts (Bhadriraju and Chen, 2002).

The general industry response as to how to increase drugs in the market that will help people to live healthier and longer has been well summarized by Macarron (2006), "... keep learning how to best select and validate targets linked with human disease and wisely exploit these recently refined technologies (in addition to and not instead of those previously established)." Two technologies probably referred to by Macarron (2006) in the above quote include high-content screening (HCS) (Proll et al., 2007; Bickle, 2008; Krausz and Korn, 2008) and pathway-based discovery (Hellerstein, 2008; Kruse et al., 2008). HCS provides spatial information of target activity within, on, and between cells. HCS is designed to extract deep biological information using advanced fluorescence protein biosensors or label-free readout. On the other hand, pathway-based discovery is being offered as an alternative paradigm relying on multiple components (or signals) of a pathway as opposed to single target–ligand interaction. Both these approaches can be characterized as information-rich and most likely low-throughput.

2.6 CONCLUDING REMARKS

Friedrich Sertürner isolated the first pharmacologically active pure compound, mor-phine, from a plant (*Papaver somnferum*) over 200 years ago (cited in Li and Vederas, 2009). This marked the beginning of an era of purifying drugs from plants and admin-istering them in prescribed dosages that did not depend on plant source and age. The Taxol case, although placed much later in time, is characteristic of that era that expanded after World War II to include massive screening of microorganisms for new antibiotics, especially after the discovery of penicillin. During the past decade, many pharmaceutical firms have eliminated their natural product research in favor of the new drug discovery approach characterized by the Gleevec case. It is not clear why.

In this chapter, elements of drug discovery and preclinical development have been introduced. Through study cases, the conventional and new target-driven drug discovery and preclinical development paradigms have been compared. The tech-nologies of combinatorial chemistry, genomics and proteomics, and HTS that have underpinned the new approach have been briefly introduced. Using data for approved drugs and biologicals since the implementation of the new approach as evidence, a case has been easily made that the promise of the new approach has not yet been realized. Several explanations of the *status quo* frequently cited in the literature have been repeated. Industry has been reacting by adopting information-rich methodolo-gies such as HCS and pathway-based technologies. Because by nature, information-rich approaches tend to be low-throughput, it is prudent to use them to complement the conventional HTS approaches.

The use of 3D culture in research is exponentially increasing as depicted by the publication trend shown in Figure 2.7. The main explanation is their superior emula-tion of the *in vivo* phenotype in comparison with their 2D counterparts (Cukierman et al., 2001; Albrecht et al., 2006; Pautot et al., 2008; Chang and Hughes-Fulford, 2009). Two applications commonly cited in 3D papers are tissue engineering and drug

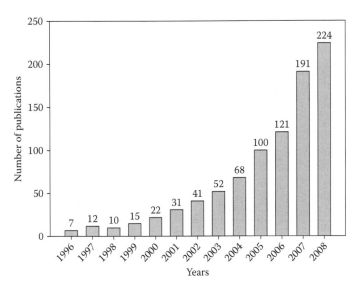

FIGURE 2.7 Publication trends obtained from the Medline database with "3D" and "culture" words/phrases. A more complete collection of 3D culture publications and information can be found at www.3dcellculture.com.

discovery. Based on the physiological relevance that 3D cultures embody, intuitively, 3D cell-based biosensors or cultures should have found widespread use in drug discovery and preclinical development. In addition to HCS and pathway-based discovery, 3D cell-based biosensors represent a third technology with potential to reduce the abnormally higher compound attrition rate. If this is the case, why are 3D cell-based biosensors or cultures not in widespread use in drug discovery and preclinical development? The rest of this book provides answers to this question and more by assembling in one place the latest—from theory to practice—about challenges and opportunities in incorporating 3D cell-based biosensors or cultures in drug discovery and preclinical development programs.

Chapters 3 and 4 (Part II) compare 3D and 2D cultures at the genomic, proteomic, and functional levels to conclusively establish how 3D cultures phenotypically better represent their *in vivo* counterparts and are physiologically more relevant. Chapters 5, 6, and 7 explore how to engineer a microenvironment *in vitro* that would support physiologically relevant phenotypes. Chapter 8 provides an approach to answering a question current knowledge has not answered—do ubiquitous three-dimensionality biomarkers exist and if they do, what are they? Chapters 9 and 10 conclude Part III by introducing readout or detection technology applicability to 3D cell-based biosensor systems as well as 3D commercial systems. In Part IV, Chapter 11 explores challenges to adopting 3D culture systems in current HTS laboratories and Chapter 12 uses study cases to highlight the opportunities. The book concludes with Chapter 13 that suggests a study for further solidifying the opportunities 3D culture may offer in drug discovery and preclinical development.

REFERENCES

Abal, M. and Andreu, J.M. 2003. Taxanes: Microtubule and centromere targets, and cell cycle dependent mechanism of action. *Curr. Cancer Targets* 3: 193–203.

Alberts, B., Bray, D., Lewis, J., Raff, M., and Roberts, K. 1989. *Molecular Biology of the Cell*, 5th edition, pp.765. New York: Garland Science/Taylor and Francis.

Albrecht, D.R., Underhill, G.H., Wassermann, T., Sah, R., and Bhatia, S. 2006. Probing the role of multicellular organization in three-dimensional microenvironments. *Nat. Methods* 3(5): 369–375.

Anonymous. 1995. Names for hi-jacking. *Nature* 373: 370.

Berrieman, H.K., Lind, M.J., and Cawkwell, L. 2004. Do β-tubulin mutations have a role in resistance to chemotherapy? *Lancent Oncology* 5: 158–164.

Bhadriraju, K. and Chen, C.S. 2002. Engineering cellular microenvironment to improve cell-based drug testing. *Drug Discov. Today* 7(11): 612–620.

Bickle, M. 2008. High-content screening: A primary screening tool? *IDRUGS* 11(11): 822–826.

Carroll, S.F. 2007. Discovery and nonclinical development. In *Drug and Biological Development*, R.P. Evens (ed.), pp. 84–106. New York: Springer.

Chang, T.T. and Hughes-Fulford, M. 2009. Monolayer culture of human liver hepatocellular carcinoma cell line cells demonstrate distinct global gene expression patterns and functional phenotypes. *Tissue Eng. Part A* 15(2): 559–567.

Chauvière, G., Guénard, D., Picot, F., Sénilh, V., and Potier, P. 1981. Analyze Structurale et étude biochimique de produits isolés de l'If.*Comptes rendus de l'Académie des Sciences de Paris, Série II* 293: 501–503.

Chesnoff, S. 1995. The use of Taxol as a trademark. *Nature* 374: 208.

Chorghade, M.S. (ed.). 2006. *Drug Discovery and Development*, Vol. 1. Hoboken, New Jersey: Wiley.

Cukierman, E., Pankov, R., Stevens, D.R., and Yamada, K. 2001. Taking cell-matrix adhesion to the third dimension. *Science* 294(5547): 1708–1712.

Daley, G.Q., McLaughlin, L., Witte, O.N., and Baltimore, D. 1987. The CML-specific P210 bcr/abl protein, unlike v-abl does not transform NIH/3T3 fibroblasts. *Science* 237: 532–535.

Dennis, J.-N., Greene, A.E., Guenard, D., Gueritte-Voegelein, F., Mangatal, L., and Potier, P. 1988. A highly efficient practical approach to natural Taxol. *J. Am. Chem. Soc.* 110: 5917–5919.

DiMasi, J.A., Hansen, R.W., and Grabowski, H.G. 2003. The price of innovation: New estimates of drug development costs. *J. Health Econ.* 22: 151–185.

Dolle, R.E., Le Bourdonnec, B., Morales, G.A., Moriarty, K.J., and Salvino, J.M. 2006. Comprehensive survey of combinatorial library synthesis: 2005. *J. Comb. Chem.* 8(5): 597–635.

Dolle, R.E., Le Bourdonnec, B., Goodman, A.J., Goodman, A.J., Morales, G.A., Salvino, J.M., and Zhang, W. 2007. Comprehensive libraries for drug discovery and chemical biology: 2006. *J. Comb. Chem.* 9(6): 855–902.

Druker, B.J., Tamura, S., Buchdunger, E., Ohno, S., Segal, G.M., Fanning, S., Zimmerman, J., and Lydon, N.B. 1996. Effects of a selective inhibitor of the Abl tyrosine kinase on the growth of Bcr-Abl positive cells. *Nat. Med.* 2(5): 561–566.

Evens, R.P. (ed.), 2007. *Drugs and Biological Development*. New York: Springer.

Franklin, J.F., Cromack, K., Jr., Denison, W., McKee, A., Maser, C., Sedell, J., Swanson, F., and Juday, G. 1981. *Ecological Characteristics of Old-Growth Duglas-fir Forests*. United States Department of Agriculture, Forest Service, Pacific Northwest Forest and Range Experiment Station, General Technical Report PNW-118.

Fuchs, D.A. and Johnson, R.K. 1978. Cytological evidence that taxol, an antineoplastic agent from *Taxus brevifolia*, acts as a mitotic spindle poison. *Cancer Treat. Rep.* 62: 750–751.

Gagné, G. 1993. Biolyse pourrait s'emparer du marché de Taxol grâce à l'if du Canada. *Le Solil (Québec)*, November 13.

Goodman, J. and Walsh, V. 2001. *The Story of Taxol: Nature and Politics in Pursuit of Anti-Cancer Drug.* New York: Cambridge University Press.

Grant, S.K. 2009. Therapeutic protein kinase inhibitors. *Cell. Mol. Life Sci.* 66: 1163–1177.

Gueritte, F. 2001. General and recent aspects of the chemistry and structure–activity relationships of taxoids. *Curr. Pharmacol. Des.* 7: 1229–1249.

Hartwell, J.L. 1969. Plants used against cancer: A survey. *Lloydia* 30: 379–436.

Hellerstein, M.K. 2008. A critique of the molecular target-based drug discovery paradigm based on principles of metabolic control: Advantages of pathway-based discovery. *Metab. Eng.* 10: 1–9.

Holton, R.A., Somoza, C., Kim, H.B., et al. 1994a. First total synthesis of taxol: 1. Functionalization of the B-ring. *J. Am. Chem. Soc.* 116: 1597–1598.

Holton, R.A., Somoza, C., Kim, H.B., et al. 1994b. First total synthesis of taxol: 2. Completion of the C-ring and D-ring. *J. Am. Chem. Soc.* 116: 1599–1600.

Hughes, I. 2006. Combinatorial chemistry in the drug discovery process. In *Drug Discovery and Development*, M.S. Chorghade (ed.), Vol. 1, pp. 129–167. Hoboken, New Jersey: Wiley.

Ideker, T., Thorsson, V., Ranish, J., Christmas, R., Buhler, J., Eng, J.K., Bumgarner, R., Good lett, D.R., Aebersold, R., and Hood, L. 2001. Integrating genomic and proteomic analyses of a systematically perturbed metabolic network. *Science* 292(5518): 929–934.

Imming, P., Sinniger, C., and Meyer, A. 2006. Drugs, their targets and the nature and number of drug targets. *Nat. Rev. Drug Discov.* 5: 821–834.

Jones, S.D. and Warren, P.G. 2006. Proteomics and drug discovery. In *Drug Discovery and Development*, M.C. Chorghade (ed.), Vol. 1, pp. 233–271. Hoboken, New Jersey: Wiley.

Jordan, M.A. and Wilson, L. 2004. Microtubules as targets for anticancer drugs. *Nat. Rev. Cancer* 4: 253–265.

Kahl, G. 2009. *The Dictionary of Genomics, Transcriptomics and Proteomics.* Hoboken, New York: Wiley-Blackwell.

Kantarjian, H., Sawyers, C., Hochhaus, A., et al. 2002. Hematologic and cytogenetic responses to imatinib mesylate in chronic myelogenous leukemia. *N. Engl. J. Med.* 346(9): 645–652.

Kola, I. and Landis, J. 2004. Can the pharmaceutical industry reduce attrition rates? *Nat. Rev. Drug Discov.* 3: 711–715.

Krausz, E. and Korn, K. 2008. High-content siRNA screening for target identification and validation. *Expert Opin. Drug Discov.* 3(5): 551–564.

Kruse, U., Bantscheff, M., Drewes, G., and Hopf, C. 2008. Chemical and pathway proteomics—powerful tools for oncology drug discovery and personalized health care. *Mol. Cell. Proteomics* 7(10): 1887–1901.

Landers, P. 2004. Human element: Drug industry's big push into technology fails short—testing machines were built to streamline research but may be stifling it—officials see payoff after 2010. *Wall Street J.* February 24.

Landry, Y. and Gies, J.-P. 2008. Drugs and their molecular targets: An updated overview. *Fundam. Clin. Pharmacol.* 22: 1–18.

Lawrence, S. 2007. Pipelines turn to biotech. *Nat. Biotechnol.* 25(12): 1342.

Li, J.W.-H. and Vederas, J. 2009. Drug discovery and natural products: End of an era or an endless frontier? *Science* 325(5937): 161–165.

Macarron, R. 2006. Critical review of the role of HTS in drug discovery. *Drug Discov. Today* 11(7/8): 277–279.

Manning, G., Whyte, D.B., Martinex, R., Hunter, T., and Sudarsanam, S. 2002. The protein kinase complement of the human genome. *Science* 298: 1912–1934.

Merrifield, R.B. 1963. Solid phase peptide synthesis. 1. The synthesis of a tetrapeptide. *J. Am. Chem. Soc.* 85: 2149–2154.

McGrogan, B.T., Gilmartin, B., Carney, D.N., and McCann, A. 2008. Taxanes, microtubules and chemoresistant breast cancer. *Biochim. Biophys.* 1785: 96–132.

Miller, J.L. 2006. Recent developments in focused library design: Targeting gene-families. *Curr. Top. Med. Chem.* 6(1): 19–29.

Neuss, N., Gorman, M., and Johnson, I.S. 1967. Natural products in cancer chemotherapy. *Methods in Cancer Research*, pp. 633–702. New York: Academic Press.

Neuss, N., Johnson, I.S., Armstrong, J.G., and Jansen, C.J. 1964. The *vinca* alkaloids. *Adv. Cancer Chemother.* 1: 133–174.

Ng, R. 2004. *Drugs: From Discovery to Approval.* Hoboken, New Jersey: Wiley.

Ng, R. 2009. *Drugs: From Discovery to Approval.* Hoboken, New Jersey: Wiley.

Noble, R.L., Beer, C.T., and Cutts, J.H. 1958. Role of chance observations in chemotherapy: *Vinca rosea. Ann. N. Y. Acad. Sci.* 76: 882–894.

Nowell, P. and Hungerford, D. 1960. A minute chromosome in human chronic granulocytic leukemia [abstract]. *Science* 132: 1497.

Nowell, P.C. and Hungerford, D.A. 1961. Chromosome studies in human leukemia. II. Chronic granulocytic leukemia. *J. Nat. Cancer Inst.* 27: 1013–1035.

Owens, J. 2007. 2006 drug approvals: finding the niche. *Nat. Rev. Drug Discov.* 6: 99–101.

Pautot, S., Wyart, C., and Isacoff, E.Y. 2008. Colloid-guided assembly of oriented 3D neuronal networks. *Nat. Methods* 5(8): 735–740.

Perdue, RE., Jr. and Hartwell, J.L. 1969. The search for plant sources of anticancer drugs. *Morris Arboretum Bull.* 20: 35–53.

Pienaar, I.S., Daniels, M.U., and Götz, J. 2008. Neuroproteomics as a promising tool in Parkinson's disease research. *J. Neural. Transm.* 115: 1413–1430.

Proll, G., Steile, L., Proll, F., Kumpf, M., Moehrle, B., Mehlmann, M., and Gauglitz, G. 2007. Potential of label-free detection in high-content-screening applications. *J. Chromatogr. A* 1161(1–2): 2–8.

Rowley, J.D. 1980. Ph1-positive leukaemia, including chronic myelogenous leukaemia. *Clin. Haematol.* 9: 55–86.

Schiff, P.B., Fant, J., and Horwitz, S.B. 1979. Promotion of microtubule assembly *in vitro* by taxol. *Nature* 277: 665–667.

Smith, L.J. 2007. Types of clinical studies. In *Drug and Biological Development*, R.P. Evens (ed.), pp. 107–122. New York: Springer.

Svoboda, G.H. 1961. Alkaloids of *Vinca rosea* (*Catharanthus roseus*). IX. Extraction and characterization of leurosidine and leurocristine. *Llyodia* 24: 173–178.

Wani, M.S., Taylor, H.L., Wall, M.E., Coggon, P., and McPhail, A.T. 1971. Plant antitumor agents. VI. The isolation and structure of taxol, a novel antileukemic and antitumor agent from *Taxus brevifolia. J. Am. Chem. Soc.* 93: 2325–2327.

Wender, P.A., Badham, N.F., Conway, S.P., et al. 1995. The pinene path to taxanes: Genesis and evolution of a strategy for synthesis. In *Taxane Anticancer Agents: Basic Science and Current Status*, G.I. Georg, T.T. Chen, I. Ojima, and D.M. Vyas (eds), pp. 326–339. Washington, DC: American Chemical Society.

Witherup, K.M., Look, S.A., Stasko, M.W., Ghiorzi, T.J., and Muschik, G.M. 1990. *Taxus* spp. Needles contain amounts of taxol comparable to the bark of *Taxus brevifolia*: Analysis and isolation. *J. Nat. Prod.* 53: 1249–1255.

Wunder, F., Kalthof, B., Müller, T., and Hüser, J. 2008. Functional cell-based assays in microliter volumes for ultra-high throughput screening. *Comb. Chem. High Throughput Screening* 11: 495–504.

Vasella, D. and Slater, R. 2003. *Magic Cancer Bullet: How a Tiny Orange Pill is Rewriting Medical History*. New York: HarperCollins.

Vidensek, N., Lim, P., Campbell, A., and Carlson, C. 1990. Taxol content in bark, wood, root, leaf, twig, and seedling from several *Taxus* species. *J. Nat. Prod.* 53: 1609–1610.

Yaish, P., Gazit, A., Gilon, C., and Levitzki, A. 1988. Blocking of EGF-dependent cell proliferation by EGF receptor kinase inhibitors. *Science* 242: 933–935.

Zhou, J. and Giannakakou, P. 2005. Targeting microtubules for cancer chemotherapy. *Curr. Med. Chem. Anticancer Agents* 5: 65–71.

Part II

3D versus 2D Cultures

3 Comparative Transcriptional Profiling and Proteomics

3.1 TRANSCRIPTIONAL PROFILING STUDIES

Transcriptional profiling in tissue engineering studies that compare *in vitro* or *in vivo* 3D to *in vitro* 2D samples can be subdivided into three categories. Studies belonging to the first category tend to focus on a specific function. For example, Olsavsky et al. (2007) cultured primary hepatocytes and hepatome-derived cell lines HepG2 and Huh7 in a 2D sandwich system, with collagen I as the substratum together with a dilute extracellular matrix (Matrigel™) overlay in a defined serum-free medium supplemented with dexamethasone at the nanomolar level. Although the study utilized the Affymetrix Human Genome U133Plus 2.0 array, only liver-specific function gene expressions were reported. Studies belonging to the second category tend to focus on profiling transcriptions only from cells cultured in 3D scaffolds without a 2D control. For example, Gelain et al. (2006) compared the expression of adult mouse neural stem cells cultured in Matrigel to those cultured in novel peptide scaffolds. The objective was to validate the new scaffolds as opposed to comparing 2D and 3D cultures. In a second example, Ikebe et al. (2007) examined the gene expression profile of an immortalized human keratinocyte, HaCaT, cultured in a neutralized type I collagen gel, with and without exposure to air for 24 h. The objective was to better understand why an air–liquid interface culture initiated the emulation of *in vivo* tissue, and as such, a 2D control was not included in the globe gene expression study. Keratinocytes make stratified epidermoid structures when cultured at an air–liquid interface and have been successfully used in 3D culture studies for more than 25 years (Boyce and Hansbrough, 1988). Studies belonging to the third category not only focused on global gene expression profiles, but also compared expressions in 2D to those in 3D formats. A summary of these studies is compiled in Table 3.1. Although the objectives are all different, a common theme seems to be better understanding the underlying reasons for morphological and functional differences between the two cultures. The purpose of this chapter is to generate insights into the same question with all the studies combined. It was possible to obtain raw data only from Li et al. (2002), Boess et al. (2003), Ghosh et al. (2005), Kenny et al. (2007), and Myers et al. (2008).

Two contemporary microarray data analysis approaches were considered: (1) gene ontology (GO) analysis (Liu et al., 2003; Sorensen et al., 2008) using the Affymetrix NetAffx website (http://www.affymetrix.com/analysis/index.affx), which assigns genes to molecular functional groups (Box 3.1), and (2) ingenuity pathway analysis (IPA)

TABLE 3.1

Summary of Comparative 3D/2D Global Gene Expression Studies

Main Study Objective	Cell Type	2D Format	3D Scaffold or Configuration	Microarray	Reference
Effect of gravity on development and function	Hepatoblastoma cell line, HepG2	Monolayer[a]	Microgravity bioreactor	UniGem V array (6144 human probe set)	Khaoustov et al. (2001)
Molecular basis of the modulation of cells phenotypes by ECM	Human aortic smooth muscle cells	Monolayer	Collagen matrix (3D geometry gel)	Nylon membrane (9600 human probe set)	Li et al. (2002)
Effect of time in culture and *in vivo* modeling	Rat liver cell lines BRL3A and NRL and primary cells	Cell lines monolayer and sandwich cultures[b]	Liver slices (primary)	Affymetrix U34A array (8799 rat probe set)	Boess et al. (2003)
Better understanding of scaffold impact	Human IMR-90 fibroblasts	Monolayer polystyrene	Collagen—glycosaminoglycan (3D geometry gel)	Affymetrix U133A array (22,200 human probe set)	Klapperich and Bertozzi (2004)
Differentiation of stem cells to osteoblasts	Human mesenchymal stem cells	Monolayer	Thermal-reversible gelation polymer (3D geometry gel)	Clonetech Atlas glass human 3.8	Hishikawa et al. (2004)
Better understanding of resistance to cancer drug treatment	NA-8 melanoma cells	Standard 2D (monolayer)	Multicellular tumor spheroids (MCTS)	Affymetrix HG-U-133A array (>20,000 probe set)	Ghosh et al. (2005)
Better understanding of the effect of 3D scaffold on hematopoietic differentiation	Mouse embryonic stem cells	Monolayer[a]	Cytomatrix™ (highly porous tantalum-based scaffold (static and in spinner flask)	Array fabricated by Professor Vishwanath Iyer, University of Texas, Austin (36,880 mouse probe set)	Liu et al. (2006)
Understanding the regulatory role of extracellular matrix in expansion of CD34+ cord blood cells in culture	CD34+ hematopoietic stem/ progenitors	In suspension	Collagen I (3D geometry gel)	Affymetrix HG-U-133A array (>20,000 probe set)	Oswald et al. (2006)

Aim	Cell type	2D culture	3D culture	Array	Reference
Gain more rigorous comprehension of cellular and molecular mechanisms underlying axon growth in physiologically relevant 3D geometry	Human neuroblastoma cell line SH-SY5Y	Monolayer of standard tissue culture dishes coated with collagen I or Matrigel	Collagen I and Matrigel matrices (3D geometry gel)	Affymetrix HG-U-133A array	Li et al. (2007)
To identify pathways relevant to cardiac tissue engineering	Primary neonatal rat heart cells	Standard tissue culture plates without rotation	On solid microcarrier surfaces within clinostatically rotated polytetrafluoroethylene vessels	Affymetrix rat genome RG-U34A array	Akins et al. (2007)
Large-scale comparison of the transcriptional profiles of a panel of human breast cancer cell lines	HCC1550, MCF-12A, MDA-MB-415, MPE-600, S1, BT-474, BT-483, HCC70, HCC1569, MCF-7, TA-2, T-47D, AU565, CAMA-1, MDA-MB-361, MDA-MB-456, MDA-MB-468, SH-BR-3, UACC-812, ZR-75-1, ZR-75-B, BT-549, Hs578T, MDA-MB-231, MDA-MB-436	Standard tissue culture dishes coated with collagen I	Laminin-rich extracellular matrix (3D geometry gel)	Affymetrix high-throughput array (22,215 probes)	Kenny et al. (2007)
To narrow the gap between neuronal cell lines and primary neurons	Human neuroblastoma SH-SY5Y and rat pheochromocytoma PC12	Standard tissue culture dishes, dishes for PC12 were coated with collagen	On Cytodex-3™ microcarrier beads in a 50 mL rotary wall vessel	Stanford Functional Genomics Facility-supplied Human Exonic Evidence-based Oligonucleotide (HEEBO) array (44,544 probe set)	Myers et al. (2008)

[a] Substrate not reported, suggesting standard polystyrene tissue culture plates.

[b] Cells were seeded on collagen-coated polystyrene plates and after an attachment period of 3 h, the medium was removed and cells were covered with a thin layer of collagen and supplemented with medium after setting.

BOX 3.1   GO ANNOTATION

GO is a major bioinformatics initiative to unify the representation of gene and gene products attributes across all species (http://www.geneontology.org). The major aims are as follows:

1. Maintain and further develop controlled vocabulary of genes and gene products
2. To annotate genes and gene products and assimilate and disseminate annotation data
3. To provide tools to facilitate access to all aspects of the data provided by the GO Project

GO molecular functions are constantly being updated. At the time of the analysis presented in this chapter, the following was the list of functions:

1. Binding
2. Catalytic activity
3. Molecular transducer activity
4. Transcription regulator activity
5. Enzyme regulator activity
6. Transporter activity
7. Structural molecular activity
8. Electron carrier activity
9. Translation regulator activity
10. Auxiliary transport protein activity
11. Chemorepellent activity
12. Antioxidant activity
13. Chemoattractant activity
14. Metallochaperone activity
15. Nutrient reservoir activity
16. Proteasome regulator activity
17. Protein tag

(Ingenuity® Systems, http://www.ingenuity.com/), which determines the molecular networks activated. Given that the raw data were pooled from different sources, GO analysis was considered more likely to yield more meaningful results.

3.2   COMPARATIVE GO ANNOTATION ANALYSIS

The molecular function distribution of the differentially expressed genes at the transcription level is presented in Figure 3.1. Figure 3.1a shows the GO annotation of the 99 probes found to be significantly upregulated or downregulated by Li et al. (2002). Of the genes included in the analysis, 77 were upregulated more than twofold and 22

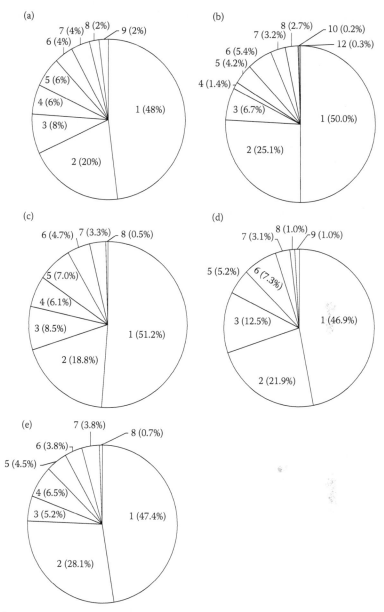

FIGURE 3.1 GO analysis of differentially expressed genes with data from (a) Li, S. et al. 2002. *FASEB J.* 16(13): 97–t (November 15) express article: doi: 10.1096/fj.02-0256fje1; (b) Boess, F. et al. 2003. *Toxicol. Sci.* 73: 386–402; (c) Ghosh, S. et al. 2005. *J. Cell. Physiol.* 204: 522–531; (d) Kenny, P.A. et al. 2007. *Mol. Oncol.* 1: 84–96; (e) Myers, T.A. et al. 2008. *J. Neurosci. Methods* 174: 31–41. Molecular functions are as follows: (1) binding, (2) catalytic activity, (3) molecular transducer activity, (4) enzyme regulator activity, (5) transcription regulator activity, (6) transporter activity, (7) structural molecule activity, (8) auxiliary transport protein activity, (9) transcription regulator activity, (10) electron carrier activity, (11) chemorepellent activity, and (12) motor activity.

were expressed less than one-half in comparison with 2D culture. Figure 3.1b shows a total of 69 probe sets found to be significantly up- or downregulated in slices at least sixfold in comparison with monolayer or sandwich cultures. Of the 69 genes, 59 were upregulated and 10 were downregulated. In Ghosh et al.'s spheroid culture study, cells of the metastatic melanoma NA8 cell line were cultured as tumor spheroids and the gene expression profile was compared with NA8 cells cultured on 2D flat surfaces. Figure 3.1c shows 107 and 74 probe sets that were upregulated and downregulated, respectively, by at least threefold, in comparison with 2D cultures. The GO annotation from the Kenny et al. (2007) study is shown in Figure 3.1d. This is a very comprehensive study in that it involved 25 breast tumor cell lines of four different 3D morphologies (round, mass, grape-like, and stellate). With a cutoff of $p = 0.00025$, 96 probe sets were found to be differentially expressed. Figure 3.1e shows the GO annotation analysis of SH-SY5Y neuroblastoma cell line cultured in a 50 mL rotating wall vessel (Synthecon). In this system 3D cell aggregates are formed in suspension with the assistance of Cytodex-3™ beads (Amersham Biosciences). Genes whose expression changed by 1.5-fold, with a corrected t-test, $p < 0.05$, were counted as differentially expressed. With this "low" cutoff (in comparison with the other studies included in this chapter), Myers et al. (2008) identified 700 genes.

All studies utilized Affymetrix U133A or U133A plus two microarray chips. Of the 17 major types of molecular functions, only 9 were found in the differentially expressed genes (2D versus 3D), and 6 of them are consistently found in all the different 3D cell culture studies. The distributions of the genes are similar in these three independent studies with different 3D cell culture systems and different cell types. Despite differences in cell types, 3D scaffold material and architecture, the top three molecular functions of binding, catalytic activity, and molecular transducer activity were almost identical. The fact that the same top three functions were found with liver slices after only 6 h in culture is significant in that liver slices can be considered to be closest to *in vivo* conditions, although 6 h is long enough for gene expression changes due to culture conditions to begin to occur.

Similar studies were extended to stem/neural progenitor cells on standard polystyrene culture plates (2D) and scaffolds (3D) as well as neural spheres in the author's laboratory (Lai et al., 2009a). Figure 3.2 shows GO annotation analysis results for 3D polystyrene scaffolds versus 2D plates, before (a) and after (b) differentiation, and *in vivo* surrogate (neural spheres) versus 2D plates (c). Hardly any differentially expressed genes were found at acceptable $q < 0.05$. The q-value is defined as the lowest "false discovery rate" at which the gene is called significantly differentially expressed; it is based on the work of Storey (2002). To generate genes for comparative purposes, the cutoff q-value was relaxed to less than 0.2. Figure 3.2b shows GO annotation analysis for neural progenitor cells cultured in polystyrene 3D scaffolds with 146 probe sets differentially expressed ($q < 0.05$). 95% of the genes were downregulated. Figure 3.2c shows GO annotation analysis results for neural spheres versus 2D plates. A high number of differentially expressed genes (1046) were found with neural spheres ($q < 0.0005$) with 65% downregulated. The neural progenitor results were consistent with previous studies highlighted in Figure 3.1.

The six common molecular functions from GO annotation are further summarized in Table 3.2. As shown, most of these genes were downregulated in 3D. The percentages

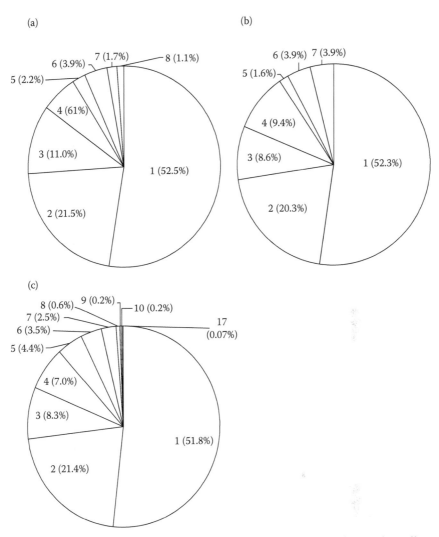

FIGURE 3.2 GO analyses of differentially expressed genes from neural progenitor cells on 2D polystyrene substrate, and in 3D polystyrene scaffold, and neural spheres: (a) 3D versus 2D before differentiation, (b) 3D versus 2D after differentiation, and (c) neural spheres versus 2D. Molecular function activities are as follows: (1) binding, (2) catalytic activity, (3) molecular transducer activity, (4) enzyme regulator activity, (5) transcription regulator activity, (6) transporter activity, (7) structural molecule activity, (8) auxiliary transport protein activity, (9) transcription regulator activity, (10) electron carrier activity, (11) chemorepellent activity, and (12) motor activity, and (17) protein tag.

of changed genes over total genes in the same molecular function category are associated with the total differentially expressed genes. Within the same study, these percentages for different molecular function are not significantly different, indicating that the dominancy of genes belonging to binding and catalytic activity is not due to the larger number of total genes in the chip belonging to these categories.

TABLE 3.2
GO Molecular Function Comparison

Molecular Function (MF)	Breast Cancer Cells (Kenny et al., 2007)		Melanoma Cells (Ghosh et al., 2005)		NP Cells (Lai et al., 2009a)		SCG Cells (Lai et al., 2009b)	
	% of Changed Genes	% of Downregulated Genes in 3D	% of Changed Genes	% of Downregulated Genes in 3D	% of Changed Genes	% of Downregulated Genes in 3D	% of Changed Genes	% of Downregulated Genes in 3D
Binding	0.29	86.7	0.70	40.4	0.26	97.0	2.3	83.4
Catalytic activity	0.31	85.7	0.57	51.2	0.22	96.2	2.1	87.6
Enzyme regulator activity	0.43	100	1.28	40.0	0.1	100	2.6	92.3
Molecular transducer activity	0.41	91.7	0.61	50.0	0.25	100	2.4	71.4
Structural molecule activity	0.28	100	0.65	42.9	0.32	100	2.3	85.2
Transporter activity	0.44	85.7	0.63	30.0	0.20	100	1.5	58.3

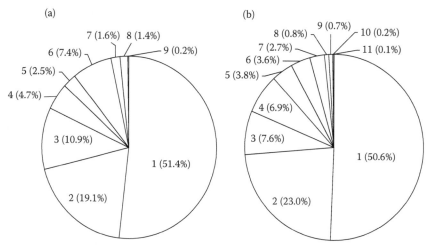

FIGURE 3.3 GO analysis of differentially expressed genes from mouse superior cervical ganglion (SCG) neurons cultured on 2D polystyrene plates, 3D polystyrene scaffold, and freshly dissected (*in vivo* surrogate): (a) 3D versus 2D and (b) freshly dissected versus 2D cells. Molecular function activities are as follows: (1) binding, (2) catalytic activity, (3) molecular transducer activity, (4) enzyme regulator activity, (5) transcription regulator activity, (6) transporter activity, (7) structural molecule activity, (8) auxiliary transport protein activity, (9) transcription regulator activity, (10) electron carrier activity, (11) chemorepellent activity, and (12) motor activity. All others below 0.1%.

To further ascertain the dominating molecular function categories and the down-regulation trends outlined above, in a system very close to *in vivo* conditions, Lai et al. (2009b) freshly dissected mouse superior cervical ganglions following published procedures (Mahantheppa and Patterson, 1998). After dissecting, half of the ganglions were enzymatically digested with collagenase and the cells were cultured in polystyrene scaffolds and control 2D plates for 5 days before RNA extraction. For the remaining half of the ganglion, RNA was extracted immediately to preserve the *in vivo* transcriptome. The GO annotation analysis results are shown in Figure 3.3, consistent with previous findings outlined before. Figure 3.4 shows the distribution of the molecular functions in the GO databases for all genes (a), human genes (b), and mouse genes (c). As expected, the dominant molecular functions correlated—it is reasonable to expect to find more differentially expressed genes from a larger pool than a small one. On the other hand, a small gene pool molecular function that yields a proportionally larger number of differentially expressed genes probably offers more interesting possibilities and is worthy of further study as discussed in Section 3.4.

3.3 PROTEOMICS STUDIES

The question needing an answer was whether the observations made in the previous section were also true at the protein level. According to Kahl (2004, p. 873), proteome refers to "the complete set of proteins in a cell or organ at a given time," while

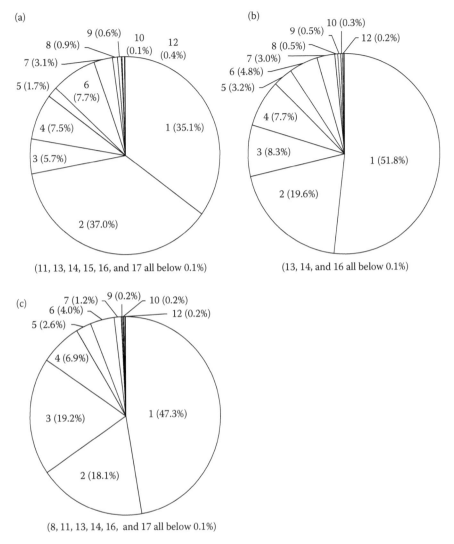

FIGURE 3.4 Molecular function distribution of genes in GO databases: (a) all genes, (b) all human genes, and (c) mouse genes.

proteomics refers to "the whole repertoire of techniques to analyze and characterize the proteome of organelle or cell," (p. 875) including protein isolation, fractionation, separation, postseparation analysis (e.g., identification of different proteins by microsequencing), analysis of posttranslation modification by matrix-assisted laser desorption ionization (MALDI) mass spectrometry, bioinformatics (storage and analysis of resulting information), and robotics (automation for high-throughput analysis). The term "proteomics" is analogous with genomics, the study of genes. The word "proteome" is a conjugation of "protein" and "genome" meaning the entire complement of proteins, including the modifications made to a particular set of proteins, produced by an organism or system. Although viral infection can add genetic material to a cell and cancer cells can undergo rapid mutation, transcriptions, and expressions as the tumor grows and develops, the genome of healthy cells is reasonably constant. This has made it relatively easy to study whole genomes of organisms including human (International Human Genome Sequencing Consortium, 2001). Proteomics confirms the presence of the protein and provides a direct measure of the quantity present, originally detected by measuring mRNA. Although the availability of the human genome sequence told researchers how many proteins are likely to be out there and the exact sequence of the amino acids they should look for, finding these proteins has been a challenge for several reasons: (1) because proteins are chemically far more heterogeneous and complex than DNA and RNA, no single machine could tell researchers everything they wanted to know about a protein; (2) abundance of proteins varies widely such that plentiful proteins can mask the signals of the rare counterparts; (3) there are more proteins than genes because proteins can be spliced together from multiple genes and, once synthesized, can later be cut down in size or modified with other chemical groups; (4) mRNA is not always translated into proteins, and the amount of protein produced by a given amount of mRNA depends on the gene it is transcribed from and on the current physiological state of the cell. This explains why the full-scale human proteome project has not yet happened almost a decade after sequencing the human genome.

A key practical application of proteomics is in drug discovery. If a certain protein is implicated in a disease (target), its structure provides information that can enable the design of drugs that can activate/inactivate or interfere with the undesirable action of the protein. A second interesting application of proteomics is using specific proteins as biomarkers for disease diagnosis or physiological state. It is in this context that the author wondered if comparative 2D–3D proteomics accomplished to date can provide insights that complement the comparative transcriptome results reported above.

A thorough search of the literature yielded only four oncology-related studies (Poland et al., 2002; Gaedtke et al., 2007; Kim et al., 2007; Kumar et al., 2008), probably due to reasons outlined above. These studies compared proteomes of multicellular tumor spheroids (MCTS) (3D) and monolayer (2D) cultures. Generally, cells from cultures were solubilized and subjected to 2D electrophoresis. Image analysis software was used to identify differentially expressed proteins that were excised and identified using various types of MALDI mass spectrometry. Differential expression criteria varied from 2.0- to 2.5-fold (for more details on proteomics procedures, see Chapter 8). The results from these studies are summarized in Table 3.3 and Figure 3.5.

3.4 CONCLUDING REMARKS

What does this chapter teach or the question can be put in a different way: what is the take-home message? One might begin to answer the above question with a question: what did one want to learn before reading the chapter. The author had two goals in writing this chapter. The first was obvious and involves informing the reader about the differences between 2D and 3D cultures at transcriptional and protein levels. The chapter is reasonably successful in reaching this goal. The second, less obvious goal,

TABLE 3.3

Results of Differentially Expressed Proteins between MCTS and Monolayer Cultures

Protein	Accession Number	Molecular Weight (kDa)	Expression in 3D/2D	Functional Association	GO Annotation	Reference
14-3-3 β	P31946	28.1	Decrease	Mediates signal transduction by binding to phosphoserine-containing proteins	Binding	Poland et al. (2002)[a]
				Regulation of neuronal development		
14-3-3 η	Q04917	28.2	Decrease	Cell growth control	Binding; transcription regulator activity	Poland et al. (2002)[a]
α-Tubulin	P68366	50	Decrease	Components of microtubules	Binding	Poland et al. (2002)[a]
				Essential for cell transport and cell division		
β-Tubulin	P07437	49.7	Decrease		Binding	Poland et al. (2002)[a]
Calreticulin precursor	P27797	46.9	Increase	Molecular chaperone	Binding	Poland et al. (2002)[a]
				Calcium-binding properties, regulation of calcium homeostasis		
				Extracellular lectin		
				Intracellular mediator of integrin function		
				Inhibitor of steroid hormone-regulated gene expression		
				C1q-binding protein		
Cytokeratin 7	P08729	51.2	Increase	Intermediate filament proteins in epithelial cells	Binding; structural molecule activity	Poland et al. (2002)[a]
				CK8 and CK18 are coordinately expressed as cytoskeletal components, involved in chemoresistance		
				Expression of CK19 is increased by vitamin A treatment		

Protein	Accession	Value	Change	Function	Activity	Reference
Cytokeratin 8	P05787	53.7	Increase		Binding; structural molecule activity	Poland et al. (2002)[a]
Cytokeratin 19	P08727	44.1	Increase		Binding; structural molecule activity	Poland et al. (2002)[a]
Cytokeratin 19 var.			Increase			Poland et al. (2002)[a]
ErbB3 binding protein EBP1	Q9UQ80	43.8	Decrease	Binds to nonphosphorylated ErbB3 Regulation by protein kinase C	Binding; transcription regulator activity	Poland et al. (2002)[a]
Glutathione S-transferase pi (isoform) GST-pi	Q28514	23.4	Decrease	Drug metabolism (conjugation of reduced glutathione to a wide number of exogenous and endogenous hydrophobic electrophiles)	Catalytic activity	Poland et al. (2002)[a]
ρ GDI, GDP dissociation inhibitor (homologous to)	P52565	23.2	Decrease	Cell signaling Proliferation Cytoskeletal organization Secretion	Binding; enzyme regulator activity	Poland et al. (2002)[a]
ρ GDI, GDP dissociation inhibitor (homologous to) var.			Increase	Cell signaling Proliferation Cytoskeletal organization Secretion		Poland et al. (2002)[a]
Thioredoxin peroxidase AO372, peroxiredoxin 4	Q13162	30.5	Increase	Regulates NFkB activation Protects thiol groups against oxidation Prevents ROS production by EGF and p53	Antioxidant activity; catalytic activity	Poland et al. (2002)[a]
Acidic calponin	JC4501	36.6	Decrease	Regulation and modulation of smooth muscle contraction The interaction of calponin with actin inhibits the actomyosin Mg–ATPase activity	Binding	Gaedtke et al. (2007)[b]
15-Hydroxyprostaglandin dehydrogenase	A35802	29.2	Increase	Inactivation of prostaglandins	Catalytic activity	Gaedtke et al. (2007)[b]

continued

TABLE 3.3 (continued)
Results of Differentially Expressed Proteins between MCTS and Monolayer Cultures

Protein	Accession Number	Molecular Weight (kDa)	Expression in 3D/2D	Functional Association	GO Annotation	Reference
Lamin A/C	Q5TCJ4	53.2	(Fragment lamin only in 3D)	Major components of the nuclear lamina	Structural molecule activity	Gaedtke et al. (2007)[b]
Acidic ribosomal protein P0	R5HUPO	34.4	Increase	Functional equivalent of *Escherichia coli* protein L10	Binding; catalytic activity	Gaedtke et al. (2007)[b]
HSP 90-β[c]	P08238	83.1	Increase	Cell stress induced	N/A	Kumar et al. (2008)[d]
HSC 71	P11142	70.8	Increase	Cell stress induced	Binding; catalytic activity	Kumar et al. (2008)[d]
Transketolase	P29401	67.8	Increase	Hexose monophosphate pathway	Binding; catalytic activity	Kumar et al. (2008)[d]
HSP 60	P10809	61	Increase	Cell stress induced	Binding; catalytic activity	Kumar et al. (2008)[d]
Pyruvate kinase, isozymes M1/M2	P14618	57.7	Increase	Glycolysis	Binding; catalytic activity	Kumar et al. (2008)[d]
Adenyl CAP-1	Q01518	51.5	Increase	Actin monomer binding protein	N/A	Kumar et al. (2008)[d]
Tubulin β-2 chain	P68371	49.8	Increase	Microtubule component	Binding	Kumar et al. (2008)[d]
α-Enolase	P06733	47	Increase	Glycolysis	Transcription regulator activity; binding; catalytic activity	Kumar et al. (2008)[d]

					Structural molecule activity; binding	Kumar et al. (2008)[d]
Actin, cytoplasmic 1	P60709	41.7	Increase	Cell motility		
Septin 2	Q15019	40.4	Increase	Cytoskeleton organization	Binding	Kumar et al. (2008)[d]
PSAT	Q9Y617	40.3	Increase	De novo serine biosynthesis	Catalytic activity	Kumar et al. (2008)[d]
PGAM1	P18669	28.7	Increase	Glycolysis	Catalytic activity; binding	Kumar et al. (2008)[d]
TPI	P60174	26.5	Increase	Glycolysis	Catalytic activity	Kumar et al. (2008)[d]
TCTP	P13693	19.5	Increase	Microtubule/calcium binding	Binding	Kumar et al. (2008)[d]
Cofilin	P23528	18.3	Increase	Actin polymerization/depolymerization	Binding	Kumar et al. (2008)[d]
Pin-1	P62937	17.9	Increase	Signal transduction	Binding	Kumar et al. (2008)[d]
CRABP1	P29762	15.4	Increase	Cytoplasmic transport	N/A	Kumar et al. (2008)[d]
Thioredoxin	P10599	11.6	Increase	Antioxidant	Binding	Kumar et al. (2008)[d]

[a] Human colorectal adenocarcinoma cell line HT-29; differential expression criteria not reported.

[b] Human colorectal cancer cell lines COGA-5 (epithelial-like) and COGA-12 (formed partly compact multicellular spheroids); differential expression was when the average normalized signals altered at least twofold between samples and the alteration was detected in at least three separate gels of at least two independent gel runs.

[c] HSP: heat shock protein, adenyl CAP-1: adenyl cyclase-associated protein 1, PSAT: phosphoserine aminotransferase, PGAM1: phosphoglycerate mutase 1, TPI: triose-phosphate isomerase, TCTP: translationally controlled tumor protein, Pin-1: peptidyl prolyl cis–trans isomerase A, and CRAB1: cellular retinoic acid binding protein 1.

[d] Human neuroblastoma cell lines SK-N-AS, SK-N-DZ, and IMR-32; differential expression was only when protein was upregulated in spheres relative to monolayers at least 2.5-fold—downregulated proteins were not reported.

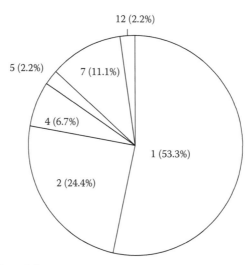

FIGURE 3.5 Differentially expressed protein distribution pooled from studies by Poland et al. (2002), Gaedtke et al. (2007), and Kumar et al. (2008). All others below 0.1%.

was to assemble preliminary evidence in support of the hypothesis that ubiquitous biomarkers for three-dimensionality exist and can be found. The need for three-dimensionality biomarkers and how to look for them are left to Chapter 8 and will not be addressed here. The success of this chapter in addressing the author's second goal is mixed. On the one hand, there is agreement on the most common molecular functions with the most differentially expressed genes and proteins. But on the other hand no set of genes/proteins have stood out among the studies examined, suggesting that it is not clear where best to look for these biomarkers. Perhaps, this lack of "standout" genes/proteins should not be surprising for a couple of reasons. First, in an excellent study, described in more detail in Chapter 8, Cukierman et al. (2001) found that novel "3D-matrix" adhesions contained the same proteins as were found in traditional 2D cell culture fibrillar and focal adhesions (Geiger and Bershadsky, 2002; Sastry and Burridge, 2000), but they were only based on the $\alpha_5\beta_1$ instead of the $\alpha_v\beta_3$ integrin receptors. Second, integrins are membrane proteins and such proteins have been difficult to detect with proteomics tools because of their poor solubility in standard protein extraction solutions in addition to their low abundance (Santoni et al., 2000). Also, membrane proteins are a class of proteins that occur in nature at a lower percentage and are typically masked by highly abundant proteins in typical analytical proteomics tools such as 2D gel electrophoresis (Vercauteren et al., 2004; Fountoulakis, 2001, 2004).

As suggested before, the correlation between differentially expressed and GO database gene distribution raised an interesting question of whether there are molecular functions that are overrepresented with respect to differentially expressed genes, relative to their GO database distribution. To answer this question, molecular function distributions from all the studies were pooled and compared with human GO database and protein percentages in Table 3.4. The only overrepresented molecular functions are electron carrier activity (gene level) and structural molecular activity

TABLE 3.4

Comparison Gene Expression Percentage Distribution

Molecular Functions (See Box 3.1 for Details)	1	2	3	4	5	6	7	8	9	10	11
Lai et al. (2009b)—freshly dissected superior cervical ganglion cells (*in vivo*)	50.6	23.0	7.6	6.9	3.8	3.6	2.7	0.8	0.7	0.2	0.1
Lai et al. (2009b)—superior cervical ganglion cells cultured *in vitro*	51.4	19.1	10.9	4.7	2.5	7.4	1.6	1.4	0.2	—	—
Lai et al. (2009a)—neural progenitor *in vitro*	52.3	20.3	8.6	9.4	1.6	3.9	3.9	—	—	—	—
Lai et al. (2009a)—neural sphere *in vitro*	51.8	21.4	8.3	7.0	4.4	3.5	2.5	0.6	0.2	0.2	—
Myers et al. (2008)	47.4	28.1	5.2	—	4.5	3.8	3.8	0.7	—	—	—
Kenny et al. (2007)	46.9	21.9	12.5	6.1	7.3	3.1	1.0	1.0	—	—	—
Ghosh et al. (2005)	51.2	18.8	8.5	1.4	7.0	4.7	3.3	0.5	—	—	—
Boess et al. (2003)	50.0	25.1	6.7	6.0	4.2	5.4	3.2	2.7	0.2	—	—
Li et al. (2002)	48.0	20.0	8.0	6.0	6.0	4.0	4.0	2.0	2.0	—	—
Average (standard deviation)	50.0 (2.0)	22.0 (3.0)	8.5 (2.2)	6.0 (2.3)	4.4 (1.9)	4.4 (1.3)	2.9 (1.0)	1.2 (0.8)	0.7 (0.8)	0.2	0.1
Human genes from GO database	51.8	19.6	8.3	7.7	3.2	4.8	3.0	0.5	0.5	0.3	—
Protein (see Table 3.3 for references)	53.3	24.4	—	6.7	2.2	—	11.1	—	—	—	—

(protein level). The structural molecular activity is consistent with numerous observations and probably offers an area for the initial search. For example, the fibronectin integrin receptor ($\alpha_5\beta_1$) is important for cell interactions with purified fibronectin, but 65–88% of the cellular response of human fibroblasts to 3D matrices is suppressed by the antibody against α_5, whereas only 27–36% inhibition was observed by the same antibody for fibroblasts on 2D fibronectin (Yamada et al., 2003). Fibronectin matrix assembly involves the binding of soluble fibronectin molecules by integrins and exposure of cryptic fibronectin self-assembly sites to promote self-association. The speculative mechanism of cryptic site exposure involves stretching by intracellular contractility (Geiger et al., 2001; Ohashi et al., 1999), which is a structural molecular activity.

REFERENCES

Akins, R.E., Gratton, K., Quezada, E., Rutter, H., Tsuda, T., and Soteropoulos, P. 2007. Gene expression profile of bioreactor-cultured cardiac cells: Activation of morphogenetic pathways for tissue engineering. *DNA Cell Biol.* 26(6): 425–434.

Boess, F., Kamber, M., Romer, S., Gasser, R., Muller, D., Albertini, S., and Suter, L. 2003. Gene expression in two hepatic cell lines, cultured primary hepatocytes, and liver slices compared to the *in vivo* liver gene expression in rats: Possible implications for toxigenomics use of *in vitro* systems. *Toxicol. Sci.* 73: 386–402.

Boyce, S.T. and Hansbrough, J.F. 1988. Biological attachment, growth, and differentiation of cultured human epidermal keratinocytes on a graftable collagen and chondroitin-6-sulfate substrate. *Surgery* 103: 421–431.

Cukierman, E., Pankov, R., Stevens, D.R., and Yamada, K.M. 2001. Taking cell-matrix adhesion to the third dimension. *Science* 294(5547): 1708–1712.

Fountoulakis, M. 2001. Proteomics: Current technologies and applications in neurological disorders and toxicology. *Amino Acids* 21: 363–381.

Fountoulakis, M. 2004. Application of proteomics technologies in the investigation of the brain. *Mass Spectrum. Rev.* 23: 231–258.

Gaedtke, L., Thoenes, L., Culmsee, C., Mayer, B., and Wagner, E. 2007. Proteomic analysis reveals differences in protein expression in spheroid versus monolayer cultures of low-passage colon carcinoma cells. *J. Proteom. Res.* 6: 4111–4118.

Geiger, B. and Bershadsky, A. 2002. Exploring the neighborhood: Adhesion-couples cell mechanosensors. *Cell* 110: 139–142.

Geiger, B., Bershadsky, A., Pankov, R., and Yamada, K.M. 2001. Transmembrane cross-talk between the extracellular matrix and the cytoskeleton. *Nat. Rev. Mol. Cell Biol.* 2: 793–805.

Gelain, F., Bottai, D., Vescovi, A., and Zhang, S. 2006. Designer self-assembling peptide nanofiber scaffolds for adult mouse neural stem cell 3-dimensional cultures. *PLoS One* (1): e119.

Ghosh, S., Spagnoli, G.C., Martin, I., Ploegert, S., Demougin, P., Heberer, M., and Reschner, A. 2005. Three-dimensional culture of melanoma cells profoundly affects gene expression profile: A high density oligonucleotide array study. *J. Cell. Physiol.* 204: 522–531.

Hishikawa, K., Miura, S., Marumo, T., Yoshioka, H., Mori, Y., Takato, T., and Fujita, T. 2004. Gene expression profile of human mesenchymal stem cells during osteogenesis in three-dimensional thermal reversible gelatin polymer. *Biochem. Biophys. Res. Commun.* 317: 1103–1107.

Ikebe, D., Wang, B., Suzuki, H., and Kato, M. 2007. Suppression of keratinocyte stratification by a dominant negative JunB mutant without blocking cell proliferation. *Genes Cells* 12: 197–207.

International Human Genome Sequencing Consortium. 2001. Initial sequencing and analysis of the human genome. *Nature* 409: 860–921.

Kahl, G. 2004. *The Dictionary of Gene Technology: Genomics, Transcription, Proteomics.* Weinheim: Wiley-VCH.

Kenny, P.A., Lee, G.Y., Myers, C.A., Neve, R.M., Semeiks, J.R., Spellman, P.T., Lorenz, K., Lee, E.H., Barcellos-Hoff, M.H., Petersen, O.W., Gary, J.W., and Bissell, M. 2007. The morphologies of breast cancer cell lines in three-dimensional assays correlate their profiles of gene expression. *Mol. Oncol.* 1: 84–96.

Khaoustov, V.I., Risin, D., Pells, N.R., and Yoffe, B. 2001. Microarray analysis of genes differentially expressed in HepG2 cells cultured in simulated microgravity: Preliminary report. *In Vitro Cell. Dev. Biol.—Anim.* 37: 84–88.

Kim, S.-W., Cheon, K., Kim, C.-H., Yoon, J.-H., Hawke, D.H., Kobayashi, R., Prudkin, L., Wistuba, I., Lotan, R., Hong, W.K., and Koo, J.S. 2007. Proteomics-based identification of proteins secreted in apical surface fluid of squamous metaplastic human tracheobronchial epithelial cells cultured by three-dimensional organotypic air–liquid interface method. *Cancer Res.* 67(14): 6565–6573.

Klapperich, C.M. and Bertozzi, C.R. 2004. Global gene expression of cells attached to a tissue engineering scaffold. *Biomaterials* 25: 5631–5641.

Kumar, H.R., Zhong, X., Hoelz, D.J., Rescorla, F.J., Hickey, R.J., Malkas, L.H., and Sandoval, J.A. 2008. Three-dimensional neuroblastoma cell culture: Proteomic analysis between monolayer and multicellular tumor spheroids. *Pediatr. Surg. Int.* 24: 1229–1234.

Lai, Y., Cheng, K., and Kisaalita, W.S. 2009a. Comparative transcriptional profiling of neuroprogenitor cells on 2D polymer substrate and in 3D polymer scaffolds. *Acta Biomater.* (submitted).

Lai, Y., Cheng, K., and Kisaalita, W.S. 2009b. Comparative transcriptional profiling of freshly dissected (*in vivo* surrogate) and 3D scaffold-cultured neurons. *Biomaterials* (submitted).

Li, G.N., Livi, L.L., Gourd, C.M., Deweerd, E.S., and Hoffman-Kim, D. 2007. Genomic and morphological changes of neuroblastoma cells in response to three-dimensional matrices. *Tissue Eng.* 13(5): 1035–1047.

Li, S., Lao, J., Chen, B.P.C., Li, Y.-S., Zhao, Y., Chu, J., Chen, K.-D., Tsuo, T.-C., Peck, K., and Chien, S. 2002. Genomic analysis of smooth muscle cells in three-dimensional collagen matrix. *FASEB J.* 16(13): 97–t (November 15) express article: doi: 10.1096/fj.02-0256fje1.

Liu, G., Loraine, A.E., Shigeta, R., Cline, M., Cheng, J., Valmeekam, V., Sun, S., Kulp, D., and Siani-Rose, M.A. 2003. NetAffx: Affymetrix probe set and annotations. *Nucleic Acids Res.* 31: 82–86.

Liu, H., Lin, J., and Roy, K. 2006. Effect of 3D scaffold and dynamic culture conditions on the global gene expression profile of mouse embryonic stem cells. *Biomaterials* 27: 5978–5989.

Mahantheppa, N.K. and Patterson, P.H., 1998. Culturing mammalian sympathoadrenal derivatives. In *Culturing Nerve Cells*, 2nd edition, G. Banker and K. Goslin (eds), pp. 289–307. Cambridge, Massachusetts: MIT Press.

Myers, T.A., Nickerson, C.A., Kaushal, D., Ott, C.M., Bentrup, K.H., Ramamurthy, R., Nelman-Gonzalez, M., Pierson, D.L., and Philipp, M. 2008. Closing the phenotypic gap between transformed neuronal cell lines in culture and untransformed neurons. *J. Neurosci. Methods* 174: 31–41.

Ohashi, T., Kiehart, D.P., and Erickson, H.P. 1999. Dynamics and elasticity of the fibronectin matrix in living cell culture visualized by fibronectin—green fluorescence protein. *Proc. Natl. Acad. Sci. USA* 96: 2153–2158.

Olsavsky, K.M., Page, J.L., Johnson, M.C., Zarbl, H., Strom, S.C., and Omiecinski, C.J. 2007. Gene expression profiling and differentiation assessment in primary human hepatocyte cultures, established hepatoma cell lines, and human liver tissues. *Toxicol. Appl. Pharmacol.* 222: 42–56.

Oswald, J., Steudel, C., Salchert, K., Joergensen, B., Thiede, C., Ehninger, G., Werner, C., and Bornhauser, M. 2006. Gene expression profiling of CD34+ hematopoietic cells expanded in collagen I matrix. *Stem Cells* 24: 494–500.

Poland, J., Sinha, P., Siegert, A., Schnolzer, M., Kort, U., and Hauptmann, S. 2002. Comparison of protein expression profiles between monolayer and spheroid cell culture of HT-29 cells revealed fragmentation of CK18 in three-dimensional cell culture. *Electrophoresis* 23: 1174–1184.

Santoni, V., Molloy, M., and Rabilloud, T. 2000. Membrane proteins and proteomics: Un amour impossible? *Electrophoresis* 21(6): 1054–1070.

Sastry, S.K. and Burridge, K. 2000. Focal adhesions: A nexus for intracellular signaling and cytoskeletal dynamics. *Exp. Cell Res.* 261: 25–36.

Sorensen, G., Medina, S., Parchaliuk, D., Phillipson, C., Robertson, C., and Booth, S. 2008. Comprehensive transcriptional profiling of prion infection in mouse models reveals networks of responsive genes. *BMC Genomics* 9: 1–14.

Storey, J.D. 2002. A direct approach to false discovery rates. *J. R. Statist. Soc.* B(64): 479–498.

Vercauteren, F.G.G., Bergeron, J.J.M., Vsndesande, F., Arckens, L., and Quirion, R. 2004. Proteomic approaches in brain research and neuropharmacology. *Eur. J. Pharmacol.* 500: 385–398.

Yamada, K.M., Pankov, R., and Cukierman, E. 2003. Dimensions and dynamics in integrin function. *Braz. J. Med. Biol. Res.* 36: 959–966.

4 Comparative Structure and Function

4.1 COMPLEX PHYSIOLOGICAL RELEVANCE

The phrase "physiological relevance" can have different meanings when used in a drug discovery context. A survey of the literature turned up three possible interpretations. The first is when a target is relevant to the disease; the cells harboring the target have been referred to as physiologically relevant. For example, Panetta and Greenwood (2008) have written about the physiological relevance of G-protein-coupled receptor (GPCR) dimers as new drug targets for schizophrenia and pre-eclampsia (development of hypertension with proteinuria or edema, or both, due to pregnancy or the influence of a recent pregnancy). Another example is voltage-gated calcium channels (VGCCs) as "physiologically relevant" targets for anxiety, epilepsy, hypertension, insomnia, and pain (Mohan and Gandhi, 2008). The second usage of the phrase is when the cell background is more relevant to the disease as the case may be for HTS methods that use primary and embryonic stem cells (Eglen et al., 2008). This is in contrast to the common practice in drug discovery where the immortalized cell employed in HTS is determined by the assay technology available; the recombinant targets are expressed in the immortalized cells followed by monitoring the drug interaction with the target using a simple quantifiable response. To illustrate further, CHO cells expressing VGCC targets are more common (John et al., 2007) than "physiologically relevant" differentiated neural progenitor cells or embryonic stem cell-derived cardiomyocytes that naturally express VGCCs (Wagner et al., 2008; Yanagi et al., 2007). The third usage of the phrase is when 3D cells are involved with the accompanying difference in structure and function. To minimize the confusion, we introduce the phrase, "complex physiological relevance," abbreviated as CPR. Hereafter, CPR is used to refer to 3D cells, *in vitro*, emulating *in vivo* structure and function that are absent in their 2D counterparts.

In the following three sections, examples that are consistent with the CPR defined above are presented. There are many systems from which to choose. For example, one can turn back to the highly successful pharmacological methods of the past, in which organ- and tissue-based systems were used (Whitby, 1987; Groneberg et al., 2002; Vickers and Fisher, 2004; Andrei, 2007; Vickers, 2009). The fact that most of these systems are low throughput makes them unattractive for our purposes here. For a sampling of a wide range of systems, see Flemming et al. (1999), Drury and Mooney (2003), Schmeichel and Bissell (2003), Kim (2005), Xu et al. (2006), Garlick (2006), Yamada and Cukierman (2007), Martínez et al. (2008), Justice et al. (2009), and Christopherson et al. (2009). Some of these systems were not investigated with 2D as control and therefore were not very useful in a comparative narrative. The selection of the three featured examples below was guided first by inclusion of 2D control in the study, second by cell type, and third by high commercialization potential (i.e., feasibly implementable in current HTS formats). In the first two example systems, hepatocytes and myocardiocytes are featured. The rationale for the first two cell types is that cardiotoxicity and hepatotoxicity remain the highest causes of drug safety liabilities and drug withdrawal during development and marketing (Mayne and Ku, 2006). In the third example system, nerve cells are featured; they are the main native source of ion channels, an emerging class of targets. For example, 15 of the top-100 selling drugs target ion channels and ion channels function is implicated in a large number of diseases such as Alzheimer's, epilepsy, cardiovascular diseases, neuromuscular diseases, and in some types of cancers. Overall, ion channels represent significant therapeutic opportunities.

4.2   CARDIOMYOCYTE CONTRACTILITY

4.2.1   CELLS AND SCAFFOLD

An interesting system that enables a quantitative evaluation of cardiomyocyte contraction based on cantilever bending principles has been proposed by a research group from Korea (Park et al., 2005; Kim et al., 2008). The group's main objective was to more accurately measure the contractile force that is more representative of the *in vivo* 3D microenvironment. The microcantilever was fabricated with polydimethylsiloxane (PDMS) substrate layer. Cardiomyocytes (isolated from the heart of a neonatal Sparague-Dawley rat at day one) were seeded on top of the substrate creating a PDMS–cardiomyocyte hybrid micro cantilever. The difference between the 2D and 3D cantilevers was that the PDMS structure in the 3D construct was grooved as shown in Figure 4.1. Grooved surfaces have been used by many investigators to induce structural emulation (rod-shaped phenotype) of *in vivo* cardiomyocyte by their *in vitro* counterparts (e.g., Yin et al., 2004; Zhao et al., 2007). Other approaches targeting tissue engineering applications have been reviewed by Akhyari et al. (2008). Any contraction in the cells causes bending of the whole cantilever and the higher the contractile force, the higher the cantilever displacement.

4.2.2   COMPARATIVE STRUCTURE

The structure was examined by staining the cytoskeleton of cells on both 2D and 3D with phalloidin and 4′,6-diamidino-2-phenylindole (DAPI) (Figure 4.2). The majority of cells on 2D cantilevers were multipolar in shape, but those on 3D cantilevers were bipolar. The actin filament and nuclei elongation from 3D cells is typical of *in vivo* cardiac cells as shown in Figure 4.3. The sample featured is from a paper that evaluated the *in vivo* calpeptin attenuation of calpain activation and cardiomyocyte in pressure-overloaded cat cardium. The purpose of Figure 4.3 in this narrative however is to make a simple *in vivo* cell morphology comparison. Overloading had no effect on the muscle cell shape.

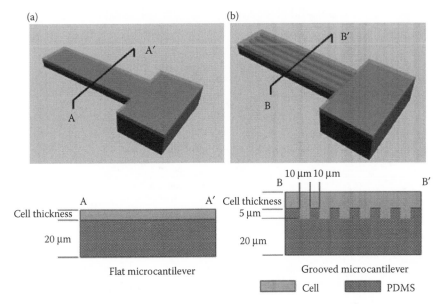

FIGURE 4.1   Cross-section schematics of 2D (a) and 3D (b) hybrid microcantilevers. (From Kim, J. et al. 2008. *J. Biomech.* 4: 2397. With permission.)

4.2.3 COMPARATIVE FUNCTION

Figure 4.4 shows still images from video recordings of the motion of the hybrid (cell–polymer) microcantilever. The motions were recorded 4 days after cell plating. The contractile force (F) can be calculated as shown in Box 4.1 (Kim et al.,

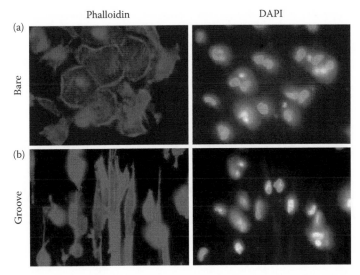

FIGURE 4.2   Immunostaining of cardiomyocytes on 2D (a) and 3D (b) with phalloidin (cytoskeleton, left) and DAPI (nucleus, right). (Adapted from Kim, J. et al. 2008. *J. Biomech.* 4: 2397.)

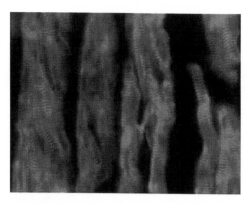

FIGURE 4.3 Perfusion-fixed ventricular sections from cats, immunostained using anti-α-actin. (Adapted from Mani, S.K. et al. 2008. *Am. J. Physiol. Heart Circ.* 295: H314–H326.)

2008). Endogenous markers (e.g., sarcomeric actin) were not different between 2D and 3D cells, ruling out the possibility that differences in contractile forces observed could be due to differences in number of cells on the cantilevers. The force on the 3D was 65–85% higher than that on 2D microcantilevers. These figures were independently confirmed using finite element analysis modeling. Other studies employing 2D culture systems (e.g., Balaban et al., 2001; Zhao and Zhang, 2005) have reported cardiomyocyte stresses of 2–5 nN/μm^2 that were verified using 2D cantilevers by the Korean group (Park et al., 2005). Also, the 65–85% force increase with 3D cantilevers translated into 4–10 nN/μm^2 stress, which is closer to stress figures obtained from skinned cardiac cells activated in solution with Ca^{2+} (Weiwad et al., 2000).

4.2.4 HTS Application Feasibility

An HTS assay utilizing the cantilever biosensing mechanism can be envisioned, probably as a label-free format, with an optical readout of the cantilever deflection.

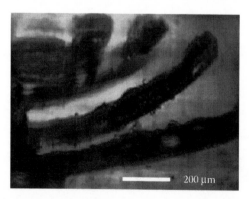

FIGURE 4.4 Vertical motion images from video recordings of the motion of the hybrid (cell–polymer) microcantilever, 4 days after cell plating. (Adapted from Park, J. et al. 2005. *Anal. Chem.* 77: 6571–6580.)

BOX 4.1 ANALYTICAL CALCULATION OF CONTRACTILE FORCE OF CARDIOMYOCYTES ON CANTILEVERS

The bending moment M is generated by shear force according to

$$M = \Sigma Fd,$$

where F is the shear force and d is the distance of F from the centroid. The bending moments M_f (flat) and M_g (grooved) caused by the shear forces applied at their surfaces are as follows:

$$M_f = Fd_f = F\frac{h_f}{2},$$

$$M_g = F\frac{d_c + d_t}{2},$$

where d_f and h_f are the distance from the centroid to the surface of the flat cantilever and the height of the flat cantilever, respectively. d_c and d_t are the distances from the centroid to the crest and trough of the cantilever, respectively. A calculation, using the above two equations and dimensions for the Kim et al. (2008) microcantilever, assuming the same shear force at the surfaces, readily shows 11.1% more bending moment with the grooved microcantilever. The difference is primarily due to differences in distances to the centroids.

The cross-section of the flat microcantilever is made up of one rectangle, whereas that of the grooved microcantilever is made up of not only the large rectangle, as in the flat structure, but also of the small rectangles. The area moments of inertia I_f (flat) and I_g (grooved) are as follows:

$$I_f = \frac{1}{12} w_1 h_1^3,$$

$$I_g = \frac{1}{12} w_1 h_1^3 + D_{y1}^2 w_1 h_1 + N\left(\frac{1}{12} w_s h_s^3 + D_{ys}^2 w_s h_s\right),$$

where w_1 and h_1 are the width and height of the large rectangle in the cross-section of the flat and grooved microcantilevers. w_s and h_s are the width and height of the small rectangles in the cross-section of the grooved microcantilever. N is the number of small rectangles in the grooved microcantilever and Dy_1 and Dy_s are the y-directional distances from the centroid to the large rectangle and the small rectangles, respectively. Using the above two equations and dimensions from the microcantilever from Kim et al. (2008), the grooved microcantilever can easily be shown to have 52.9% larger moment of inertia. Area moment of inertia indicates stiffness of the structure to the bending moment.

Kim et al. (2008) considered the above two factors and showed that to yield the same amount of bending moment in the two types of microcantilevers, the grooved type needed 34.8 higher contractile forces in comparison with the flat type. This figure was confirmed by finite element analysis.

A number of cantilevers can be chemically "welded" into wells of a standard high-density plate (e.g., 1536). With such a system, the 2D embryonic stem cell assay for cardiotoxicity prediction proposed by Chaudhary et al. (2006) can be transformed to a 3D cell-based biosensor system. Compounds against targets that affect contractility such as ion channels would be suitable for screening in the envisioned assay. There may be technical hurdles such as uniform chemical "welding" of grooved cantilevers in high-density well plates, but they are not insurmountable.

4.3   LIVER CELL BILE CANALICULI *IN VITRO*

4.3.1   Cells and Scaffold

The second example comes from a research group at Durham University, UK, led by Stefan Przyborski. This group developed a thin polystyrene membrane that is porous enough to allow the entry of cells into the internal structure (Bokhari et al., 2007). The polystyrene scaffolds were fabricated using high internal phase emulsion (HIPE) polymerization (Akay et al., 2004; Hayman et al., 2005). Thin wafers (120 μm) were produced by sectioning the monolith on an automated microtome. The membranes were mounted ("welded") on the bottom of well inserts as shown in Figure 4.5. The group evaluated the structure and function of human HepG2 cells cultured in the membrane in the presence of a hepatotoxic drug, methotrexate. HepG2 cell is well recognized and frequently used in assays for liver function (Liguori et al., 2008; Ye et al., 2007; Miret et al., 2006). HepG2 was originally derived from a hepatic carcinoma. Bokhari et al. (2007) chose methotrexate because it is widely used in the treatment of malignant tumors and other diseases and hepatotoxicity is a commonly associated complication that in some cases has led to irreversible liver damage (Wu et al., 1983).

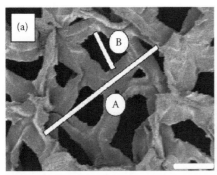

FIGURE 4.5   Polystyrene scaffolds. (a) SEM images of microscale architecture consisting of voids (A) and interconnects (B). (b) Photomicrograph showing 6- and 12-well inserts with polystyrene scaffolds "welded to the bottom." Scale bars are 20 μm (a) and 1 cm (b). (From Bokhari, M. et al. 2007. *Biochem. Biophys. Res. Commun.* 354: 1097. With permission.)

4.3.2 COMPARATIVE STRUCTURE AND FUNCTION

The structural difference between HepG2 cells cultured on 2D and 3D substrates is presented in Figure 4.6. Unlike 2D, 3D cultures formed tight junction between cells and exhibited bile canaliculi-like structures into which microvilli projected. Such structures have been linked to the establishment of membrane polarity (Peshwa et al., 1996). Tight junctions and bile canaliculi structures have been reported in rat primary hepatocytes that self-assemble into spheroids (Abu-Absi et al., 2002). In this spheroid 3D model, a fluorescent bile analogue, fluorescein isothiocyanate-labeled glycocholate, was taken up into spheroids and excreted into bile canalicular channels, demonstrating functional polarity. This function has not been demonstrated in 2D cultures.

Albumin production is generally recognized as an indicator of liver-specific activity (Kane et al., 2006). Rat hepatocyte spheroids have shown stable expression of more than 80% of the 242 liver-related genes including those of albumin synthesis (Brophy et al., 2009). Albumin secretion in the presence of methotrexate in the Bokhari et al. (2007) study is presented in Figure 4.7. *In vivo* albumin synthesis was

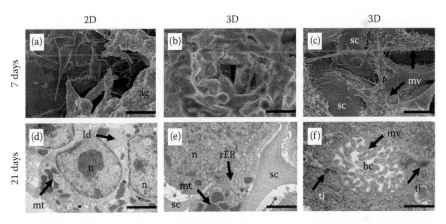

FIGURE 4.6 SEM (a–c) and TEM (d–f) micrographs showing examples of HepG2 cells cultured on 2D (a, d) and 3D (b, c, e, and f) polystyrene scaffolds for either 7 (a–c) or 21 days (d–f), 2D cells were significantly more heterogeneous in structure (a) in comparison with 3D cells (b). Cells seeded at lower density allowed visualization of individual cell structure on 3D scaffolds (sc) (c). HepG2 cells developed complex 3D shapes and interactions with neighboring cells. Higher magnification images revealed the expression of large numbers of microvilli (mv) on the surface of cells (c). TEM showed that cells grown on 2D plastics contained typical cellular structures, for example, nuclei (n), mitochondria (mt), and lipid droplets (ld) (d). Images of cells cultured on 3D scaffolds showed how the cells grow in close association with the polymer, completely surrounding struts of the scaffold (sc) (e). Hepatocytes in 3D scaffolds also displayed an array of typical cellular organelles such as nuclei (n), mitochondria (mt), and rough endoplasmic reticulum (rER) (e). Higher magnification TEM showed the formation of tight junction (jt) complexes between adjacent cells (f). The space formed in between cells resembles bile canaliculi (bc) into which microvilli (mv) are projected. Bars are 25 μm (a, b); 12 μm (c); 4 μm (d); 2 μm (e); and 500 nm (f). (From Bokhari, M. et al. 2007. *J. Anat.* 211: 570. With permission.)

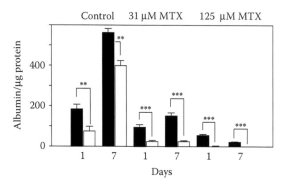

FIGURE 4.7 HepG2 albumin secretion into the culture medium by HepG2 cells cultured in 3D (open bar) and 2D (solid bars) substrates. Albumin levels were normalized to the total protein per well. Significance is denoted by **$P < 0.01$ and ***$P < 0.001$ using the Mann-Whitney U test. MTX is the abbreviation of methotrexate. (From Bokhari, M. et al. 2007. *J. Anat.* 211: 571. With permission from Wiley.)

estimated by Morgan and Peters (1971) to be 4 µg (min mg DNA)$^{-1}$. In contrast, albumin production by 2D cultures has been reported to be several folds lower (Ise et al., 1999). The results presented in Figure 4.7 are consistent with these previous studies; HepG2 cells cultured in polystyrene scaffolds synthesized a significantly higher albumin concentration than their 2D counterparts. The higher resistance to methotrexate (concentrations used are in the range used in clinical settings) demonstrated in Figure 4.7 is a consistent theme in 3D culture toxicological studies. The result's implication here, that absence of 3D cell organization in toxicological screening may lead to overestimation of drug toxicity, is further explored in one of the case studies presented in Chapter 12.

4.3.3 HTS APPLICATION FEASIBILITY

The author is not aware of a commercial product based on this technology. But a patent was applied for by the scientists involved and a start-up company, Relnnervate Limited, was put in place to commercialize the technology (http://www.sciencedaily.com/release/2007/09/070919073020.htm, accessed on September 27, 2007).

4.4 NERVE CELL VOLTAGE-GATED CALCIUM SIGNALING

4.4.1 CELLS AND SCAFFOLD

The third example comes from the author's Cellular Bioengineering Laboratory. As in the previous example, polystyrene scaffolds were used, but the fabrication process incorporates the 3D scaffolds into standard cell culture dishes and multiwell plates in a precise and rapid manner (Cheng et al., 2008; US patent pending). A schematic of the process is presented in Figure 4.8. Generally, a viscous polymer solution was prepared by dissolving polystyrene in chloroform. Sieved ammonium bicarbonate

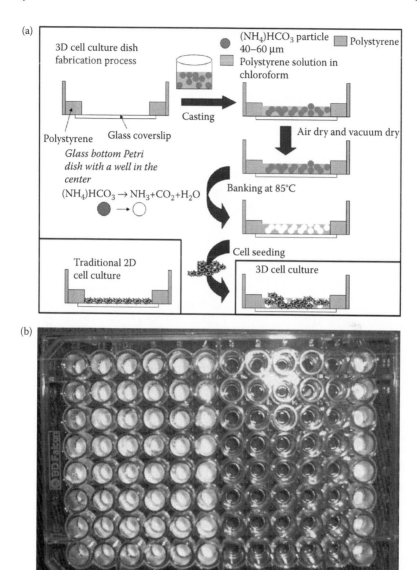

FIGURE 4.8 Fabrication and characterization of 3D cell culture vessels. (a) Schematic of the fabrication process. (b) Outlook of a 3D cell culture 96-well plates with 3D polystyrene scaffolds in the left half of the columns. (From Cheng, K., Lai, Y., and Kisaalita, W.S. 2008. *Biomaterials* 29: 2804. With permission.)

particles in the range of 40–60 μm were added to the polymer solution and mixed thoroughly. The paste mixture of polymer/salt/solvent was cast into the wells of a standard cell culture vessel with a single- or multichannel pipette. The mixture was viscous and as such the vessels were gently rocked to obtain even layers. MatTek® glass bottom Petri dishes (Cat#: P35G-0-14; well diameter: 14 mm) and 96-well plates (Cat#: P96G-0-5-F; well diameter: 5 mm) were used for 3D vessel fabrication. The casting mixture was composed of 0.1 g polystyrene, 2 g ammonium bicarbonate, and 4 mL chloroform. Usually, 70–150 and 10–30 μL of mixture were cast into each well of Petri dishes and 96-well plates, respectively. After casting, the vessel was immediately covered to control the evaporation rate of chloroform. This step was crucial as chloroform also served as an adhesive to "weld" the scaffold by partially dissolving the polystyrene wall of the well. After chloroform was completely evaporated, the dishes and plates were baked in an oven overnight. At any temperature above 36°C, ammonium bicarbonate decomposed to ammonia, carbon dioxide, and water and left pores, creating a porous polymer scaffold. Although a higher temperature can achieve removal of ammonium bicarbonate faster, the baking temperature cannot exceed the glass temperature (T_g) of the polymer in use, as well as the maximum temperature the cell culture vessels can tolerate.

Human neural stem (NS) cells used were isolated by the Regenerative Bioscience Center at the University of Georgia and are now commercially available as ENStem-A™ from Millipore (Billerica, MA). The cells were maintained in neural basal media (Life Technologies/Invitrogen, Carlsbad, CA) supplemented with penicillin/streptomycin, L-glutamine, recombinant human leukemia inhibitory factor (hLIF), basic fibroblast growth factor (bFGF), and B-27 (a serum-free supplement). The composition of differentiation media was similar to the subculture media described above but without bFGF. The NS cells were incubated at 37°C in a 5% CO_2 humidified atmosphere. Before cell seeding, both the scaffolds in Petri dishes and 96-well plates were prewetted and sterilized in 70% ethanol overnight. To achieve better cell attachment and rule out any difference caused by the polymer materials, both the scaffolds and flat coverslips were coated with polyornithine (Sigma-Aldrich, St. Louis, MO; molecular weight is 30,000–70,000) and laminin (Sigma-Aldrich, St. Louis, MO; from Engelbreth-Holm-Swarm murine sarcoma basement membrane). The scaffolds were submerged in 20 μg/mL polyornithine water solution overnight and then in 5 μg/mL laminin water solution overnight for achieving complete coating. Before cell seeding, the laminin solution was aspirated from the scaffolds. For both 2D and 3D cell cultures, NS cells were seeded with a uniform density of 50,000 cells/cm².

Neural spheres and freshly dissected mouse superior cervical ganglion (SCG) were used as *in vivo* surrogate and *in vivo* controls, respectively. For neural sphere cell culture, 1×10^6 cells were seeded onto a 35-mm Petri dish without coating of polyornithine and laminin. Uncoated surfaces prevented cells from adhesion but encouraged them to form neural spheres. In some experiments, NS cells were induced to differentiate by replacing the growth media with differentiation media.

4.4.2 COMPARATIVE STRUCTURE

Only morphological studies were conducted in the 3D scaffolds with freshly dissected mouse SCG; the detailed results will be appearing in a forthcoming paper (Lai et al., 2010). The cells in 3D scaffolds developed shorter neurites and were less spread than the 2D cultured cells. The 3D cultured cell morphology more closely resembled the cell morphology found in freshly dissected intact SCG tissue. The above differences between 2D versus 3D cultured cells suggested that polymer scaffolds are a unique substrate that promotes cell attachment and differentiation that differs from what was observed with 2D substrates, which are consistent with conclusions from previous studies (Desai et al., 2006; Mao and Kisaalita, 2004; Wu et al., 2006) in the author's laboratory.

4.4.3 COMPARATIVE FUNCTION

To further test the value of using these 3D cell culture vessels in drug discovery programs, the studies were extended to the functionality validation level. VGCC functionality on 2D substrates and 3D scaffolds was compared. VGCC was chosen because it is an emerging drug target—there is a strong link between diseases of the nervous and cardiovascular systems and channel dysfunction (e.g., Bear et al., 2009; Lorenzon and Beam, 2008). Previous studies in the author's laboratory and by other investigators had shown differences in calcium currents between intact and dissociated adult mouse SCG cells (Mains and Patterson, 1973), and differences in VGCC function between 2D and 3D cultured human neuroblastoma cells on collagen hydrogels and Cytodex microbead scaffolds (Mueller-Klieser, 1997; Weaver et al., 1997; Powers et al., 2002). The effect of an agonist (drug) was simulated by using high K^+ (50 mM) depolarization.

For 3D Petri dishes, single cell bodies were selected as the region of interest (ROI) and the intracellular calcium concentration was recorded continuously in time by the membrane-permeable dye Calcium Green-1 acetoxymethyl ester (AM), with a confocal laser scanning microscope. Figure 4.9a–d shows typical changes and the time course in terms of Calcium Green-1 AM fluorescence intensity for responsive NS cells after 2 days' culture in 3D polystyrene scaffolds. A cell was considered responsive only when it showed an increase in fluorescence intensity of 15% or higher over the basal fluorescence intensity level. To extend the study to the HTS level, Fluo-4, as a calcium indicator was utilized and both the 3D and standard 2D 96-well vessels were read on a Molecular Devices FlexStation® fluorescence multiwell plate reader. In this case, each well in the plate consisted of the ROI, and the time course of fluorescence changes was recorded (Figure 4.9e). In both cases, the magnitudes of the response from each cell/well were expressed as a peak fractional increase over basal fluorescence intensity $(F - F_o)/F_o$, where F is the peak fluorescence intensity and F_o is the basal fluorescence intensity.

For NS cells on 3D dishes, VGCC functionality was characterized before differentiation, 1 week into differentiation and 2 weeks into differentiation. Before differentiation, 87.1% and 50.4% of the NP cells on 2D and 3D substrates were responsive to high K^+ buffer, respectively. As shown in Figure 4.9f, the response magnitude of 3D cultured cells (0.63 ± 0.08) was much lower than that of 2D cultured cells (2.37 ± 0.44) with a p-value of 5.33e-7. After 1 week into differentiation, the percentage of responsive cells on 2D substrates increased to 90.2% with no significant difference in a response magnitude of 2.29 ± 0.39 at a p level of 0.05, whereas 70.2% of cells on 3D substrate were responsive with an average response magnitude increase to 0.72 ± 0.07 $(p = 0.042)$. The 3D cultured cells' response magnitude was still much lower than that of 2D cultured cells $(p = 4.88e-6)$. After

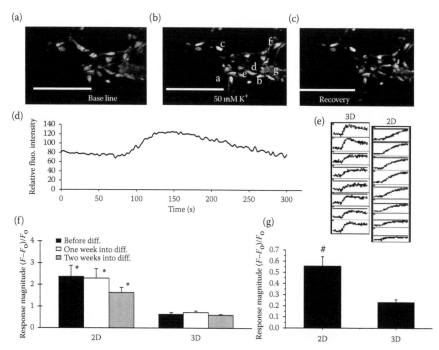

FIGURE 4.9 VGCC functionality. Cells on 2D/3D Petri dishes and 2D/3D 96-well plates were loaded with calcium indicator Calcium Green-1 AM and Fluo-4, respectively. The cells were exposed to 50 mM high K$^+$ depolarization for calcium imaging. (a–c) Confocal micrographs of cells on a 3D scaffolds within a Petri dish showing changes in [Ca^{2+}] levels following addition of high K$^+$ buffer. The elapsed time between (a), (b), and (c) was 30 and 150 s. (d) Plot of relative fluorescent intensity versus recording time for a cell labeled "d" in (b). The increase in fluorescence intensity is proportional to the increase in intracellular [Ca^{2+}] concentration. (e) Typical [Ca^{2+}] time courses of cells from eight wells in both 2D and 3D culture 96-well plates. (f) High K$^+$ buffer stimulated VGCC response magnitudes from NS cells cultured on 2D surface and in 3D vessels. Single cell body was selected as the ROI and fluorescence changes were recorded by a time course of laser confocal scanning microscopy. Fifty cells from each specimen across two experiments were selected for calcium imaging ($n = 100$). (g) High K$^+$ buffer stimulated VGCC response magnitudes from NS cells cultured in 2D and 3D 96-well plates. Time course of fluorescence changes was recorded on FlexStation$^\circledR$. Assays were performed in triplicate for both 2D and 3D plates ($n = 3$). * indicates that 2D and 3D response magnitude means compared were significantly different at $p < 0.00001$. # indicates that 2D and 3D response magnitude means compared were significantly different at $p < 0.01$. Error bars are standard deviations. (From Cheng, K., Lai, Y., and Kisaalita, W.S. 2008. *Biomaterials* 29: 2810. With permission.)

2 weeks into differentiation, 99.1% of the 2D cultured cells were responsive to high K$^+$ buffer with a response magnitude decrease to 1.63 ± 0.24 ($p = 0.024$). Meanwhile, 60.3% of the 3D cultured cells were responsive with the response magnitude of 0.6 ± 0.032. The 3D cultured cells' response magnitude was still much lower than that of 2D cultured cells ($p = 5.17e-4$). In sum, the 3D culture's VGCC response magnitudes were significantly lower than those from 2D culture at all the three time

points. This finding was consistent with previous studies (Weaver et al., 1997; Powers et al., 2002; Webb et al., 2003; Bear et al., 2009), suggesting that cellular responses observed in 2D are probably exaggerations of *in vivo* functionality. In other words, NS cells in 3D vessels more closely emulated cells *in vivo* with respect to the VGCC functionality.

4.4.4   HTS Application Feasibility

To further establish the suitability of the 3D 96-well plates for a Calcium FLIPR® Assay (a popular HTS assay format used in the pharmaceutical industry), undifferentiated NS cells were cultured onto 3D plates, loaded with the calcium indicator Fluo-4 and the assay was performed on a FlexStation® I. Figure 4.9e shows typical time courses for a group of wells of traditional 2D and 3D 96-well plates. A similar pattern of calcium dynamic courses within 3D dishes was observed. The differences in response magnitude between 2D and 3D plates were consistent with those between 2D and 3D dishes (Figure 4.9g). The average response magnitude of cells on 3D scaffolds was 0.23 ± 0.025, which is lower than that of 2D plates: 0.49 ± 0.19 ($p = 0.00443$). It is also notable that the standard deviation of 3D plates was in an acceptable range (10% of the mean), suggesting that with further optimization, the quality of HTS assays performed on this new 3D cell-based platform will feasibly meet acceptable industrial standards. A start-up company, Spatiumgen LLC (http://www.spatiumgen.com), is developing the technology toward commercial production of the plate for HTS in drug discovery and preclinical studies.

4.5   CONCLUDING REMARKS

It is now an axiom that the context in which a cell is grown matters and changing its microenvironment can radically alter the resulting phenotype. The goal in this and the previous chapter was to "walk" the reader through some evidence that underpins this axiom at all levels—genomic, proteomic, structural, and functional. The factors that contribute to a specific microenvironment are many. It is therefore necessary to organize them into a few categories to enable a more general discussion and probably systematic study. The literature (e.g., Griffith and Swartz, 2006; Green and Yamada, 2007) has provided guidance that leads to three main categories or MEFs, hereafter: (1) chemical or biochemical composition, (2) spatial (geometric 3D) and temporal dimensions, and (3) force and substrate physical properties. Interestingly, the three factors can be graphically visualized as space described by 3D Cartesian coordinates (Figure 4.10) (Yamada et al., 2003; Yamada and Cukierman, 2007). Within this space, regions exist that give rise to given phenotypes that accurately emulate their *in vivo* counterparts. It is feasible that with some cell types, one of the coordinates may not be necessary (e.g., some cells may be insensitive to force), in which case the space is reduced to a plane. The next three chapters are devoted to a discussion of each of these MEFs. It should be pointed out that although each MEF is considered in its own chapter, in practice the effect on cells is a combination of the three.

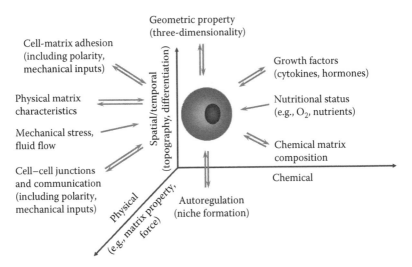

FIGURE 4.10 Three major categories of microenvironment factors (MEFs) that regulate cellular phonotype outcomes. The biochemical coordinate includes short range (e.g., substrate coatings) and long range (e.g., growth factors and nutrients) factors. The physical/force coordinate includes substrate material property (e.g., rigidity and mechanical inputs like stress). The spatial/temporal coordinate includes geometry (topography, aspect ratio, etc.) and proliferation, migration, and differentiation. Taken together, cells respond to environmental cues resulting from information mediated by these factors. (Adapted from Yamada, K.M., Pankov, R., and Cukierman, E. 2003. *Braz. J. Med. Biol. Res.* 36(8): 959–966.)

REFERENCES

Abu-Absi, S.F., Friend, J.R., Hansen, L.K., and Hu, W.S. 2002. Structural polarity and functional bile canaliculi in rat hepatocytes spheroids. *Exp. Cell. Res.* 274: 56–67.

Akay, G. and Birch, M.A. 2004. Microcellular polyHIPE polymer supports osteoblasts growth and bone formation in vitro. *Biomaterials* 25: 3991–4000.

Akhyari, P., Kamiya, H., Haverich, A., Karck, M., and Lichterberg, A. 2008. Myocardial tissue engineering: The extracellular matrix. *Eur. J. Cardiothorac. Surg.* 34: 229–241.

Andrei, G. 2007. Three-dimensional culture model for human viral diseases and antiviral drug development. *Antiviral Res.* 71: 96–107.

Balaban, N.Q., Schwarz, U.S., Riveline, D., Goichberg, P., Tzur, G., Sabanay, I., Mahalu, D., et al. 2001. Force and focal adhesion assembly: A close relationship studied using elastic micropatterned substrates. *Nat. Cell Biol.* 3: 466–472.

Bear, B., Asgian, J., Termin, A., and Zimmermann, N. 2009. Small molecule s targeting sodium and calcium channels for neuropathic pain. *Curr. Opin. Drug Discov. Devel.* 12(4): 543–561.

Bokhari, M., Carnachan, R.J., Cameron, N.R., and Przyborski, S.A. 2007. Culture of HepG2 liver cells on three dimensional polystyrene scaffolds enhances cell structure and function during toxicological challenge. *J. Anat.* 211: 567–576.

Brophy, C.M., Luebke-Wheeler, J.L., Amiot, B.P., Khan, H., Remmel, R.P., Rinaldo, P., and Nyberg, S.L. 2009. Rat hepatocyte spheroids formed by rocked technique maintained differentiated hepatocyte gene expression and function. *Hepatology* 49(2): 578–586.

Chaudhary, K.W., Barrezuelta, N.X., Bauchmann, M.B., Milici, A.J., Beckius, G., Stedman, D.B., Hambor, J.E., et al. 2006. Embryonic stem cell in predictive cardiotoxicity: Laser capture microscopy enables assay development. *Toxicol. Sci.* 90(1): 149–158.

Cheng, K., Lai, Y., and Kisaalita, W.S. 2008. Three-dimensional polymer scaffolds for high throughput cell-based assay systems. *Biomaterials* 29: 2802–2812.

Christopherson, G.T., Song, H., and Mao, H.-Q. 2009. The influence of fiber diameter of electrospun substrate on neural stem cell differentiation and proliferation. *Biomaterials* 30: 556–564.

Desai, A., Kisaalita, W.S., Keith, C., and Wu, Z.-Z. 2006. Human neuroblastoma (SH-SY5Y) cell culture and differentiation in 3-D collagen hydrogels for cell-based biosensing. *Biosens. Bioelectron.* 21: 1483–1492.

Drury, J.L. and Mooney, D.J. 2003. Hydrogels for tissue engineering: Scaffold design variables and applications. *Biomaterials* 24: 4337–4351.

Eglen, R.M., Gilchrist, A., and Reisine, T. 2008. An overview of drug screening using primary and embryonic stem cells. *Comb. Chem. High Throughput Screening* 11(7): 566–572.

Flemming, R.G., Murphy, C.J., Abrams, G.A., Goodman, S.L., and Nealey, P.F. 1999. Effects of synthetic micro- and nano-structured surfaces on cell behavior. *Biomaterials* 20: 573–588.

Garlick, J.A. 2006. Engineering skin to study human disease—tissue models for cancer biology and wound repair. *Adv. Biochem. Eng./Biotechnol.* 103: 207–239.

Green, J.A. and Yamada, K.M. 2007. Three-dimensional microenvironments moderate fibroblast signaling responses. *Adv. Drug Deliv. Rev.* 59: 1293–1298.

Griffith, L.G. and Swartz, M.A. 2006. Capturing complex 3D tissue physiology *in vitro*. *Nat. Rev. Mol. Biol.* 7: 211–224.

Groneberg, D.A., Grosse-Siestrup, C., and Fisher, A. 2002. *In vitro* models to study hepatotoxicity. *Toxicol. Pathol.* 30(3): 394–399.

Hayman, M.W., Smith, K.H., Cameron, N.R., and Przyborski, S.A. 2005. Growth of human stem cell-derived neurons on solid three-dimensional polymers. *J. Biochem. Biophys. Methods* 62: 231–240.

Ise, H., Takashima, S., Nagaoka, M., Ferdous, A., and Akaike, T., 1999. Analysis of cell viability and differential activity of mouse hepatocytes under 3D and 2D culture in agarose gel. *Biotechnol. Lett.* 21: 209–213.

John, V.H., Dale, T.J., Hollands, E.C., Chen, M.X., Partington, L., Downie, D.L., Meadows, H.J., and Trezie, D.J. 2007. Novel 384-well population patch clamp electrophysiology assays for Ca^{2+}-activated K^+ channels. *J. Biomol. Screening* 12(1): 50–60.

Justice, B.A., Badr, N.A., and Felder, R.A. 2009. 3D cell culture opens new dimensions in cell-based assays. *Drug Discov. Today* 14(1/2): 102–107.

Kane, B.J., Zinner, M.J., Yarmush, M.L., and Toner, M. 2006. Liver-specific functional studies in microfluidic array of primary mammalian hepatocytes. *Anal. Chem.* 78: 4291–4298.

Kim, J., Park, J., Na, K., Yang, S., Baek, J., Yoon, E., Choi, S., Lee, S., Chun, K., Park, J., and Park, S. 2008. Quantitative evaluation of cardiomyocyte contractility in a 3D microenvironment. *J. Biomech.* 41: 2396–2401.

Kim, J.B. 2005. Three-dimensional tissue culture models in cancer biology. *Semin. Cancer Biol.* 15: 365–377.

Lai, Y., Cheng, K., and Kisaalita, W.S. 2010. Comparative transcriptional profiling of freshly dissected (in vivo surrogate) and 3D scaffold-cultured neurons. *Biomaterials.* Submitted.

Liguori, M.J., Blomme, E.A.G., and Waring, J.F. 2008. Trovafloxacin-induced gene expression changes in liver-derived *in vitro* systems: Comparison of primary human hepatocytes to HepG2 cells. *Drug Metab. Dispos.* 36(2): 223–233.

Lorenzon, N.M. and Beam, K.G. 2008. Diseases causing mutations of calcium channels. *Channels* 2(3): 163–179.

Mains, R.E. and Patterson, P.H. 1973. Primary cultures of dissociated sympathetic neurons: I. Establishment of long-term growth in culture and studies of differentiated properties. *J. Cell. Biol.* 59: 329–345.

Mani, S.K., Shiaishi, H., Balasubramania, S., Yamane, K., Chellaiah, M., Cooper, G., Banik, N., Zile, M.R., and Kuppuswamy, D. 2008. *In vivo* administration of calpetin attenuates calpain activation and cardiomyocyte loss in pressure-overloaded felin myocardium. *Am. J. Physiol. Heart Circ.* 295: H314–H326.

Mao, C. and Kisaalita, W.S. 2004. Characterization of 3-D collagen gels for functional cell-based biosensing. *Biosens. Bioelectron.* 19: 1075–1088.

Martínez, E., Engel, E., Planell, J.A., and Samitier, J. 2008. Effects artificial micro- and nanostructured surfaces on cell behavior. *Ann. Anat.* 191(1): 126–135.

Mayne, J.T. and Ku, W.W. 2006. Informed toxicity assessment in drug discovery: Systems-based toxicology. *Curr. Opin. Drug Discov. Dev.* 9: 75–83.

Miret, S., Groene, E.M., and Klaffke, W. 2006. Comparison of *in vitro* assays of cellular toxicity in the human hepatic cell line HepG2. *J. Biomol. Screening* 11(2): 184–193.

Mohan, C.G. and Gandhi, T. 2008. Therapeutic potential of voltage gated calcium channels. *Mini Rev. Med. Chem.* 8(12): 1285–1290.

Morgan, E.H. and Peters, Jr., T. 1971. The biosynthesis of rat serum albumin. V. Effect of protein depletion and refeeding on albumin and transferring synthesis. *J. Biol. Chem.* 246: 3500–3507.

Panetta, R. and Greenwood, M.T. 2008. Physiological relevance of GPCR oligomerization and its impact on drug discovery. *Drug Discov. Today* 13(23–24): 1059–1066.

Park, J., Ryu, J., Choi, S.K., Seo, E., Cha, J.M., Ryu, S., Kim, J., Kim, B., and Lee, S.H. 2005. Real-time measurement of the contractile forces of self-organized cardiomyocytes on hybrid biopolymer microcantilevers. *Anal. Chem.* 77: 6571–6580.

Peshwa, M.V., Wu, F.J., Sharp, H.L., Cerra, F.B., and Hu, W.S. 1996. Mechanistics of formation and ultrastructural evaluation of hepatocytes spheroids. *In Vitro Cell. Dev. Biol. Anim.* 32: 197–203.

Powers, M.J., Domansky, K., Kaazempur-Mofrad, M.R., Kalezi, A., Capitano, A., Upadhyaya, A., Kurzawski, P. et al. 2002. A microfabricated array for perfused 3D liver culture. *Biotechnol. Bioeng.* 78(3): 257–269.

Schmeichel, K.L. and Bissell, M.J. 2003. Modeling tissue-specific signaling and organ function in three dimensions. *J. Cell Sci.* 116(2): 2377–2388.

Vickers, A.E. 2009. Tissue slices for the evaluation of metabolism-based toxicity with the example of diclofenac. *Chem. Biol. Interact.* 179(1): 9–16.

Vickers, A.E. and Fisher, R.L. 2004. Organ slices for the evaluation of human toxicity. *Chem. Biol. Interact.* 150(1): 87–96.

Wagner, F., Kraft, R., Härtig, W., Schaarschmidt, G., Schwartz, S.C. Schwarz, J., and Hevers, W. 2008. Functional and molecular analysis of GABA receptors in human midbrain-derived neural progenitor cells. *J. Neurochem.* 107(4): 1056–1069.

Weaver, V.M., Petersen, O.W., Wang, F., Larabell, C.A., Briand, P., Damsky, C., and Bissell, M.J. 1997. Reversion of the malignant phenotype of human breast cells in three-dimensional culture and *in vivo* by integrin blocking antibodies. *J. Cell Biol.* 137(1): 231–245.

Webb, K., Li, W.H., Hitchcock, R.W., Smeal, R.M., Gray, S.D., and Tresco, P.A. 2003. Comparison of human fibroblast ECM-related gene expression on elastic three-dimensional substrates relative to two-dimensional films of the same material. *Biomaterials* 24(25): 4681–4690.

Weiwad, W.K.K., Linke, W.A., and Wussling, M.H.P. 2000. Sarcomere length–tension relationship of rat cardiac myocytes at lengths greater than optimum. *J. Mol. Cell. Cardiol.* 32: 247–259.

Whitby, K.E. 1987. Teratological research using *in vitro* systems. III. Embryonic organ culture. *Environ. Health Perspect.* 72: 221–223.

Wu, G.Y., Wu, C.H., and Stocker R.J. 1983. Model for specific rescue of normal hepatocytes during methotrexate treatment of hepatic malignancy. *Proc. Natl. Acad. Sci. USA* 80: 3078–3080.

Wu., Z.-Z., Zhao, Y.-P., and Kisaalita, W.S. 2006. A packed Cytodex microbead array for three-dimensional cell-based biosensing. *Biosens. Bioelectron.* 22: 685–693.

Xu, M., Kreeger, P.A., Shea, L.D., and Woodruff, T.K. 2006. Tissue-engineered follicles produce live, fertile offspring. *Tissue Eng.* 12(10): 2739–2746.

Yamada, K.M. and Cukierman, E. 2007. Modeling tissue morphogenesis and cancer in 3D. *Cell* 130: 601–610.

Yamada, K.M., Pankov, R., and Cukierman, E. 2003. Dimensions and dynamics in integrin function. *Braz. J. Med. Biol. Res.* 36(8): 959–966.

Yanagi, K., Takano, M., Narazaki, G., Uosaki, H., Hoshino, T., Ishii, T., Misaki, T., and Yamashita, J.K. 2007. Hyperpolarization-activated cyclic nucleotide-gated channels and T-type calcium channels confer automaticity of embryonic stem cell-derived cardiomyocytes. *Stem Cells* 25(11): 2712–2719.

Ye, N.N., Qin, J.H., Shi, W.W., Liu, X., and Lin, BC. 2007. Cell-based high content screening using an integrated microfluidic device. *Lab on a Chip* 7(12): 1696–1704.

Yin, L., Bien, H., and Entcheva, E. 2004. Scaffold topography alters intracellular calcium dynamics in cultured cardiomyocytes networks. *Am. J. Physiol. Heart Circ. Physiol.* 287: H1276–H1285.

Zhao, Y., Lim, C.C., Sawyer, D.B., Liao, R., and Zhang, X., 2007. Simultaneous orientation and cellular force measurements in adult cardiac myocytes using three-dimensional polymeric microstructures. *Cell Motil. Cytoskeleton* 64: 718–725.

Zhao, Y. and Zhang, X. 2005. Contractile forces measurement in cardiac myocytes using PDMS pillar arrays. In *Proceedings of IEE MicroElectroMechanical Systems*, Florida, USA, pp. 843–837.

Part III

Emerging Design Principles

5 Chemical Microenvironmental Factors

5.1 CELL ADHESION MOLECULES

Cell adhesion molecules (CAMs) are cell-surface molecules grouped into four subclasses based on their structural characteristics: (1) cadherins, (2) the integrin family, (3) selectins, and (4) the immunoglobulin (Ig)-domain containing superfamily of CAMs (Paschos et al., 2009). CAMs are in most cases transmembrane proteins responsible for mediating adhesion of cells to other cells and/or the extracellular matrix (ECM) via their extracellular domains. CAMs provide the "bridge" between the extracellular and intracellular scaffolds; while the extracellular domain mediates adhesion, the intracellular domain interacts with the cytoskeleton (Hulpiau and van Roy, 2009). CAMs are not merely "molecular glue," as some have been associated with numerous signaling events with major physiological implications (Wheelock and Johnson, 2003; Resink et al., 2009; Takeichi, 2007). CAMs interact with like molecules (hemophilic interactions) and nonlike molecules (heterophilic interaction) on neighboring cells or ECM. This enables cell position and dynamic interactions with other cells through mechanisms that depend on intracellular signaling, ultimately resulting in contact-mediated attraction or repulsion and in chemoattraction or chemorepulsion (Maness and Schachner, 2007). More details are provided below for each of the four CAM subclasses.

5.1.1 CADHERINS

Nearly 30 years ago, cadherins were identified as calcium-dependent hemophilic cell CAMs (Yoshida and Takeichi, 1982; Gallin et al., 1983; Peyrieras et al., 1983). Cadherins are glycoproteins, identified in vertebrates as well as in invertebrates, representing a superfamily of more than 100 members in vertebrates. They have been subdivided into subfamilies depending on their amino acid sequence. As shown in Figure 5.1, the subfamilies are designated as classic Types I and II (e.g., E-, N-, P-, R- and VE-), desmosomal (e.g., desmogleins and desmocollins), protocadherins (e.g., R-, CNR-, and ARCADLIN-), Flamingo/CELSR (e.g., Celsr1, Celsr2, and Celsr3), and FAT (e.g., Fat1) (Resink et al., 2009). However, some cadherins do not fit in the above subfamilies. Examples include T-cadherin, LI-cadherin, and RET proto-oncoproteins (Koch et al., 2004). A key feature that qualifies proteins as members of

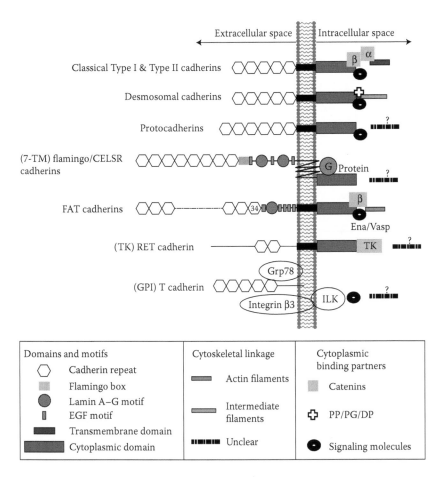

FIGURE 5.1 Schematic of basic differences in the molecular features of members of cadherin superfamily. A common characteristic of all the members of the superfamily is the reserved sequences involved in calcium binding, also referred to as "cadherin repeats" of varying numbers. With 34 repeats, FAT turns out to be the largest cadherin. Nonclassical cadherins have additional extracellular motifs such as laminins A–G and EGF domains and flamingo boxes. There is a lot of variation in the cytoplasmic domain among the subfamilies. For example, Types I and II cadherins associate with actin cytoskeleton through the β-catenin complex, whereas desmosomal cadherins associate with the intermediate filaments through plankophilin (PP), plankoglobin (PG), and desmoplakin (DP). FAT cadherins interact with β-catenin and the Ena/Vasp family of actin regulators. How the other subfamilies interact with the cytoskeleton is not yet well understood. The "bridging" between the extracellular and intracellular scaffolds is achieved through interaction with signaling molecules by the cytoplasmic domain. Examples of these molecules include cytoplasmic and transmembrane proteins that participate in cellular signaling such as GFRs, receptor tyrosine kinases and phosphatases, PI3-kinases, Shc, and small GTPases. Another mode of interaction involves control of cytoskeletal dynamics through formin-1, Arp2/3, and dynein. These interactions may be both cadherin and cell-specific. Abbreviations: 7-TM = seven-transmembrane domain, TK = tyrosine kinase, GPI = glycosylphosphatidylinositol anchor, ILK = integrin-linked kinase. (From Resink, T., et al. 2009. *Swiss Med. Wkly.* 139: 122–124. With permission.)

the cadherin family is the repetitive conserved sequence in the extracellular domain (cadherin repeat) involved in calcium binding. An overview of cadherin-related 3D structures is available in the Worldwide Protein Data Bank (http://www.wwpdb.org). It is through this repeat that cadherins establish intercellular adhesion and other biological functions including but not limited to cell recognition and sorting, boundary formation in tissues, induction and maintenance of structural and functional cell and tissue polarity, cytoskeletal organization, cellular phenotype modulation, cell migration, cell proliferation, and cell survival (Suzuki and Takeichi, 2008; Resink et al., 2009). Defects in cadherin expression lead to disruption of normal tissue, which can lead to disease; cadherins are just beginning to be explored as targets in drug discovery (Mascini et al., 2007; Vaidya and Welch, 2007; Luk and Wong, 2006).

5.1.2 Selectins

Members of the selectin family, like the cadherins, are calcium-dependent transmembrane glycoproteins, with the exception that all selectins exhibit an extracellular lectin-like domain. Selectins comprise three CAMs: E-selectins, present exclusively in endothelial cells, P-selectins in platelets and endothelial cells, and L-selectins in leukocytes. Figure 5.2 shows a schematic of selectin structures. Their cytoplasmic tail mediates their action as signaling receptors, while they require carbohydrates to

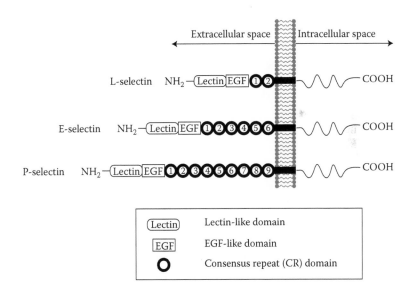

FIGURE 5.2 Schematic of selectin structures. In all cases, the extracellular N-terminus carries the lectin-like domain followed by an EGF-like domain and "consensus repeats" in various numbers. Membrane anchorage is achieved with a single transmembrane domain (TMD) and the cytoplasmic domain is made up of a small tail. With nine repeats, P-selectin turns out to be the longest in humans. (Adapted from Kneuer, C., et al. 2006. *Drug Discov. Today* 11(21/22): 1034–1040.)

mediate adhesion. Selectin ligands with high affinity include primarily oligosaccharides and sulfopolysaccharidess. These ligands usually represent the binding epitope of physiological selectin counter receptors, including glycoproteins and glycolipids. Table 5.1 presents a sample of selectin–ligand interaction parameters (Hoke et al., 1995) to illustrate the high binding constants. For a definition of the equilibrium dissociation constant (K_D), see Box 5.1. Selectins are expressed when certain mediators are present. Examples include interleukins, tumor necrosis factor, or toxins (Witz, 2008; Garcia et al., 2007; Borsig, 2004). Selectins mediate the initial tethering of circulating blood cells to endothelial cells or among each other to deliver immune cells to the site of inflammation or injury (Springer, 1994; Worthylake and Burridge, 2001). Because circulating cells in the blood stream are subjected to high shear stresses, the immigration of leukocytes from the circulation requires a sophisticated mechanism. One of the recent discoveries in cellular adhesion is biological adhesive bonds, called catch bonds that are enhanced by tensile mechanical force. When a receptor–ligand bond is regulated by tensile mechanical force to become longer lived at higher forces, it is called a catch bond (Marshall et al., 2003; Isberg and Barnes, 2002). Selectins have been shown to form catch bonds and Thomas (2008) has

---

TABLE 5.1
Selectin–Counter Receptor (or Ligand) Interaction Parameters

Selectin	Ligand	K_D (μM)	IC_{50} (μM)
P-selectin	PSGL-1 (human)	0.2–0.32	
	rPSGL-IgG	0.06	520–1300
	sLex	7800	220
	6-sulfo-sLex		
E-selectin	ESL-1 (mouse)	56–62	
	Recombinant ESL-1	66	100–750
	sLex	100–2000	
L-selectin	GlyCAM-1 (mouse)	108	
	sLex	3900	2300
	6-sulfo-sLex		800

Abbreviations: K_D, equilibrium dissociation constant; IC_{50}, concentration that yields 50% inhibition; PSGL-1, P-selectin glycoprotein ligand-1 (CD162), a mucin-like transmembrane protein that forms homodimers by linking two 120 kDa chains via a disulfide bridge; ESL-1, E-selectin ligand; sLex, sialy-Lewis x epitopes present on *O*-glycans of various glycoproteins in high endothelial venules; 6-sulfo-sLex, sulfated sLex; and GlyCAM-1, glycosylation-dependent CAM-1, secreted and primarily transducer signals into leukocytes rather than support adhesion.

Source: Adapted from Hoke, D., et al. 1995. *Curr. Biol.* 5(6): 670–678.

BOX 5.1 SIMPLE RECEPTOR-COUNTER RECEPTOR (OR LIGAND) BINDING

Consider reversible equilibrium binding of free receptor [R] to a free ligand [L], forming a receptor–ligand complex [C], expressed in equation form as shown below:

$$R + L \leftrightarrow C$$

with association and dissociation constants of k_f and k_r, respectively.

Typically, the units of [R] and [L] are number/cell and moles/liter, or M, respectively. However, in some cases, units of moles/volume solution or density (number/cell surface area) may be used. The units of k_f and k_r are M^{-1} time^{-1} and time^{-1}, respectively.

The equilibrium dissociation constant, K_D, is defined as

$$K_D = \frac{[R][L]}{[C]}.$$

The equilibrium association constant is defined as

$$K_A = \frac{k_f}{k_r} = \frac{1}{K_D}.$$

K_D values vary widely depending on the system; values less than 10^{-12} M represent high affinity, such as the avidin–biotin bond ($K_D = 10^{-15}$ M) (Green, 1975). Values around 10^{-6} M represent low affinity such as fibronectin–fibronectin receptor bond ($K_D = 0.86 \times 10^{-6}$ M) (Akiyama and Yamada, 1985).

Source: Adapted from Lauffernburger, D.A. and Liderman, J.J. 1993. *Receptors: Models for Binding, Trafficking and Signaling.* New York: Oxford University Press.

reviewed the literature on how catch bonds mediate shear-enhanced adhesion of cells or artificial surfaces. The frequent involvement of selectins in serious illnesses such as cancer, asthma, allergy and autoimmune reactions and transplant failures have made them potential targets for drug discovery (Kneuer et al., 2006). However, only a few selectin-targeting drugs have progressed to clinical trial and only a limited number have proven their *in vivo* potential.

5.1.3 The Integrin Superfamily

Recognized a little over two decades ago as a receptor family (Haynes, 1987), integrins are now well established as a cell-surface receptors system, constituting the major component of cell–cell adhesion receptors. Also, cells use integrins to assemble and recognize a functional ECM. In addition to mediating cell adhesion, integrins make transmembrane connections to the cytoskeleton, as shown in the ECM schematic in Figure 5.3, and activate many intracellular signaling pathways (Zaidel-Bar et al., 2007; Streuli and Akhtar, 2009). Integrins are unusual transmembrane receptors; they are bidirectional, meaning that they mediate information flow from extracellular stimuli to induce intracellular changes as well as intracellular stimuli to induce extracellular changes (Zaidel-Bar et al., 2007; Legate et al., 2009).

Integrins are composed of α- and β-dimers, which combine to form heterodimeric receptors (Figure 5.4). Each subunit crosses the membrane once, with most of each polypeptide (>1600 amino acids in total) in the extracellular space and two short peptides (20–50 amino acids) in the cytoplasmic domain (Haynes, 2002; Moser et al.,

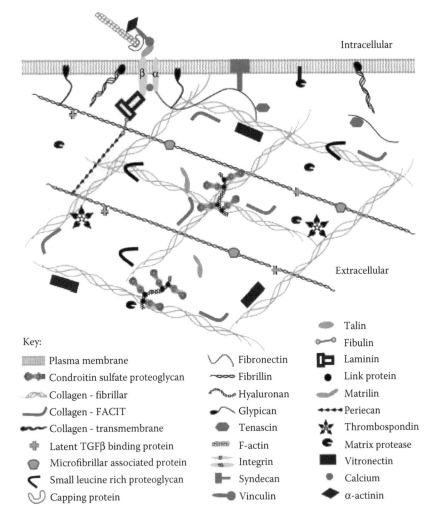

FIGURE 5.3 Schematic representation of the ECM depicting major classes of extracellular molecules. (From Huxley-Jones, J. et al. 2008. *Drug Discov. Today* 13(15/16): 685–694. With permission.)

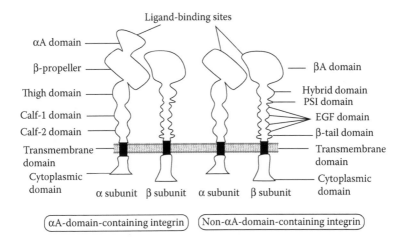

FIGURE 5.4 Integrin domain structure. (Adapted from Humphries, J.D., Byron, A., and Humphries, M.J. 2006. *J. Cell Sci.* 119: 3901–3903.)

2009). To date, there are 18 α and 8 β subunits known to assemble into 24 distinct integrins, each with its own ligand specificity (Humphries et al., 2006). Table 5.2 depicts the mammalian α–β subunit associations and their ligands (Humphries et al., 2006).

5.1.4 THE IG-DOMAIN-CONTAINING SUPERFAMILY OF CAMS

This Ig superfamily is one of the most ancient and diverse groups of CAMs, which are surface and soluble proteins. These proteins are involved in intercellular recognition, binding, and adhesion. Subgroups include antigen receptors, growth factors, and cytokine receptors (e.g., IL-6 receptor), antigens of tumor cells (e.g., carcinoembryonic antigen, CEA), T-lymphocytes and natural killer cell receptors (e.g., CD4), and intercellular adhesion molecules (e.g., ICAM-1). Known CAMs of the superfamily are presented in Table 5.3. All these CAMs present an Ig-like structure in their extracellular part as shown in Figure 5.5. The ICAMs are transmembrane glycoproteins with 2–9 Ig domains, a hydrophobic transmembrane region and a short intracellular domain.

5.2 SHORT-RANGE CHEMISTRY

It is well established that CAMs do not only "glue" cells together or on noncellular substrates but are also associated with numerous signaling events. It is not therefore surprising that the surface chemistry (short-range chemistry) plays major roles in the resultant cellular behavior and/or function. To be able to reproduce the *in vivo* phenotype as accurately as possible, the approach adopted by most cell culturists is to present a substrate to the cell that resembles the ECM, chemically. The question therefore to answer is what is the practice, but before answering this question, it is important to have a good understanding of the ECM and its chemistry.

TABLE 5.2

Major Integrin–Ligand Combinations

Ligand	Integrin Receptors
Collagen	$\alpha10\beta1$, $\alpha2\beta1$, $\alpha2\beta1$, $\alpha11\beta1$, **$\alpha X\beta2$**
Laminin	$\alpha10\beta1$, $\alpha2\beta1$, $\alpha2\beta1$, _$\alpha7\beta1$_, _$\alpha6\beta4$_, _$\alpha6\beta4$_, _$\alpha6\beta1$_, _$\alpha3\beta1$_
Thrombospondin	_$\alpha3\beta1$_, $\alpha2\beta1$ _**$\alpha4\beta1$**_, $\alpha V\beta3$, $\alpha IIb\beta4$
Fibronectin	$\alpha IIb\beta3$, $\alpha V\beta3$, $\alpha V\beta6$, $\alpha V\beta1$, $\alpha5\beta1$, $\alpha8\beta1$, _**$\alpha4\beta1$**_, _**$\alpha4\beta7$**_
VCAM-1	_$\alpha D\beta2$_, _$\alpha9\beta1$_, _**$\alpha4\beta7$**_, _**$\alpha4\beta1$**_
ICAM	_**$\alpha X\beta2$**_, _**$\alpha M\beta2$**_, _**$\alpha L\beta2$**_, _**$\alpha D\beta2$**_
MAdCAM-1	_**$\alpha4\beta7$**_, _**$\alpha4\beta1$**_
E-cadherin	_**$\alpha E\beta7$**_
iC3b	_**$\alpha X\beta2$**_, _**$\alpha M\beta2$**_
Factor X	_**$\alpha M\beta2$**_
Fibrinogen	_**$\alpha M\beta2$**_, _**$\alpha X\beta2$**_, $\alpha IIb\beta3$, $\alpha V\beta3$
Fibrillin	$\alpha V\beta3$
LAP-TGF-β	$\alpha V\beta3$, $\alpha V\beta6$, $\alpha V\beta8$, $\alpha V\beta1$
Tenascin	$\alpha8\beta1$, _**$\alpha9\beta1$**_, $\alpha V\beta3$
vWF	$\alpha V\beta3$, $\alpha IIb\beta3$
Vitronectin	$\alpha IIb\beta3$, $\alpha8\beta1$, $\alpha V\beta3$, $\alpha V\beta5$
MFG-E8 Del-1	$\alpha V\beta5$, $\alpha V\beta3$
BSP	$\alpha V\beta5$, $\alpha V\beta3$
Osteopontin	$\alpha V\beta5$, $\alpha V\beta3$, $\alpha V\beta6$, $\alpha V\beta1$, $\alpha5\beta1$, $\alpha8\beta1$, _**$\alpha9\beta1$**_, _**$\alpha4\beta1$**_, _**$\alpha4\beta1$**_
PECAM-1	$\alpha V\beta3$

Key to integrin receptors: bold, RGD-binding; underlined bold, LDV binding; underlined bold italics, αA-domain-containing LDV binding; italics, αA-domain-containing β1; underlined italic, non-αA-domain-containing laminin binding.

Ligand abbreviations: BSP, bone sialoprotein; Del-1, developmental endothelial locus-1; EGF, epidermal growth factor; ICAM, intercellular cell adhesion molecule; iC3b, inactivated complement C3b; LAP-TGF-β, latency-associated peptide transforming growth factor β; MAdCAM-1, mucosal addressin cell adhesion molecule 1; MFG-E8, milk fat globule EGF factor 8; PECAM-1, platelet endothelial cell adhesion molecule 1 (CD31); VCAM-1, vascular cell adhesion molecule 1; and vWF, von Willebrand factor.

Source: Adapted from Humphries, J.D., Byron, A., and Humphries, M.J. 2006. *J. Cell Sci.* 119: 3901–3903.

5.2.1   ECM COMPOSITION

The ECM is composed of a great variety of organized structural and functional proteins including the collagen family, elastin fibers, glycosoaminoglycans (GAGs) and proteoglycans, and adhesive glycoproteins. Figure 5.3 depicts the major classes of these molecules in an ECM schematic. By combining spatial and temporal availability of a wide variety of chemical species to choose from, nature is able to give rise to distinct tissues and organs.

Collagen is the most abundant protein in vertebrates accounting for about 30% of the body's total proteins and is essentially present in all tissues and organs of the

TABLE 5.3
CAMs of the Ig Superfamily

Subfamily	Members/Groups	Reference
ICAMs	ICAM-1 (CD54), -2, -3, -4, and -5	Jimenez et al. (2005); Yang et al. (2004)
Vascular cell adhesion molecule (VCAMs)	VCAM-1 (CD106)	Wu (2007)
Platelet endothelial cell adhesion molecule (PECAMs)	PECAM-1 (CD31)	Newman (1997); Gong and Chatterjee (2003)
Neural adhesion molecule (NCAM; CD56)	NCAM-120, -140, -180, PSA (polysialylated)-NCAM	Nielsen et al. (2008)
L1	L1CAM, NrCAM, neurofascin, CHL1	Maness and Schachner (2007)
Carcinoembryonic antigen cell adhesion molecule (CEACAMs)	CEACAM1, CEACAM3–8, CEACAM16, CEACAM18–21	Kuespert et al. (2006); Hammarstrom (1999)

body. Collagens are the major component of the ECM (Canty and Kadler, 2002). At least 210 different collagens have been characterized and shown to be tissue specific and have different functional properties (Table 5.4). For example, Type III is found only in cartilage (Zhang et al., 2003) and types IV, VII, IX, X, and XII are found associated with collagen fibrils or organized in the network of basement membranes. Collagen is usually organized in fibrils that resist tensile, shear, or compressive forces in tendons, arteries, bone, skin, and other force-baring structures (Silver et al., 2000). Types I, II, and III are the most abundant in humans and form fibrils responsible for tensile strength of tissues (Hulmes, 1992). Other types of collagens are present in varying amounts in different tissues. These collagens perform important functions, for example, initiating fibril formation and limiting fibril diameter (Aigner and Stove, 2003). Besides mechanical and structural functions, collagens are critical in cell attachment and spreading and as such influence cell migration and differentiation (Rosso et al., 2004).

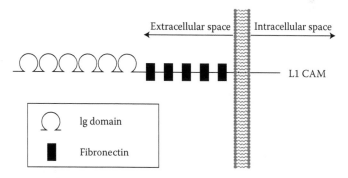

FIGURE 5.5   Representative schematic structure of L1 CAM, displaying six Ig-like domains at the N-terminal and five fibronectin III (FN-III) repeats. Other CAMs in this superfamily display differing numbers of Ig-like domains and FN-III repeats. (Adapted from Zhang, Y., et al. 2008. *Restor. Neurol. Neurosci.* 26: 81–96.)

TABLE 5.4
Collagen Superfamily

Type	Classification by Assembly Mode	Comment
I	Fibril-forming (F-F)	Heterotrimeric protein. It is the major fibrillary collagen of bone (reinforced with calcium hydroxyapatite), tendon (rope-like structure), and skin (sheet-like structure); in the fibril, the neighboring collagen molecules are packed into a quarter-staggered array
II	F-F	Homotrimer; exists in two forms resulting from alternative splicing; abundant in cartilage, intervertebral disc and vitreous humor; supports chondrocyte adhesion
III	F-F	Homotrimer abundant in hollow organs, dermis, placenta, and uterus where it occurs as fine banded, D-periodic fibrils. It is often found as a partially processed pNcollagen that contains the N-propeptides but not the C-propeptides
V	F-F	Frequently occurs as heterotypic fibrils with type I collagen in cornea and placenta. Can be homotrimers of $proα1(V)$, or heterotrimers of two $proα1(V)$ and one $proα2(V)$ chains, or heterotrimers of $proα1(V)$, $proα2(V)$, and $proα3(V)$
XI	F-F	Heterotrimeric procollagen of three distinct chains
IX	Fibril-associated collagen with interrupted triple (FACIT) helices	These do not form fibrils themselves but attach to the surfaces of preexisting collagen fibrils such as types I and II. It is a heterotrimers of $α1(XI)$, $α2(XI)$, and $α3(XI)$. It has an antiparallel orientation relative to the collagen II molecules within the fibrils
XII	FACIT	Homologous to type XIV. Homotrimeric molecules. Alternative splicing generates different lengths of molecules XIIA and XIIB
XIV	FACIT	Homologous to type XII. NC3 domain is smaller than in collagen XIIB. The same as type XII, type XIV are found in most dense connective tissues
XVI	FACIT	Contain multiple triple-helical and non-triple-helical domains. Have a wide range of expressing tissues
XIX	FACIT	Discovered through cDNA cloning with RNA from human cell line (CCL136). Contains five triple-helical domains, interspersed with and flanked by six non-triple-helical domains
XX	FACIT	Most prevalent in corneal epithelium and also detectable in embryonic skin, sterna cartilage, and tendon
XXI	FACIT	An ECM component of the blood vessel walls, secreted by smooth-muscle cells
IV	Non-F-F—network-forming	Basement membrane-associated collagen, exclusively exists in basement membranes; composed of three α chains selected from the translated products of six genetically distinct genes; critical in the separation of cell sheets from underlying or surrounding connective tissue or from different types of cell sheets

TABLE 5.4 (continued)
Collagen Superfamily

Type	Classification by Assembly Mode	Comment
VIII	Non-F-F—network-forming	A specialized basement membrane elaborated by corneal endothelial cells. Also synthesized by vascular endothelial cells and epithelia. Major component of Descemet's membrane; short-chain collagen with only about 60 kDa subunits. Hexagonal network
X	Non-F-F—network-forming	Hypertrophic and mineralizing cartilage
VII	Anchoring fibril	Forms anchoring fibrils in dermal epidermal junctions
VI	Beaded microfibril	Most interstitial tissue, associated with type I collagen
XIII	Transmembrane	Transmembrane collagen interacts with integrin $\alpha 1\beta 1$, fibronectin, and components of basement membranes like nidogen and perlecan
XVII	Transmembrane	Also known as BP180, a 180 kDa protein
XV	Multiplexin	Expressed in specialized basement membranes. Source of endostatin
XVIII	Multiplexin	Expressed in specialized basement membranes. Source of restin

Source: Adapted from Kadler, K. 1995. *Protein Profile* 2(5): 491–619; Kreis, T. and Vale, R. (eds), 1999. *Guide Book to Extracellular Matrix Anchor, and Adhesion Proteins*, 2nd edition. New York: Oxford University Press.

GAGs are linear polysaccharides formed by repeating disaccharide units that may contain sulfate groups. The most common sulfated GAG chains are chondroitins, keratins, and dermatans. The most common nonsulfated GAG is hyaluronan. Hyaluronan provides matrix structure and lubrication and is involved in cell locomotion (Toole, 2004). Heparans form another GAG class that is mainly associated with basement membranes. Proteoglycans are GAGs associated with a protein backbone. Proteoglycans are responsible for the spatial organization of the ECM in tissues such as skin. They are also present on the cell surface and act as receptors for various hormones and growth factors and adhesive glycoproteins (Ruoslahti, 1988). One class of well-characterized proteoglycans is cartilage aggrecans whose structure consists of several chondroitinsulfate and keratinsulfate GAGs covalently attached to a linear polypeptide. Aggregates of these aggrecans are assembled by hyaluronan, to which the individual aggrecan core protein binds, forming huge water-containing pockets that enable the cartilage to withstand the compression generated during joint movement (Kiani et al., 2002). Another class of interesting proteoglycans is the syndecan family of four (in mammals). Syndecans are type I transmembrane heparin sulfate proteoglycans that modulate cell adhesion, cell–cell interactions, and ligand–receptor interactions. One of the members (Syndecan-3) is a novel regulator of feeding behavior and body weight (Reizes et al., 2008). Another member (Syndecan-2) has been implicated in wound healing (Verderio et al., 2009).

Elastin is an amorphous protein that makes up the elastic fibers. Large quantities of elastic fibers are found in arterial walls, dermis, and the lungs. The elastic fiber function is to provide elasticity. With regard to creating *in vivo*-like microenvironments *in vitro*, elastins have been ignored mainly because the protein lacks the traditional binding domains. However, elastin has been considered in fabricating tunable hydrogels that incorporate fibronectin-derived GRGDS domains (Jia and Kiick, 2009).

Key members of the adhesive glycoprotein class of ECM molecules include fibronectin, laminin, vitronectin, thrombosponin, and tenascin (Rosso et al., 2004). Fibronectins are abundant in the ECM (insoluble form) and in body fluids (soluble form or long range) (Haynes, 1990). Fibronectin is a dimer made up of two not necessarily identical subunits linked by a pair of disulfide bonds at their carboxyl terminals. Functional distinctive properties are displayed by monomer folds (type I, II, and II repeats), which may contain binding sites for other ECM proteins like collagen and GAGs, cell-surface receptors. There are variant forms of fibronectin displaying multiple RGD, RGDS, LDV, and REDV sequences that are responsible for cell adhesion (Kao, 1999). In addition to cell adhesion, other critical roles that fibronectin plays include migration, maintenance of normal cell morphology, homeostasis, thrombosis, wound healing, differentiation, and proliferation (Kaspar et al., 2005, 2006). Due to its broad binding properties, fibronectin is widely used in cell culture systems to promote cell adhesion and spreading.

The laminin family of glycoproteins was first discovered as a product of mouse Engelbreth–Holme–Swarm (EHS) sarcoma cells (Timpl et al., 1979). Laminins are extracellular heterotrimeric glycoproteins composed of various combinations of α, β, and γ subunits. Currently, five α, four β, and three γ subunits as well as their splice variants have been identified (Aumailley et al., 2005) and these subunits are known to form 16 known laminins in mammals (Tzu and Marinkovich, 2008) (Table 5.5). These adhesive proteins are characterized by high binding affinity for the cell surface as well as for heparin and type IV collagen. They exhibit the RGD and other sequences such as PDSGR, YIGSR, and IKVAV known to recognize and bind to cell-surface receptors (Timpl et al., 2000). Laminins play a significant role in basement membrane assembly, architecture, and regulation of cellular differentiation, adhesion, and migration. Other basement membrane components, besides laminin, are collagen IV, perlecan, and nidogen (Hohenester and Engel, 2000). Detailed presentation of the relationship between structure and function as well as the roles in pathological processes is presented in an excellent review by Tzu and Marinkovich (2008). Due to its high binding affinity, laminin is widely use either alone or in combination with other ECM molecules in cell culture systems to promote attachment and spreading (El-Ghannam et al., 1998).

Like fibronectin, vitronectin can be found in fibrillar form in the ECM and in soluble form in the blood. Through its RGD sequence, it can bind to integrin receptors, cell surfaces, as well as collagen, heparin. For detailed discussions of the other ECM glycoproteins not mentioned above, the following reviews can be consulted: tenascin (Jones and Jones, 2000) and thrombospondin (Tucker, 2004).

5.2.2 SUBSTRATE SURFACE CHEMISTRY

A schematic of cell-surface adhesion via a receptor (integrin)-adhesive protein (fibronectin) binding is shown in Figure 5.6. For cell adhesion to occur, to unmodified surfaces, a layer of adsorbed protein is necessary and this adsorption is affected by

TABLE 5.5
Laminin Superfamily

Type	α	β	γ	Comments
Laminin 1	1	1	1	Associated with integrins, dystroglycan, etc.; involved in neurite outgrowth, cell migration, milk protein production, and AchR clustering on myotubes
Laminin 2	2	1	1	Associated with integrins, dystroglycan, heparin sulfates, etc.; involved in neurite outgrowth, receptor/cytoskeletal clusters in myotubes, and survival for myotubes
Laminin 3	1	2	1	Major basement membrane component and major sites of expression include myotendinous junction
Laminin 4	2	2	1	Receptor interaction, presumed to be similar to those of laminin 2
Laminin 5	3A	3	2	Found in basement membranes underlying specialized epithelia; associated with integrins. Highly adhesive for epithelial cells or promotes migration. Involved in hemidesmosome formation, skin regeneration, assembly of gap junctions in keratinocytes, etc.
Laminin 5B	3B	3	2	Known to have strong activity to promote cell adhesion and migration
Laminin 6	3A	1	1	Ψ-shaped structure. Formation of anchoring filaments in the epidermis and some other organs that stabilizes the junction between cells and basement membranes. Do not participate in the formation of networks
Laminin 7	3A	2	1	Binds nidogen with high affinity
Laminin 8	4	1	1	Have affinity for heparin and a similar Schwannoma laminin stimulates neurite outgrowth
Laminin 9	4	2	1	Major basement membrane component and major sites of expression include endothelium and smooth muscle
Laminin 10	5	1	1	Drosophila laminin 10 was shown to bind heparin and to be cell adhesive
Laminin 11	5	2	1	Stop signal for Schwann cells. Major sites of expression include mature epithelium, mature endothelium, smooth muscle, neuromuscular junction, and glomerular basement membrane
Laminin 12	2	1	3	Major site of expression is ciliated epithelia
Laminin 13	3	2	3	Major site of expression is central nervous system/retina
Laminin 14	4	2	1	Major site of expression is central nervous system/retina
Laminin 15	5	2	3	Major site of expression is central nervous system/retina

Sources: Adapted from Kreis, T. and Vale, R. (eds), 1999. *Guide Book to Extracellular Matrix Anchor, and Adhesion Proteins*, 2nd edition. New York: Oxford University Press; Rousselle, P. 2002. *Med. Sci.* 18(10): 989–994; Hallman, R. et al. 2005. *Physiol. Rev.* 85: 979–1000.

the surface hydrophobicity/hydrophilicity. The substrate has to exhibit a chemistry that facilitates the binding of fibronectin or for that matter any other adhesive protein. This has given rise to a subfield of scaffold surface engineering involving surface treatments aimed at increasing the hydrophilicity so that adsorbed proteins can largely retain their normal functionality (Ma et al., 2007; Kikuchi and Okano, 2005). The synthetic materials of interest in drug discovery HTS systems are mainly polymers. Place et al. (2009) have summarized scaffold surface modification techniques to

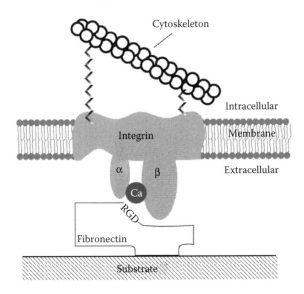

FIGURE 5.6 Schematic of integrin–fibronectin-mediated cell–substrate adhesion. The α and β subunits of the integrin are operative only in the presence of calcium.

include (1) plasma treatment by air, nitrogen, oxygen, methane, or other gases, which introduces charged groups to the surface; (2) chemical etching such as NaOH on poly(lactic-co-glycolic acid) (PLGA) and NHO_3 on poly(etherurethane); (3) following the above two treatments with covalent linking with a second polymer or proteins; (4) entrapment of a hydrophilic polymer following swelling of the scaffold primary polymer with a solvent that entraps the second hydrophilic polymer on shrinking; and (5) depositing polymer layers non-covalently by techniques such as electrostatic deposition.

Peptide motifs derived from binding regions of ECM proteins, for example, RGD (which is found in all the adhesive proteins discussed above), have been incorporated into a wide range of surfaces, hydrogels, and scaffolds, as a means to improve adhesion. The simple presence of such motifs affects cell adhesion, and the spacing between the groups (packing density) is important. For example, Massia and Hubbell (1991) found that an average distance of 440 nm between RGD peptides was sufficient for fibroblast attachment and spreading, but a lower distance of 140 nm was required for focal contact formation and cytoskeletal reorganization into stress fibers. A second design feature with peptide motifs involves the use of propeptide sequences. For example, Benoit and Anseth (2005) found that PEG-based hydrogels containing the sequence $RGD_{13}PHSRN$ improved osteoblast adhesion and spreading in comparison to just RDG, suggesting the importance of spatial relationships in designing surfaces or scaffolds for cell culture.

5.3 LONG-RANGE CHEMISTRY

Long-range chemistry refers to diffusible chemical species that are found in the ECM that contribute to the maintenance or control of the cell behavior and/or phenotype outcome. The list includes basic nutrients, growth factors, cytokines, and other morphogens, as well as metalloproteinases. Owing to its relatively low solubility in culture medium, oxygen is usually the first component to become rate-limiting. This is why many studies have been conducted on the effects of hypoxia in different cell culture systems (e.g., Glicklis et al., 2004; Gao et al., 2001; Alsat et al., 1996). The relationship between oxygen concentration profiles and size of 3D cell aggregate will be further explored in Chapter 6. Effects of other basic nutrients, amino acids, and glucose that have not been studied in similar contexts as gradients in these nutrients are typically negligible in most culture systems. A brief discussion of the key diffusible chemical species identified above as microenvironmental factors follows.

5.3.1 Cytokines, Chemokines, Hormones, and Growth Factors

A definition of the subtitle terms is in order. Cytokines are regulatory peptides or proteins (e.g., interleukin-2, IL-2) secreted by white blood cells and a variety of other cells in the body. Cytokines exhibit pleiotrophic regulatory effects on hematopoietic and many other cells that participate in host defense and repair processes. Chemokines (e.g., monocyte chemoattractant protein-1, MCP-1) are a family of small cytokines with their name derived from their ability to induce chemotaxis (*chemo*tactic/cyto*kines*) in nearby responsive cells. Polypeptide hormones are generally produced by specialized endocrine glands and are present in the circulatory system. These hormones serve to maintain homeostasis. On the other hand, most cytokines act over short distances as autocrine or paracrine intercellular signals in local tissues, with a few exceptions. Cytokines are generally produced in response to danger signals to mitigate challenges to the integrity of the host.

Growth factors (e.g., fibroblast growth factor, FGF) refer to naturally occurring protein (tends to be produced constitutively as opposed to tightly regulated production of cytokines) capable of stimulating cellular growth, proliferation, and differentiation. Despite the differences outlined above, many of the characteristics of cytokines are shared by hormones and growth factors. For example, they all function as extracellular signaling molecules featuring fundamentally similar mechanisms of action (Vilček, 2003).

Many growth factors and cytokines regulate the synthesis of the ECM. For example, TGF-β, which exists in three isoforms in mammals (TGF-β1, TGF-β2, and TGF-β3), is known to stimulate the production of ECM proteins (MacKenna et al., 2000). Others examples include various interleukins (e.g., IL-1, IL-6, and IL-8), insulin-like growth factor-I (IGF-I), FGFs, vascular endothelial growth factor (VEGF), platelet-derived growth factor (PDGF), tumor necrosis factor (TNF-α and -β) (Wang and Thampatty, 2006). The ECM has for a long time been thought of as a static rather than a dynamic structure, but this paradigm is changing particularly when it comes to the regulation of growth factors and cytokine activities. The ECM acts as a reservoir for these signaling molecules; sequestered in latent forms that are activated upon appropriate stimulation. For example, heparin-bound VEGF controls blood vessel branching (e.g., Ruhrberg et al., 2002). Also, members of the fibrillin superfamily bind TGF-β and have been implicated in its regulation (Chaudhry et al.,

2007). This property may greatly slow the diffusion of these signaling molecules and therefore fine-tune local concentrations as well as gradients. A second reservoir mode is degradation of other ECM components into components with growth factor-like sequences that act on cells. Laminin is a good example that possesses EGF-like peptide domains that can stimulate cell proliferation and differentiation (Schenk et al., 2003).

Given their role in ECM, it is easy to imagine them at the center of some diseases when they are not tightly regulated. High concentrations of TGF-β have been found in several pathological conditions including cardiac fibrosis, chronic tendinosis, and osteoarthritis (Kim et al., 1994). Many interleukins function as key mediators in many diseases such as inflammatory and autoimmunity disorders (Sivakumar and Das, 2008; Andereakos and Feldmann, 2003), cancer (Shim et al., 2009; Cooper and Caligiuri, 2003), and asthma (Takatsu et al., 2009; Oriss and Ray, 2003), to name a few. This has made these molecules attractive targets in drug discovery efforts.

5.3.2 MATRIX METALLOPROTEINASES (MMPs)

MMPs are a family of zinc-dependent enzymes that share a common functional domain structure. As shown in Figure 5.7, most MMPs have a common domain structure with a signal peptide (directs secretion from the cell), a catalytic domain incorporating a propeptide (contains a cysteine residue that binds zinc in the active site to form the cysteine switch—necessary to maintain enzyme latency), a hinge region, and in a majority of cases, a C-terminal hemopexin domain (which contributes to substrate specificity and to interactions with endogenous inhibitors (Overall, 2002). MMPs were first described by Gross and Lapiere (1962), and since then, 28 MMPs have been described and 24 of these have been identified in vertebrates (Table 5.6). In the past, MMPs taxonomy has been based on substrate specificity; however, another classification based on structural similarity has been proposed by Egeblad and Werb (2002) to include collagenases, gelatinases, stromelysins, and membrane-type MMPs (Table 5.6). Most MMPs, with the exception of membrane types, are secreted and act in the extracellular space. MMPs are produced in latent form, but once activated (by free radicals or enzymes that free the cysteine bond or cleave the propeptide region), regulate many physiological processes such as migration, invasion, proliferation, and apoptosis (Sternlicht and Werb, 2001) as well as branching, morphogenesis, angiogenesis, and wound healing (Vu and Werb, 2000).

5.4 CONCLUDING REMARKS

Advances are being made in our understanding of how the biochemical or chemical microenvironmental factors influence cellular phenotypic outcomes. The question that is appropriate to ask is whether a critical point has been reached at which it is possible to translate this knowledge into a set of design principles to facilitate the building of cell-based biosensors or tissues that emulate the *in vivo* situation as closely as possible. To begin to answer this question requires examining and

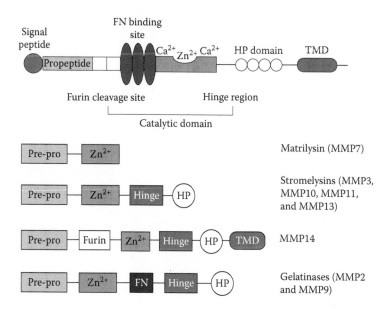

FIGURE 5.7 Schematic of the MMP protein structure. The cysteine switch is made up of a signal peptide and the propeptide region. The switch folds over the zinc in the catalytic site and maintains the latent state. A cleavage site enables the proconvertase furin to activate MMP by cleaving the propeptide. A fibronectin (FN) binding site is present in MMP-2 and MMP-9, connecting them with basal lamina. The catalytic zinc site is present in all MMPs. A hemopexin (HP) domain is joined to the catalytic site by the hinge region. MMP-14 (a membrane-type MMP) has a TMD. (Adapted from Rosenberg, G. 2009. *Lancet Neurol.* 8: 205–216.)

integrating how functional components of the *in vivo* ECM are being used in engineering *in vitro* microenvironments, which is a difficult task. An attempt in this direction has recently been published by Roach et al. (2007), who reviewed literature on the effect of surface functional groups on adhesion, proliferation, and differentiation. It should be pointed out that such examinations are not complete without integrating the other three microenvironmental factors. At this point, it is sufficient to point out the dominant approaches to engineering physiologically relevant microenvironments: (1) recapitulating the ECM with synthetic or naturally derived hydrogels to the greatest extent possible and/or (2) providing only either the initial set of cues (e.g., geometry, short-, and long-range chemistry) or a minimal set of dominant cues and let the cells complete the establishment of the physiologically relevant microenvironment (Griffith and Swartz, 2006; Lutolf and Hubbell, 2005). A recent excellent example of the latter involves programmed connectivity of multiple cell types by functionalizing the cells with short oligonucleotides that imparted specific adhesive properties. Using this approach, Gartner and Bertozzi (2009) have demonstrated the ability to design microtissues with defined cell–cell composition. By providing a single connective cue, a CHO cell line engineered to express IL-3 and a hematopoietic progenitor cell line (FL5.12), whose survival and replication depend on the presence of IL-3, formed 3D multicellular aggregates and established a paracrine

TABLE 5.6

The Matrix Metalloproteinases (MMPs) Family

Name	Latent/Active Molecular Weight (kDa)	Substrates
	Interstitial collagenases	
MMP-1 (collagenase-1)	52/41	Collagen I, II, III, VII, VIII, X, aggrecan, gelatin, pro-MMP-9
MMP-8 (collagenase-2)	85/64	Collagen I, II, III, VII, VIII, X, aggrecan, gelatin
MMP-13 (collagenase-3)	65/55	Collagen I, II, III, aggrecan, gelatin
MMP-18 (collagenase-4)	53/42	Missing—look in the literature
	Gelatinases	
MMP-2 (gelatinase-A)	72/66	Collagen, I, II, III, IV, V, VII, X, XI, XIV, gelatin, elastin, fibronectin, aggrecan
MMP-9 (gelatinase-B)	92/85	Collagen IV, V, VII, X, XIV, gelatin, pro-MMP-9, pro-MMP-13, elastin, aggrecan
	Stromelysins	
MMP-3 (stromelysin-1)	57/45, 28	Collagen II, III, IV, IX, X, XI, elastin, pro-MMP-1, pro-MMP-7, pro-MMP-8, pro-MMP-9, pro-MMP-13
MMP-10 (stromelysin-2)	56/47, 24	Collagen III, IV, V, gelatin, fibronectin
MMP-11 (stromelysin-3)	58/28	Fibronectin, laminin, gelatin, aggrecan
	Membrane-type (MT) MMPs	
MMP-14 (MT1-MMP)	66/60	Pro-MMP2, pro-MMP-13, collagen I, II, III, gelatin, aggrecan, fibronectin, laminin
MMP-15 (MT2-MMP)	68/62	Pro-MMP-2, gelatin, fibronectin, laminin
MMP-16 (MT3-MMP)	64/55	Pro-MMP-2
MMP-17 (MT4-MMP)	57/53	Unknown
MMP-18 (MT5-MMP)	63/45	Pro-MMP-2
MMP-19 (MT6-MMP)	Unknown	Gelatin
	Others	
MMP-7 (matrilysin-2)	28/19	Collagen II, III, IV, IX, X, XI, elastin, pro-MMP-1, pro-MMP-7, pro-MMP-8, pro-MMP-9, pro-MMP-13, gelatin, aggrecan, fibronectin, laminin
MMP-12 (metalloelastase)	54.45, 22	Elastin
MMP-19	57/45	Tenascin, gelatin, aggrecan
MMP-20 (enamelysin)	54/22	Enamel, gelatin
MMP-21	70/53	Unknown
MMP-23	Unknown	Unknown
MMP-26 (matrilysin)	28/unknown	Collagen IV, gelatin, fibronectin
MMP-27	Unknown	Unknown
MMP-28 (epilysin)	Unknown/58, 55	Unknown

Source: From Gueders, M.M., et al. 2006. *Eur. J. Pharmacol.* 533: 135. With permission.

signaling network *in vitro*. Singly, in the absence of IL-3, the FL5.12 cell cycle is arrested and the onset of apoptosis occurs within 12–36 h.

A good understanding of spatiotemporal cell–ECM interactions will move the *in vitro* development of cultures that more accurately emulate their *in vivo* counterparts from art to engineering, where scientists/engineers desiring a particular phenotype have available to them a set of principles, which can be followed to achieve desired outcomes.

REFERENCES

Aigner, T. and Stove, J. 2003. Collagens—major component of the physiological cartilage matrix, major target of cartilage degeneration, major tool in cartilage repair. *Adv. Drug Deliv. Rev.* 55(12): 1569–1593.

Alsat, E., Wyplosz, P., Malassine, A., Guibourdenche, J., and Evain-Brion, D. 1996. Hypoxia impairs cell fusion and differentiation process in human cytotrophoblast *in vitro. J. Cell. Physiol.* 168: 346–353.

Akiyama, S.K. and Yamada, K.M. 1985. The interaction of plasma fibronectin with fibroblastic cells in suspension. *J. Biol. Chem.* 260: 4492–4500.

Andereakos, E. and Feldmann, M. 2003. Autoimmunity and cytokines: Pathogenesis and therapy. In *The Cytokine Handbook*, A.W. Thomson and M.T. Lotze (eds), 4th edition, pp. 1189–1211. San Diego: Academic Press.

Aumailley, M., Bruckner-Tuderman, L., Carter, W.G., Deutzmann, R., Edgar, D., Ekblom, P., Engel, J., et al. 2005. A simplified laminin nomenclature. *Matrix Biol.* 24(5): 326–332.

Benoit, D.W.S. and Anseth, K.S. 2005. Heparin functionalized PEG gels that modulate protein adsorption for hMSC adhesion and differentiation. *Acta Biomater.* 1(4): 461–470.

Borsig, L. 2004. Selectins facilitate carcinoma metastasis and heparin can prevent them. *News Physiol. Sci.* 19: 16–21.

Canty, E.G. and Kadler, K.E. 2002. Collagen fibril biosynthesis in tendon: A review and recent insights. *Comp. Biochem. Physiol. A Mol. Integr. Physiol.* 133(4): 979–985.

Chaudhry, S.S., Cain, S.A., Morgan, A., Dallas, S.L., Shuttleworth, C.A., and Kielty, C.M. 2007. Fibrillin-1 regulates the bioavailability of TGF-β1. *J. Cell Biol.* 176: 355–367.

Cooper, M.A. and Caligiuri, M.A. 2003. Cytokines and cancer. In *The Cytokine Handbook*, A.W. Thomson and M.T. Lotze (eds), 4th edition, pp. 1213–1232. San Diego: Academic Press.

Egeblad, M. and Werb, Z. 2002. New functions for the matrix metalloproteinases in cancer progression. *Nat. Rev. Cancer* 2(3): 161–174.

El-Ghannam, A., Starr, L., and Jones, J. 1998. Laminin-5 coating enhances epithelial cell attachment, spreading and hemidesmosome assembly on Ti-6A1-4V implant material *in vitro. J. Biomed. Mater. Res.* 41: 30–40.

Gallin, W.J., Edelman, G.M., and Cunningham, B.A. 1983. Characterization of L-CAM, a major cell adhesion molecule from embryonic liver cells. *Proc. Natl Acad. Sci. USA* 80: 1038–1042.

Gao, H., Ayyaswamy, S.P.S., Ducheyne, P., and Radin, S. 2001. Surface transformation of bioactive glass in bioreactors simulating microgravity conditions. Part II: Numeric simulations. *Biotechnol. Bioeng.* 75: 379–385.

Garcia, J., Callewaert, N., and Borsig, L. 2007. P-selectin mediates metastatic progression through binding to sulfatides on tumor cells. *Glycobiology* 17(2): 185–196.

Gartner, Z.J. and Bertozzi, C.R. 2009. Programmed assembly of 3-dimensional microtissues with defined cellular connectivity. *Proc. Natl. Acad. Sci. USA* 106(12): 4606–4610.

Glicklis, R., Merchuk, J.C., and Cohen, S. 2004. Modeling mass transfer in hepatocyte spheroids via cell viability, spheroid size, and hepatocellular functions. *Biotechnol. Bioeng.* 86: 672–680.

Gong, N. and Chatterjee, S. 2003. Platelet endothelial cell adhesion molecule in cell signaling and thrombosis. *Mol. Cell Biochem.* 253(1–2): 151–158.

Green, N.M. 1975. Avidin. *Adv. Protein Chem.* 29: 85–133.

Griffith, L.G. and Swartz, M.A. 2006. Capturing complex 3D tissue physiology *in vitro*. *Nat. Rev. Mol. Cell Biol.* 7: 211–224.

Gross, J. and Lapiere, C.M. 1962. Collagenolytic activity in amphibian tissues: A tissue culture assay. *Proc. Natl. Acad. Sci. USA* 48: 1014–1022.

Hallman, R., Horn, N., Selg, M., Wendler, O., Pausch, F., and Sorokin, M. 2005. Expression and function of laminins in the embryonic and mature vasculature. *Physiol. Rev.* 85: 979–1000.

Hammarstrom, S. 1999. The carcinoembryonic antigen (CEA) family: Structures, suggested functions and expression in normal and malignant tissues. *Semin. Cancer Biol.* 9(2): 67–81.

Haynes, R.O. 1990. *Fibronectins*. New York: Springer.

Hohenester, E. and Engel, J. 2000. Domain structure and organization in extracellular matrix proteins. *Matrix Biol.* 21: 115–128.

Hoke, D., Mebius, R.E., Dybdal, N., Dowbenko, D., Gribling, P., Kyle, C., Baumhueter, S., and Watson, S.R. 1995. Selective modulation of the expression of L-selectin ligands by an immune-response. *Curr. Biol.* 5(6): 670–678.

Hulmes, D.J.S. 1992. The collagen superfamily—diverse structures and assemblies. *Essays Biochem.* 27: 49–67.

Hulpiau, P. and van Roy, F. 2009. Molecular evolution of the cadherin superfamily. *Int. J. Biochem. Cell Biol.* 41: 349–369.

Humphries, J.D., Byron, A., and Humphries, M.J. 2006. Integrins ligands at a glance. *J. Cell Sci.* 119: 3901–3903.

Huxley-Jones, J., Foord, S.M., and Barnes, M.R. 2008. Drug discovery in the extracellular matrix. *Drug Discov. Today* 13(15/16): 685–694.

Haynes, R.O. 1987. Integrins: A superfamily of cell surface receptors. *Cell* 48: 549–554.

Haynes, R.O. 2002. Integrins: Bidirectional, allosteric signaling machines. *Cell* 110: 673–687.

Isberg, R.R. and Barnes, P. 2002. Dancing with the host; flow-dependent bacteria adhesion. *Cell* 110(1): 1–4.

Jia, X. and Kiick, K.L. 2009. Hybrid multicomponent hydrogels for tissue engineering. *Macromol. Biosci.* 9(2): 140–156.

Jimenez, D., Roda-Navarro, P., Springer, T.A., and Casasnovas, J.M. 2005. Contribution of N-linked glycans to the conformation and function of intercellular adhesion molecules (ICAMs). *J. Biol. Chem.* 280(7): 5854–5861.

Jones, F.S. and Jones, P.L. 2000. The tenascin family of ECM glycoproteins: Structure, function, and regulation during embryonic development and tissue remodeling. *Dev. Dyn.* 218: 235–259.

Kadler, K. 1995. Extracellular matrix 1: Fibril-forming collagens. *Protein Profile* 2(5): 491–619.

Kao, W.Y.J. 1999. Evaluation of protein-modulated macrophage behavior on biomaterials: Designing biomimetic materials for cellular engineering. *Biomaterials* 20(23–24): 2213–2221.

Kaspar, M., Zardi, L., and Neri D. 2005 Fibronectin as target for tumor therapy. *Int. J. Cancer* 118: 1331–1339.

Kaspar, M., Zardi, L., and Neri, D. 2006. Fibronectin as target for tumor therapy. *Int. J. Cancer* 118: 1331–1339.

Kuespert, K., Pils, S., and Hauck, C.R. 2006. CEACAMs: Their role in physiology and pathophysiology. *Curr. Opin. Cell Biol.* 18(5): 565–571.

Kiani, C., Wu, Y.J., and Yang, B.B. 2002. Structure and function of aggrecan. *Cell Res.* 12(1): 19–32.

Kikuchi, A. and Okano, T. 2005. Nanostructured designs of biomedical materials: Applications of cell sheet engineering to functional regenerative tissues and organs. *J. Control Release* 101(1–3): 69–84.

Kim, S.J., Romeo, D., Yoo, Y.D., and Park, K. 1994. Transforming growth-factor-beta-expression in normal and pathological conditions. *Hormone Res.* 42(1–2): 5–8.

Kneuer, C., Ehrardt, C., Radomski, M.W., and Bakowsky, U. 2006. Selectins—potential pharmacological targets. *Drug Discov. Today* 11(21/22): 1034–1040.

Koch, A.W., Manzur, K.L., and Shan, W. 2004. Structure-based models of cadherin-mediated cell adhesion: The evolution continues. *Cell. Mol. Life Sci.* 61: 1884–1895.

Kreis, T. and Vale, R. (eds), 1999. *Guide Book to Extracellular Matrix Anchor, and Adhesion Proteins*, 2nd edition. New York: Oxford University Press.

Lauffernburger, D.A. and Liderman, J.J. 1993. *Receptors: Models for Binding, Trafficking and Signaling*. New York: Oxford University Press.

Legate, K.R., Wickstrom, S.A., and Fassler, R. 2009. Genetic and cell biological analysis of integrin outside-in signaling. *Genes Dev.* 23(4): 397–418.

Luk, J.M. and Wong, K.F. 2006. Monoclonal antibodies as targeting and therapeutic agents: Prospects for liver transplantation, hepatitis and hepatocellular carcinoma. *Clin. Exp. Pharmacol. Physiol.* 33(5–6): 482–488.

Lutolf, M.P. and Hubbell, J.A. 2005. Synthetic biomaterials as instructive extracellular microenvironments for morphogenesis in tissue engineering. *Nat. Biotechnol.* 23(1): 47–55.

Ma, Z.W., Mao, Z.W., and Gao, C.Y. 2007. Surface modification and property analysis of biomedical polymers used for tissue engineering. *Colloids Surf. B* 60(2): 137–157.

MacKenna, D., Summerour, S.R., and Villarreal, F.J. 2000. Role of mechanical factors in modulating cardiac function and extracellular matrix synthesis. *Cardiovasc. Res.* 46(2): 257–263.

Maness, P.F. and Schachner, M. 2007. Neural recognition molecules of the immunoglobulin superfamily: Signaling transducers of axon guidance and neuronal migration. *Nat. Neurosci.* 10(1): 19–26.

Marshall, B.T., Long, M., Piper, J.W., Yago, T., McEver, R.P., and Zhu, C. 2003. Direct observation of catch bonds involving cell-adhesion molecules. *Nature* 423: 1990–1993.

Mascini, M., Guibault, G.C., Monk, I.R., Hill, C., Del Carlo, M., and Campagnone, D. 2007. Screening of rationally designed oligopeptide for *Listeria monocytogenes* detection by means of high density colorimetric microarray. *Microchim. Acta* 163(3–4): 227–235.

Massia, S.P. and Hubbell, J.A. 1991. An RGD spacing of 440 nm is sufficient for integrin alpha V beta 3-mediated fibroblast spreading and 140 nm for focal contact and stress fiber formation. *J. Cell. Biol.* 114(5): 1089–1100.

Moser, M., Legate, K.R., Zent, R., and Fässler, R. 2009. The tail of integrins, talin, and kindlins. *Science* 324(5929): 895–899.

Newman, P.J. 1997. The biology of PECAM-1. *J. Clin. Invest.* 99(1): 3–8.

Nielsen, J., Kulahin, N., and Walmod, P.S. 2008. Extracellular protein interactions mediated by the neural cell adhesion molecule, NCAM: Heterophilic interactions between NCAM and cell adhesion molecules, extracellular matrix proteins, and viruses. *Neurochem. Res.* 29(36): 11360–11376.

Oriss, T.B. and Anuradha, R. 2003. Cytokines and asthma. In *The Cytokine Handbook*, A.W. Thomson and M.T. Lotze (eds), 4th edition, pp. 1314–1334. San Diego: Academic Press.

Overall, C.M. 2002. Molecular determinants of metalloproteinase substrate specificity: Matrix metalloproteinase substrate binding domains, modules, and exocites. *Mol. Biotechnol.* 22: 51–86.

Paschos, K.A., Canovas, D., and Bird, N.C. 2009. The role of cell adhesion molecules in the progression of colorectal cancer and the development of liver metastasis. *Cell. Signal.* 21: 665–674.

Peyrieras, N., Hyafil, F., Louvard, D., Ploegh, H.L., and Jacob, F. 1983. Uvomorulin: A non-integral membrane protein of early mouse embryo. *Proc. Natl. Acad. Sci. USA* 80: 6274–6277.

Place, E.S., George, J.H., Williams, C.K., and Stevens, M.M. 2009. Synthetic scaffolds tissue engineering. *Chem. Soc. Rev.* 38(4): 1139–1151.

Reizes, O., Benoit, S.C., and Clegg, D.J. 2008. The role of syndecans in the regulation of body weight and synaptic plasticity. *Int. J. Biochem. Cell Biol.* 40(1): 28–45.

Resink, T., Philippova, M., Joshi, M.B., Kyriakakis, E., and Erne, P. 2009. Cadherins in cardiovascular disease. *Swiss Med. Wkly.* 139: 122–134.

Roach, P., Eglin, D., Rohde, K., and Perry, C.C. 2007. Modern biomaterials: A review—bulk properties and implications of surface modifications. *J. Matr. Sci.: Mater. Med.* 18: 1263–1277.

Rosenberg, G. 2009. Matrix metalloproteinases and their multiple roles in neurodegenerative diseases. *Lancet Neurol.* 8: 205–216.

Rosso, F., Giordano, A., Barbarisi, M., and Barbarisi, A. 2004. From cell–ECM interactions to tissue engineering. *J. Cell. Physiol.* 199: 174–180.

Rousselle, P. 2002. Cell migration and cancer. *Med. Sci.* 18(10): 989–994.

Ruhrberg, C., Gerhardt, H., Golding, M., Watson, R., Ioannidou, S., Fujisawa, H., Betsholtz, C., and Shima, D.T. 2002. Spatially restricted patterning cues provided by heparin-binding VEGF control blood vessel branching morphogenesis. *Genes Dev.* 16: 2684–2698.

Ruoslahti, E. 1988. Structure and biology of proteoglycans. *Annu. Rev. Cell Biol.* 4: 229–255.

Schenk, S., Hintermann, E., Bilban, M., Koshikawa, N., Hojilla, C., Khokha, R., and Quaranta, V. 2003. Binding to EGF receptor of laminin-5 EGF-like fragment liberated during MMP-dependent mammary gland involution. *J. Cell Biol.* 161: 197–209.

Shim, H., Oishi, S., and Fulii, N. 2009. Chemokine receptor CXCR4 as a therapeutic for neuroectodermal tumors. *Sem. Cancer Biol.* 19(2): 123–134.

Silver, F.H., Christiansen, D.L., Snowhill, P.H., and Chen, Y. 2000. Role of storage on changes in the mechanical properties of tendon and self-assembled collagen fibers. *Connect. Tissue Res.* 41(2): 155–164.

Sivakumar, P. and Das, A.M. 2008. Fibrosis, chronic inflammation and new pathways for drug discovery. *Inflamm. Res.* 57: 410–418.

Springer, T.A. 1994. Traffic signals for lymphocyte recirculation and leukocyte emigration: The multistep paradigm. *Cell* 76: 301–314.

Sternlicht, M.D. and Werb, Z. 2001. How matrix metalloproteinases regulate cell behavior. *Annu. Rev. Cell. Dev. Biol.* 17: 463–516.

Streuli, C.H. and Akhtar, N. 2009. Signaling co-operation between integrins and other receptor systems. *Biochem. J.* 418: 491–506.

Suzuki, S.C. and Takeichi, M. 2008. Cadherins in neuronal morphogenesis and function. *Develop. Growth Differ.* 50: S119–S130.

Takatsu, K., Kouro, T., and Nagai, Y. 2009. Interleukin 5 in the link between the innate and acquired immune response. *Adv. Immunol.* 101: 191–236.

Takeichi, M. 2007. The cadherin superfamily in neuronal connections and interactions. *Nat. Rev. Neurosci.* 8: 11–20.

Timpl, R., Rohde, H., Robey, P.G., Rennard, S.I., Foidart, J.M., and Martin, G.R. 1979. Laminin—a glycoprotein from basement membranes. *J. Biol. Chem.* 254: 9933–9937.

Timpl, R., Tisi, D., Talts, J.F., Andac, Z., Sasaki, T., and Hohenester, E. 2000. Structure and function of laminin LG molecules. *Matrix Biol.* 19: 309–317.

Thomas, W. 2008. Catch bonds in adhesion. *Annu. Rev. Biomed. Eng.* 10: 39–57.

Toole, B.P. 2004. Hyaluronan: From extracellular glue to pericellular cue. *Nat. Rev. Cancer* 4(7): 528–539.

Tucker, R.P. 2004. The thrombospondin type 1 repeat superfamily. *Int. J. Biochem. Cell Biol.* 36: 969–974.

Tzu, J. and Marinkovich, M.P. 2008. Bridging structure with function: Structural, regulatory, and developmental role of laminins. *Int. J. Biochem. Cell Biol.* 40: 199–214.

Vaidya, K.S. and Welch, D.R. 2007. Metastasis suppressors and their roles in breast carcinoma. *J. Mammary Gland Neoplasia* 12(2–3): 175–190.

Verderio, E.A.M., Scarpellini, A., and Johnson, T.S. 2009. Novel interactions of TG2 with heparin sulfate proteoglycans: Reflection of physiological implications. *Amino Acids* 36(4): 671–677.

Vilček, J. 2003. The cytokines: An overview. In *The Cytokine Handbook*, A.W. Thomson and M.T. Lotze (eds), 4th edition, pp. 3–18. San Diego: Academic Press.

Vu, T.H. and Werb, Z. 2000. Matrix metalloproteinases: Effectors of development and normal physiology. *Genes Dev.* 14(17): 2123–2133.

Wang, J.H.-C. and Thampatty, B.P. 2006. An introductory review of cell mechanobiology. *Biomechan. Model. Mechanobiol.* 5: 1–16.

Wheelock, M.J. and Johnson, K.P. 2003. Cadherins as modulators of cellular phenotype. *Annu. Rev. Cell. Dev. Biol.* 19: 207–235.

Witz, I.P. 2008. The selectin–selectin ligand axis in tumor progression. *Cancer Metastasis Rev.* 27(1): 19–30.

Worthylake, R.A. and Burridge, K. 2001. Leukocyte transendothelial migration: Orchestrating the underlying molecular machinery. *Curr. Opin. Cell Biol.* 13: 569–577.

Wu, T.C. 2007. The role of vascular cell adhesion molecule-1 in tumor immune evasion. *Cancer Res.* 67(13): 6003–6006.

Yang, Y., Jun, C.D., Liu, J.H., Zhang, R., Joachimiak, A., Springer, T.A., and Wang, J.H. 2004. Structural basis for dimerization of ICAM-1 on the cell surface. *Mol. Cell* 14(2): 269–276.

Yoshida, C. and Takeichi, M. 1982. Teratocarcinoma cell adhesion: Identification of cell-surface protein involved in calcium dependent cell aggregation. *Cell* 28: 217–224.

Zaidel-Bar, R., Itzkovitz, S., Ma'ayan, A., Iyengar, R., and Geiger, B. 2007. Functional atlas of the integrin family. *Nat. Cell Biol.* 9(8): 858–867.

Zhang, H., Marshall, K.W., Tang, H., Hwang, D.M., Lee, M., and Liew, CC. 2003. Profiling genes expressed in human fetal cartilage using 13,155 expressed sequence tags. *Osteoarthritis Cartilage* 11(5): 309–319.

Zhang, Y., Yeh, J., Richardson, P.M., and Bo, X. 2008. Cell adhesion molecules of the immunoglobulin superfamily in axonal regeneration and neural repair. *Restor. Neurol. Neurosci.* 26: 81–96.

6 Spatial and Temporal Microenvironmental Factors

The language used to describe physical surface modifications can be misleading. For example, Flemming et al. (1999) have used "micro- and nanostructured surfaces" to include surfaces with grooves up to 120 µm deep (Brunette, 1986a, 1986b). For the purposes of this book, micro- and nano-structured surfaces are reserved for physical modification with aspect ratios such that cells cultured thereon remain in a monolayer and the structure imparts its effect by interaction with only the bottom side of the cells. For example, nanotube surfaces present wells with nanoscale dimensions such that cells with micron-scale sizes interact with the surfaces as monolayers (Oh et al., 2009). The word "scaffold" is used to refer to structures with high aspect ratio architecture that allow cells to be embedded into the structure pores and in some cases to form microtissues. For example, Cheng et al. (2008) fabricated polystyrene structures with random pores of more than 60 µm that supported growth and differentiation of neural aggregates of progenitor cells.

When cells interact with most surfaces, they attach and move; as they migrate, they "feel" or sense the environment by extending and retracting filopodia that is endowed with a high concentration of integrin receptors. Vogel and Sheetz (2006) have defined geometry sensing as "the formation of signaling complexes by changes in spacing of molecular recognition sites [on] the geometrical shape of the substrate." Astrocytes have been shown to preferentially adhere and proliferate on carbon nanofibers of 100 nm in diameter (McKenzie et al., 2004), suggesting that fibers are sensed by cells differently in comparison to plane 2D surfaces. Further evidence in support of curvature sensing includes studies that have demonstrated inward curvature of the plasma membrane to the release of *Rac* (small gTP-binding protein involved in regulating actin cytoskeleton) (Habermann, 2004; Zimmerberg and McLaughlin, 2004) and outward curvature to the opening of an ion channel (Patel et al., 2001).

The effect of surface topography or scaffold architecture on cell behavior is one of the most investigated topics with mainly tissue engineering or regenerative medicine applications. Three excellent reviews by Flemming et al. (1999), Yim and Leong (2005), and Martínez et al. (2009) have summarized the work done in the past three decades. In the next three sections, engineering of structures and their effects on biological outcomes are presented; temporal effects are considered in the fourth section, followed by a first-step proposal toward establishing design principles from current and possibly future results.

6.1 NANO- AND MICROSTRUCTURED SURFACES

Microstructuring has been justified on the basis of recreating the 3D geometry found *in vivo* and the rationale for nanostructuring is based on recreating *in vivo* dimensions such as collagen fibers, ranging in diameter between 50 and 500 nm (Kadler et al., 1996). In Tables 6.1 through 6.3, summaries of the effects of nano- and microstructuring on different cell fates are provided. The literature (Table 6.1) shows that there were as many references reporting enhanced adhesion as those reporting less adhesion or no change. What is clear is that alignment and migration can be easily controlled by the kind of surface modifications reported. The same is true for morphology (Table 6.2) and differentiation (Table 6.3). The majority of these changes are also reflected in changes in the cell cytoskeleton.

Nanostructuring of surfaces to mimic the architecture of the natural ECM is a more recent undertaking and has mainly been accomplished by phase separation (Yang et al., 2004), electrospinning (Mathews et al., 2002), and self-assembly (Yang et al., 2009; Gelain et al., 2007; Zhang et al., 2005) techniques. Typical architectures of nanostructured surfaces are shown in Figure 6.1. Jayaraman et al. (2004) compared the techniques and came up with the results reported in Table 6.4, identifying the electrospinning technique as the only one meeting industrial applications. This explains the relatively large body of literature for the electrospinning technique in comparison to phase separation and self-assembly approaches (Barnes et al., 2007). Like nanostructuring at low aspect ratios (Tables 6.1 through 6.3), the nanofibrous architecture has been shown to modulate cell adhesion and function in comparison to smooth surfaces (Woo et al., 2003; Chen et al., 2006; Hu et al., 2008; Nisbet et al., 2009). However, the underlying mechanisms are not well understood. By using easily reproducible surfaces, such as nanotubes, it is possible to conduct experiments that will provide insights into the underlying mechanisms as explored in Section 6.5.

6.2 SCAFFOLDS

Scaffolds have been produced from either natural or synthetic materials. Elsdale and Bard (1972) pioneered 3D or scaffold culture with a model system for fibroblastic cells in the body using collagen I matrices polymerized *in vitro* to form a 3D fibrous network. Collagen hydrogels have been used widely since then for studying cell motility (Friedl and Brocker, 2000) and the roles of the physical 3D state (Grinnell, 2000). In addition to collagen, other natural polymers used for fabricating scaffolds have included alginate, agarose, chitosan, fibrin, and hyaluronan (Cheung et al., 2007). Collagen is a fibrous protein that is the major component in connective tissue. Alginate is a polysaccharide extracted from algae. Agarose is also a polysaccharide, but in this case, it is extracted from seaweed. Chitosan is derived from chitin, a naturally occurring polysaccharide extracted from crab shells or shrimps, or from fungal fermentation processes.

Synthetic polymers used in scaffold fabrication include poly(ethylene oxide), poly(vinyl alcohol), poly(acrylic acid), and poly(propylene furmarate-co-ethylene glycol). For excellent reviews, see Drury and Mooney (2003) and Lozinsky and

TABLE 6.1

Summary of Effects on Nano- and Microstructuring on Cell Adhesion, Alignment, and Migration

Feature Type	Material	Feature Dimension	Aspect Ratio Range (Depth/Width)	Cell Type	Effect	Reference
Ridges	Silicon oxide on polystyrene	4 μm	0.0125	Murine neuroblastoma cells	Cells adhered to lines and processes aligned along the lines; processes grew in bipolar manner	Cooper et al. (1976)
Groove	Ti-coated silicon	15 μm	0.2	Human gingival fibroblasts	Microtubules were the first element to become aligned; microtubules aligned at the bottom of grooves after 20 min; actin was observed first at wall-ridge edges after 40–60 min; after 3 h, a majority of cells exhibited aligned focal contacts	Oakley and Brunette (1993)
Groove	Titanium-coated silicon	15 μm	0.2	Porcine epithelial cells	Oriented in the direction of grooves; actin filaments and microtubules aligned along walls and edges; single cells showed less variability of aligned cytoskeletal arrangements than cell clusters	Oakley and Brunette (1995)
Groove	Photo-responsive PMMA	1 μm	0.25	Human astrocytes	Improved adhesion, strong alignment	Baac et al. (2004)
Groove	PDLA	10 μm	0.3	Schwann cells (nerve cells)	Strong alignment	Miller et al. (2001)
Groove	PS	10 μm	0.3	Rat astrocytes	Less adhesion, strong alignment	Recknor et al. (2004)

continued

TABLE 6.1 (continued)
Summary of Effects on Nano- and Microstructuring on Cell Adhesion, Alignment, and Migration

Feature Type	Material	Feature Dimension	Aspect Ratio Range (Depth/Width)	Cell Type	Effect	Reference
Pits	PCL	150 nm	0.53	Fibroblasts	Less focal contacts and vinculin pattern	Gallagher et al. (2002)
Groove	Epoxy replica of silicon original	17 μm	0.59	Porcine periodontal ligament epithelial cells, rat parietal implant model	Epithelial cells attached to grooved surfaces more and were oriented by grooves; shorter length epithelial attachment and longer connective tissue attachment in grooved parts of implant; grooves impeded epithelial down growth on implants	Chehroudi et al. (1988)
Groove	PDMS cast of silicon original	1 μm	1	Human gingival fibroblasts	Vinculin-positive attachment sites were observed; cells aligned to grooves in PDMS, which had been made hydrophilic by glow discharge treatment; focal adhesion contacts also aligned to grooves	Meyle et al. (1994)
Groove	Serum-coated glass	2 μm	1	Human neutrophil leukocytes	When cells moving across the plane of glass encountered a groove, they were highly likely to migrate along the groove rather than cross it	Wilkinson et al. (1982)
Groove	Polyimide	4 μm	1.25	Osteoblasts	Strong alignment, no changes in adhesion	Charest et al. (2004)
Groove	Epoxy replica of silicon original	0.5 μm	2	Human gingival fibroblasts	Cells showed strong alignment to topography; cells bridged or conformed to features	Meyle et al. (1993b)

Groove	Epoxy replica of silicon original	0.5 μm	2	Human gingival fibroblasts	Cells grew mostly in monolayers; some cells extended processes into grooves; inner corners of grooves not occupied by cellular processes; some cells bridged grooves; cytoskeletal elements oriented parallel to the long axis of grooves	Meyle et al. (1993a)
Groove	Silicon dioxide	0.5 μm	2	Human fibroblasts, gingival keratinocytes, neutrophils, monocytes, macrophages	100% of fibroblasts and 20% of macrophages aligned; no orientation or alignment was observed with keratinocytes or neutrophils; some macrophages extended processes parallel to the long axis of grooves after 2 h	Meyle et al. (1995)
Groove	Silicon dioxide	0.5 μm	2	Fibroblasts	Strong alignment	Meyle et al. (1995)
Groove	Silicon dioxide	0.5 μm	2	Keratinocytes	No alignment	Meyle et al. (1995)
Ridges	Polystyrene cast of silicon original	0.5–100.0 μm	0.0003–10	Uromyces appendiculatus fungus	Ridge spacing of 0.5–6.7 μm caused a high degree of orientation of the fungus	Hoch et al. (1987)
Groove and chemical pattern	Amino-silane and methyl-silane-coated quartz	2.5, 6.0, 12.5, 25.0, and 50.0 μm	0.002–2.4	BHK cells	Cells aligned most to 25 μm aminosilane tracks and 5 μm wide, 6 μm deep grooves; alignment increased when adhesive tracks and grooves were parallel; adhesive cues dominant	Britland et al. (1996)
Groove	Quartz and protein-coated quartz	2.0, 10.0 μm	0.003–0.141	P3881D1 macrophages, rat peritoneal macrophages	Cells spread more on 282 nm deep grooves than on plain substrate; degree of orientation of cells increased with increasing depth and decreasing width of grooves; cells on grooves showed higher phagocytic activity	Wojciak-Stothard et al. (1996)

continued

TABLE 6.1   (continued)
Summary of Effects on Nano- and Microstructuring on Cell Adhesion, Alignment, and Migration

Feature Type	Material	Feature Dimension	Aspect Ratio Range (Depth/Width)	Cell Type	Effect	Reference
Groove	Quartz, poly-L-lysine-coated quartz and polystyrene replicas	1.0, 2.0, and 4.0 μm	0.0035–1.1	Embryonic Xenopus spinal cord neurons, rat hippocampal neurons	Orientation of Xenopus and hippocampal neurites was unaffected by cytochalasin B, which eliminated filopodia; taxol, and nocodazole disrupted hippocampal microtubules, but did not affect orientation or turning toward grooves; perpendicular alignment of hippocampal neurites was not inhibited by several calcium channel, G protein, protein kinase, and protein tyrosine kinase inhibitors; some calcium channel and protein kinase inhibitors did inhibit alignment	Rajnicek and McCaig (1997)
Groove	Quartz, poly-L-lysin-coated and polystyrene replicas	1.0, 2.0, and 4.0 μm	0.0035–1.1	Embryonic Xenopus spinal cord neurons, rat hippocampal neurons	Xenopus neurites grew parallel to all groove sizes; hippocampal neurites grew perpendicular to narrow, shallow grooves and parallel to wide, deep grooves; Xenopus neurites emerged from soma regions parallel to grooves; rat hippocampal presumptive axons emerged perpendicular to grooves, but presumptive dendrites emerged parallel to grooves; neurites turned to align to grooves	Rajnicek et al. (1997)
Grooves	PS	0.02–1 μm	0.005–26.5	Fibroblasts	No alignment for shallow structure or widths <100 nm	Loesberg et al. (2007)

Wells and nodes	Polycarbonate, polyetherimide	7.0, 25.0, and 50.0 μm	0.01–0.36	Human neutrophils, fibroblasts	Neutrophil movement was greater on some of the textured surfaces than on an untextured surface; chemical signal is greater on neutrophil movement than texture signal. No effects on fibroblast orientation, spreading, or elongation	Hunt et al. (1995)
Wells	PC	7, 25, 50 μm	0.01–0.36	Fibroblasts	No orientation	Hunt et al. (1995)
Groove	PMMA	2.0, 3.0, 6.0, and 12.0 μm	0.017–0.95	BHK cells, MDCK cells, chick embryo cerebral neuron	Alignment of BHK cells increased with depth but decreased with increasing width; alignment of MDCK cells increased with depth but not affected by width; response of MDCK cells depended on whether or not the cells were isolated or not; alignment of chick embryo cerebral neurons also increased with depth	Clark et al. (1990)
Groove	PMMA	2.0, 3.0, 6.0, 12.0 μm	0.017–0.95	BHK cells	Alignment increased with d. and decreased with w.	Clark et al. (1990)
Groove	Quartz	25 μm	0.02–0.2	Murine P388D1 macrophage	More elongation; cells spread faster for shallow grooves, but elongated faster on deeper grooves; orientation dependent on depth during first 30 min; 60% more F-actin in cells in grooves; LPS-activation enhanced orientation	Wojciak-Stothard et al. (1995a)
Groove	Quartz	0.5, 5.0, 10.0, 25.0 μm	0.02–10.0	Murine P388D1 macrophage	Smaller groves induced stronger orientation	Wojciak-Stothard et al. (1995b)
Groove	Poly-D-lysine-coated chrome-plated quartz	0.13–4.01 μm	0.025–9	Rat optic nerve oligodendrocytes, optic nerve astrocytes, hippocampal cerebellar neurons	Oligodendrocytes were highly aligned by features as small as 100 nm depth and 260 nm repeat spacing: astrocytes were also aligned while hippocampal and cerebellar neuron cells were not; oligodendrocyte alignment induced by pattern corresponding to diameter of axon in 7-day optic nerve	Webb et al. (1995)

continued

TABLE 6.1 (continued)
Summary of Effects on Nano- and Microstructuring on Cell Adhesion, Alignment, and Migration

Feature Type	Material	Feature Dimension	Aspect Ratio Range (Depth/Width)	Cell Type	Effect	Reference
Groove	PDMS cast of silicon oxide original	1.0–10.0 μm	0.045–1	Rat dermal fibroblasts	Cells oriented and elongated along grooves with ridge widths 4.0 μm or less; protrusions contacting ridges observed on oriented cells; cells randomly oriented on grooves with ridges more than 4.0 μm wide; groove width and depth did not affect orientation	den Braber et al. (1996a)
Groove	PDMS cast of silicon original	2.0, 5.0, and 10.0 μm	0.05–0.25	Rat dermal fibroblasts	Microfilaments and vinculin aggregates oriented along 2 μm grooves after 1, 3, 5, and 7 days, less oriented on 5 and 10 μm grooves; bovine and endogeneous fibronectin and vitronectin were oriented along grooves	den Braber et al. (1998)
Groove	PDMS cast of silicon original	2.0, 5.0, and 10.0 μm	0.05–0.25	Rat dermal fibroblasts	Cells on 2 and 5 μm grooves were elongated and aligned parallel to grooves; cells on 10 μm grooves were similar to those on smooth substrate	den Braber et al. (1996b)
Groove	PDMS cast of silicon original	2.0, 5.0, and 10.0 μm	0.05–0.25	Rat dermal fibroblast	2, 5 μm grooves induced stronger orientation than 10 μm	Weiss (1958)
Groove	PS	1–10 μm	0.05–1.5	Rat bone cells	Focal adhesions all over the surface for large grooves but only on the edges for narrow grooves	Matsuzaka et al. (2003)
Groove	Quartz	0.5, 5, and 10 μm	0.05–10	Murine P388D1 macrophage	Cells spread faster on shallow grooves, but elongated faster on deeper grooves; orientation dependent on depth during first 30 min; 60% more F-actin in cells in grooves; LPS-activation enhanced orientation	Wojciak-Stothard et al. (1995b)

Feature	Material	Size	Range	Cell type	Observations	Reference
Groove	Quartz	1.65–8.96 μm	0.077–0.42	Chick heart fibroblasts	Ridge width more important than groove width in determining cell alignment; alignment of cells inversely proportional to ridge width	Dunn and Brown (1986)
Groove	Quartz	1.65–8.96 μm	0.077–0.42	Chick heart fibroblasts	Ridge width more important than groove width in determining cell alignment; alignment of cells inversely proportional to ridge width	Dunn and Brown (1986)
Groove	Quartz	1.0, 4.0 μm	0.28–1.1	Mesenchymal stem cells (MSCs)	Aligned better when feature dimension is bigger	Wood (1988)
Groove	Quartz	0.98–4.01 μm	0.28–1.20	Mesenchymal tissue cells	Cells migrated along grooves; highly polarized; highest alignment on widest repeat spacing	Wood (1988)
Grooves	Silicon	0.33–2.1 μm	0.29–1.82	Human corneal epithelial cells	Perpendicular alignment for small pitch, parallel for large pitch	Teixeira et al. (2006)
Groove	Titanium-coated epoxy replica of silicon original	30 μm repeat spacing with 3, 10, or 22 μm depth or 7 and 39 μm spacing with 3 or 10 μm depth	0.43–1.43	Rat parietal implant model	Endothelial cells bridged 22 μm horizontal grooves, and attached to other sizes grooves; fibroblasts encapsulated smooth and 3, 10 μm horizontal grooves; fibroblasts inserted obliquely into 22 μm horizontal grooves; epithelial down growth greatest on vertical and smooth surface while least on 10, 22 μm horizontal grooves	Chehroudi et al. (1990)
Groove	Quartz and poly-L-lysine-coated quartz	0.13 μm	0.77–3.08	BHK, MDCK, chick embryo cerebral neurons	BHK cells aligned on all groove patterns, alignment increased with increasing depth; MDCK aligned and elongated to grooves, only elongation increased with depth; MDCK cells in groups and chick embryo cerebral neurons not affected by grooves	Clark et al. (1991)
Protein tracks	Quartz, hydrophobic silane, laminin	25.0 μm	ND	Embryonic xenopus laevis neurites	65% of neurites aligned to tracks after 5 h	Britland and McCaig (1996)

continued

TABLE 6.1 (continued)
Summary of Effects on Nano- and Microstructuring on Cell Adhesion, Alignment, and Migration

Feature Type	Material	Feature Dimension	Aspect Ratio Range (Depth/Width)	Cell Type	Effect	Reference
Pores	Uncoated and silicon coated filters	0.2–10.0 μm	ND	*In vivo* canine model	Nonadherent, contracting capsules around implants with pores smaller than 0.5 μm; implants with 1.4–1.9 μm pores showed adherent capsules but no inflammatory cells; pores bigger than 3.3 μm were infiltrated with inflammatory tissue; pores 1–2 μm allowed for fibroblast attachment	Campbell and von Recum (1989)
Groove	Titanium-coated silicon	ND (3 μm depth)	ND	Human gingival fibroblasts	Cells elongated and oriented along grooves; cell height 1.5-fold greater on grooves	Chou et al. (1995)
Groove	Ti-coated silicon	ND width, 3 –μm depth	ND	Human gingival fibroblasts	Cells oriented along grooves by 16 h; cells on grooves showed altered matrix metalloproteinase-2 mRNA time-course expression and levels	Chou et al. (1998)
Steps	PMMA	1–18 μm steps	ND	BHK cells, chick embryonic neural, chick heart fibroblast, rabbit neutrophils	Cells exhibited decrease in frequency of crossing steps and increased alignment at steps with increasing step height regardless of direction of approach; rabbit neutrophils showed twice the crossing frequency over 5 μm steps as did the other cells	Clark et al. (1987)
General roughness	Ti, Ti/Al/V alloy, Ti30Ta alloy	0.04, 0.36, and 1.36 μm peak-to-valley height	ND	Human gingival fibroblasts	Cells aligned to grinding marks: 10% of cells oriented on the surface with 0.04 μm roughness, 60% on 0.36 μm roughness, and 72% on 1.36 μm roughness	Eisenbarth et al. (1996)

Spheres	Poly(NIPAM) particles on polystyrene surface	0.86–0.63 μm	ND	Neutrophil-like induced HL-60 cells	Cells loosely adhered but did not spread on sphere-coated surface and could roll easily	Fujimoto et al. (1997)
Roughness	Nitro-cellulose, PVDF	Smooth and rough surfaces, feature size ND	ND	Rat sciatic nerve implant model	Bell-shaped tissue adhered to rough tube implants; free-floating nerve cables, containing myelinated and unmyelinated axons and Schwann cells grouped in microfascicles and surrounded by an epineurial layer observed in smooth tubes; macrophages comprised initial cell layer on rough polymers	Guenard et al. (1991)
General roughness	PMMA	Peak-to-valley heights from 0.07–3.34 μm	ND	Chick embryo vascular and corneal cells	Cells aligned to grinding marks: 10% of cells oriented on surface with 0.04 μm roughness, 60% on 0.36 μm roughness, and 72% on 1.36 μm roughness	Lampin et al. (1997)
General roughness	Titanium	1–2 μm pits, 1 μm pits, 10 μm craters, 10–20 globules, and sharp features of <0.1 μm	ND	MG63 osteoblast	Electropolished surface had more cells, while TI plasma-sprayed had less than TCPS; sandblasted surfaces had the same as TCPS	Martin et al. (1995)
Groove	Polystyrene, epoxy replicas	2 and 10 μm	ND	Chick heart fibroblasts, murine epithelial cells	75% of cells aligned on 5 μm grooves; 60% of cells aligned on 30 μm grooves; cytoplasmic extensions not related to surface features; alignment of cells not guided by lamellae or filopodia; cells bridged 2 and 10 μm grooves without touching surface	Ohara and Buck (1979)

continued

TABLE 6.1 (continued)
Summary of Effects on Nano- and Microstructuring on Cell Adhesion, Alignment, and Migration

Feature Type	Material	Feature Dimension	Aspect Ratio Range (Depth/Width)	Cell Type	Effect	Reference
General roughness	Polystyrene- and $H_2SO_4^-$-treated polystyrene	ND	ND	Murine peritoneal macrophages fibroblasts	Macrophages accumulated preferentially on roughened surfaces, while fibroblasts preferred smooth surfaces	Rich and Harris (1981)
Waves	PDMS gels of varying softness	ND	ND	Human dermal fibroblasts and keratinocytes	Fibroblasts proliferated equally on all substrates; keratinocytes spread more and secreted more ECM on soft gels than on hard gel	Rosdy et al. (1991)
Cylinders	Fused quartz	12–13 or 25 μm	ND	Primary mouse embryo fibroblasts and rat epithelial cell lines	Cells in the polarization stage of spreading with straight actin bundles became elongated, oriented along the cylinder, and resisted bending around cylinders; cells in the radial stage of spreading with circular actin bundles or cells with no actin bundles tended to bend around the cylinder and exhibited less elongation and orientation to the long axis of the cylinder	Rovensky Yu and Samoilov (1994)
Protein tracks	Glass coated with fibronectin	0.2–5.0 μm	ND	BHK cells, rat tendon fibroblasts, rat dorsal root ganglia cells, P388D1 macrophages	Fibers increased spreading and alignment in the direction of fiber; fibers increased speed and persistence of cell movement and rate of neurite outgrowth; macrophages had actin-rich microspikes and became polarized and migratory	Wojciak-Stothard et al. (1997)
Groove	Titanium-coated silicon, epoxy replicas, photoresist	ND (5–120 μm depth)	ND	Human gingival fibroblasts	Alignment observed in grooves and on rat ridges; cells oriented preferentially to major grooves; minor grooves caused orientation of cells in the absence of major grooves or when discontinuity existed in major groove pattern	Brunette (1986a), Brunette et al. (1983)

Steps	PMMA	1–18 μm	ND	BHK	Alignment at steps	Clark et al. (1987)
Pits	PCL, PMMA	35, 75, and 120 nm	ND	Fibroblasts	Reduced adhesion, orientation, and distinction of symmetries	Curtis et al. (2001), Curtis et al. (2004), Dalby et al. (2004)
Pillars and pores	PMMA	1.0, 5.0, 10.0, and 50.0 μm	ND	Human osteoblasts and amniotic epithelial cells	Cells attached to edges of pores, especially on 10 μm pores; increase in cell adhesion on all materials but PMMA; greatest increase in adhesion was on 50 μm PET pillars; 10 μm pores caused 5% increase in resistance to shear force	Fewster et al. (1994)
Random	PLGA, PU, PCL	206, 370 nm	ND	Bladder smooth muscle cells	Enhanced adhesion	Thapa et al. (2003a)
Protein tracks	Oriented collagen or fibrin	ND	ND	Human neutrophil leukocytes	Cells tended to move in the direction of fiber axis alignment; movement was bidirectional; no chemotaxis was evident	Wilkinson et al. (1982)
General roughness	Areas of rougher reative ion etched silicon and smoother, wet etched silicon	Reactive ion etched: 57 nm average diameter, 230 nm height; wet etched: 115 nm peak-to-valley roughness	Reactive ion etched: 4.04; wet etched: 0.87–2.17	Transformed rat astrocytes, primary rat cortical astrocytes	Transformed cells attached preferentially to wet-etched regions rather than reactive ion etched columnar structures; primary cells preferred columnar structures of reactive ion etched areas and did not spread on wet etched areas	Turner et al. (1997)

Abbreviations: ND = not determinable, PDMS = poly(methacrylate) siloxane, PC = polycarbonate, NPAM = *N*-isopropyl-acrylamide, TCPS = tissue culture polystyrene, PCL = polycaprolactone, PLGA = polylactic-co-glycolic-acid, PU = poly-ether-urethane, PMMA = poly(methylmethacrylate), PDLA = poly(D,L-lactic acid), BHK = baby hamster kidney, and MDCK = Madin Darby canine kidney.

TABLE 6.2
Summary of Effects on Nano- and Microstructuring on Cell Morphology

Feature Type	Material	Feature Dimension	Aspect Ratio Range (Depth/Width)	Cell Type	Effect	Reference
Ridges	Silicon oxide on polystyrene	4 μm	0.0125	Murine neuroblastoma cells	Cells adhered to lines and processes aligned along the lines; processes grew in bipolar manner	Cooper et al. (1976)
Groove	Titanium-coated silicon	15 μm	0.2	Porcine epithelial cells	No significant elliptical morphology	Oakley and Brunette (1995)
Groove	Epoxy replica of silicon original	17 μm	0.59	Porcine periodontal ligament epithelial cells, rat parietal implant model	Shorter length epithelial attachment and longer connective tissue attachment in grooved parts of implant compared to smooth parts	Chehroudi et al. (1988)
Groove	Quartz and protein-coated quartz	2.0, 10.0 μm	0.003–0.141	P3881D1 macrophages, rat peritoneal macrophages	Cells on grooves had increased number of protrusions extending perpendicular to grooves; grooves caused increase in F-actin; F-actin and vinculin accumulated along groove/ridge boundaries; cells on grooves showed higher phagocytic activity	Wojciak-Stothard et al. (1996)

continued

Groove	Quartz	0.02–0.2	25 µm	Murine P388D1 macrophage	More elongation; cells spread faster on shallow grooves, but elongated faster on deeper grooves; orientation dependent on depth during first 30 min; 60% more F-actin in cells in grooves; LPS-activation enhanced orientation	Wojciak-Stothard et al. (1995a)
Groove	Quartz	0.02–1	5, 10, and 25 µm	BHK cells	F-actin condensations observed at topographic discontinuities; condensations often at right angles to groove edge with periodicity of 0.6 lm; vinculin organization similar to that of actin; microtubules observed after 30 min	Wojciak-Stothard et al. (1995a)
Groove	PDMS cast of silicon oxide original	0.045–1	1.0–10.0 µm	Rat dermal fibroblasts	Protrusions contacting ridges observed on oriented cells; cells randomly oriented and were more circular on grooves with ridges more than 4.0 µm wide; groove width and depth did not affect cell size or shape	den Braber et al. (1996a)
Groove	PDMS cast of silicon original	0.05–0.25	2, 5, 10 µm	Rat dermal fibroblasts	Vinculin located primarily on surface ridges; groove-spanning filaments also observed	den Braber et al. (1998)

TABLE 6.2 (continued)
Summary of Effects on Nano- and Microstructuring on Cell Morphology

Feature Type	Material	Feature Dimension	Aspect Ratio Range (Depth/Width)	Cell Type	Effect	Reference
Wells and nodes	PDMS replicas of silicon original	2.0, 5.0, and 8.0 µm diameter round	0.071–0.23	Rabbit implant model, murine macrophages	2 and 5 µm textured implants had fewer mononuclear cells and thinner fibrous capsules than did smooth and 8 µm textured implants; cells on smooth PDMS were round with few pseudopods, but cells on 2 and 5 µm textures were elongated with pseudopods	Schmidt and von Recum (1991)
Groove	Quartz and poly-L-lysine-coated quartz	0.13 µm	0.77–3.08	BHK, MDCK, chick embryo cerebral neurons	MDCK elongation increased with depth; MDCK cells in groups, and chick embryo cerebral neurons not affected by grooves	Clark et al. (1991)
Protein tracks	Quartz, hydrophobic silane, laminin	25.0 µm	ND	Embryonic *Xenopus laevis* neurites	Neuritogenesis not affected	Britland and McCaig (1996)
Groove	Titanium-coated silicon	ND (3 µm depth)	ND	Human gingival fibroblasts	Cell height 1.5-fold greater on grooves; fibronectin mRNA and secreted fibronectin increased in cells on grooves; GAPD mRNA not affected; half-lives of fibronectin mRNA altered	Chou et al. (1995)

Groove	Ti-coated silicon	ND width, 3 μm depth	ND	Human gingival fibroblasts	Cells on grooves showed altered matrix metalloproteinase-2 mRNA time-course expression and levels compared to cells on smooth Ti or tissue culture plastic	Chou et al. (1998)
Protein tracks	Quartz, hydrophobic silane, laminin	2.0, 3.0, 6.0, 12.0, 25.0 μm	ND	Chick embryo neurons, murine dorsal root ganglia neurons	Isolated 2 μm tracks strongly guided neurite extension, while 2 μm repeat tracks did not; growth cone morphology was simpler on narrower single tracks; growth cones spanned many tracks on narrow repeats; neurite branching decreased on 25 μm tracks	Clark et al. (1993)
Micro-textured surface	Polyurethane positive cast of PMMA negative	Micro- and nanometer-scale topography	ND	Bovine aortic endothelial cells	Cells grown on replicas of ECM had 3D appearance and spread areas at confluence which appeared more like cells in their native arteries than cells grown on untextured control surfaces	Goodman et al. (1996)
Roughness	Nitro-cellulose, PVDF	Smooth and rough surfaces, feature size ND	ND	Rat sciatic nerve implant model	Free-floating nerve cables, containing myelinated and unmyelinated axons and Schwann cells grouped in microfascicles and surrounded by an epineurial layer observed in smooth tubes; epineurial layer was thinner on rough PVDF than on rough nitrocellulose; smooth PVDF showed more myelinated axons than did smooth nitrocellulose	Guenard et al. (1991)

continued

TABLE 6.2 (continued)

Summary of Effects on Nano- and Microstructuring on Cell Morphology

Feature Type	Material	Feature Dimension	Aspect Ratio Range (Depth/ Width)	Cell Type	Effect	Reference
General roughness	Titanium	0.14, 0.41, and 0.80 μm peak-to-valley heights	ND	Human gingival fibroblasts	Cells on smooth, electropolished surfaces showed a flat morphology and grew in layers; cells on sandblasted Ti grew in clusters; round and flat cells found on etched and sandblasted Ti; actin bundles and vinculin-containing focal adhesions were observed in spreading cells on electropolished and etched Ti, but not in spreading cells on sandblasted Ti	Kononen et al. (1992)
General roughness	Titanium	1–2 μm pits, 1 μm pits, 10 μm craters, 10–20 globules, and sharp features of <0.1 μm	ND	MG63 osteoblast	Thymidine incorporation inversely related to roughness; proteoglycan synthesis decreased on all surfaces; alkaline phosphatase production decreased with increasing roughness except on coarse blasted Ti; correlation observed between roughness and RNA and CDP production	Martin et al. (1995)

Cylinders	Fused quartz	12–13 or 25 μm	ND	Primary mouse embryo fibroblasts and rat epithelial cell lines	Cells in the polarization stage of spreading with straight actin bundles became elongated, oriented along the cylinder, and resisted bending around cylinders; cells in the radial stage of spreading with circular actin bundles or cells with no actin bundles tended to bend around the cylinder and exhibited less elongation and orientation to the long axis of the cylinder	Rovensky Yu and Samoilov (1994)
Wells and nodes	PDMS cast of silicon original	2.0, 5.0, 8.0, and 10.0 μm diameter	ND	Murine peritoneal macrophages	Cells on 5 μm textures had smallest dimensions while cells on smooth silicone and glass had largest dimensions; mitochondrial activity highest on cells on 5 and 8 μm variable pitch surfaces and on polystyrene; PMA-stimulated cells on smaller textures were less active than unstimulated cells	Schmidt and von Recum (1992)
Protein tracks	Glass coated with fibronectin	0.2–5.0 μm	ND	BHK cells, rat tendon fibroblasts, rat dorsal root ganglia cells, P388D1 macrophages	Actin aligned in fibroblasts; increased polymerization of F-actin; macrophages had actin-rich microspikes and became polarized and migratory	Wojciak-Stothard et al. (1997)

continued

TABLE 6.2 (continued)
Summary of Effects on Nano- and Microstructuring on Cell Morphology

Feature Type	Material	Feature Dimension	Aspect Ratio Range (Depth/Width)	Cell Type	Effect	Reference
Pits	PCL, PMMA	35, 75, and 120 nm	ND	Fibroblasts	Reduced adhesion, orientation, and distinction of symmetries	Curtis et al. (2001), Curtis et al. (2004), Dalby et al. (2004)
General roughness	Areas of rougher reactive ion etched silicon and smoother, wet etched silicon	Reactive ion etched: 57 nm average diameter, 230 nm height; wet etched: 115 nm peak to valley roughness	Reactive ion etched: 4.04 wet etched: 0.87–2.17	Transformed rat astrocytes,	Transformed cells on wet-etched areas spread in an epithelial-like manner and were smooth; transformed cells on columnar regions were rounded, loosely attached, and exhibited complex surface projections	Turner et al. (1997)

Abbreviations: ND = not determinable, PDMS = polydimethyl siloxane, BHK = baby hamster kidney, MDCK = Madin Darby canine kidney, GAPD = glyceraldehyde-3-dehydrogenase, PMA = phorbol 12-mystrate, ECM = extracellular matrix, PVDF = poly(vinylidene fluoride), PMMA = poly(methyl methacrylate).

TABLE 6.3

Summary of Effects on Nano- and Microstructuring on Cell Apoptosis, Replication, and Differentiation

Feature Type	Material	Feature Dimension	Aspect Ratio Range (Depth/ Width)	Effect	Cell Type	Reference
Ridges	Polystyrene cast of silicon original	0.5–100.0 µm	0.0003–10	Maximum cell differentiation observed for ridges or plateaus 0.5 µm high; ridges higher than 1.0 µm or smaller than 0.25 µm were not effective signals	Uromyces appendiculatus fungus	Hoch et al. (1987)
Wells and nodes	PDMS cast of silicon original	2.0, 5.0, and 10.0 µm diameter	0.05–0.25	Cells on 2 and 5 µm nodes showed increased rate of proliferation and increased cell density compared to cells on 2 and 5 µm wells; 10 µm nodes and wells did not differ statistically from smooth surfaces	ATCC human abdomen fibroblasts	Green et al. (1994)
Groove	PDMS cast of silicon original	2, 5, 10 µm	0.05–0.25	Cells on smooth PDMS entered S phase of cell cycles faster than cells on textured PDMS; cells on 10 µm texture proliferated less than those on 2 and 5 µm textures	Human skin fibroblasts	van Kooten et al. (1998)
Groove	Polystyrene cast of silicon original	0.5 µm	1.0–10.0	Multiple layer protein adsorption from serum; cells grew to confluence in 4 days and produced ECM after 7 days	Sprague-Dawley rat calavarial cells	Chesmel and Black (1995)
Protein tracks	Quartz, hydrophobic silane, laminin	25.0 µm	ND	Neutritogenesis not affected	Embryonic Xenopus laevis neurites	Britland and McCaig (1996)

Abbreviations: ND = not determinable, CEM = extracellular matrix, PDMS = poly-dimethyl siloxane, ATCC = America Type Culture Collection.

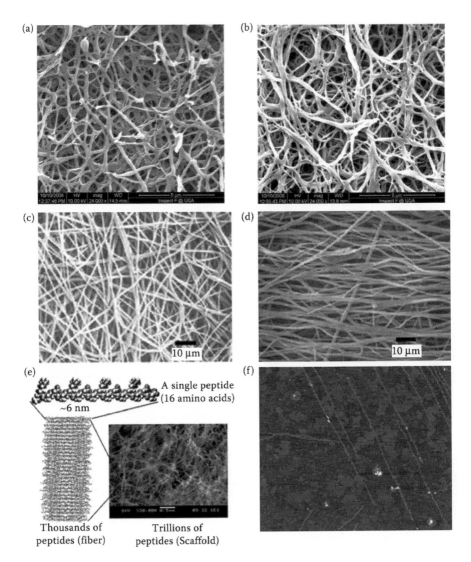

FIGURE 6.1 Examples of nanostructured surfaces. (a) and (b): Scanning electron micrograph nanofibrous surfaces prepared by the phase separation method with PLLA in THF (wt/v) of 1.25% (a) and 2.85% (b). ((b) adapted from Wang, L. and Kisaalita, W.S. 2010. *J. Biomed. Mater. Res. Part B* (accepted).) (c) and (d): Scanning electron micrographs of nanofibrous surfaces of electrospun blends of 75:25 PLA: PGA, randomly oriented (c) and aligned (d). ((d) adapted from Barnes, C.P., et al. 2007. *Adv. Drug Deliv. Rev.* 59: 1413–1433.) (e): Molecular model of self-assembling peptide (RADA16-I) amino acid (~6 nm long, 1.3 wide and 0.8 nm thick that self-assemble into nanofibers to create the surface as shown in the SEM image, and (f) peptide nanofibers aligned by water flowing through a microfluidic device. ((f) from Yang, Y., et al. 2009. *Nano Today* 4: 194. With permission.)

TABLE 6.4

Comparison of Nanofiber Processing Techniques

Techniques	Scale	Level of Difficulty	Advantages	Limitations
Self-assembly	Lab	Difficult	Can achieve fiber diameters at the lowest ECM scale (5–8 nm)	Only short fibers can be created (<1 μm); low yield; matrix directly fabricated
Phase separation	Lab	Easy	Can manipulate mechanical properties, pore size, and interconnectivity; batch-to-batch consistence	Low yield; matrix directly fabricated; limited to a few polymers
Electrospinning	Lab and industrial	Easy	Cost effective; long continuous fibers; can produce aligned fibers; can manipulate mechanical properties, size, and shape; can use a plethora of polymers	Large nanometer to microscale fibers; use of organic solvents; no control of pore structure

Source: Adapted from Jayaraman, K. et al. 2004. *J. Nanosci. Nanotechnol.* 4(1–2): 52–65.

Plieva (1998). The main application of both natural and synthetic scaffold research has been either drug delivery or engineered tissue replacement or regenerative medicine (Friess, 1998; Woerly et al., 1996; Sherwood et al., 2002; Deng et al., 2002; Hutmacher et al., 2003).

Relative to 2D substrates, scaffold matrix interactions have displayed differences in structure, localization, function, shape (Friedl and Brocker, 2000; Krewson et al., 1994), and proliferation (Senoo et al., 1996). One limitation with these structures is that their fabrication processes do not lend themselves easily to precise aspect ratio control (Freyman et al., 2001).

Micromachining techniques mainly from the microelectronic industry have been used to engineer scaffolds. The fabrication of high aspect ratio microstructures is a standard technique for microelectrical–mechanical systems (MEMS) (Petronis et al., 2003; Madou, 2002; Zhao et al., 2003). Powers et al. (2002a, 2002b) described a scaffold fabricated by deep reactive ion etching of silicon wafers to create an array of channels with cell-adhesive walls (Figure 6.2) and revealed significantly greater functional activity and morphological stability in comparison to 2D primary rat liver cell cultures. The effect of different patterned microstructures such as hexagonal microstructure and micropillar arrays on the neurite outgrowth was successfully conducted by Craighead and Turner's group (Dowell et al., 2000; Turner et al., 2000; Kam et al., 1999). A little more than a decade ago, Hockberger et al. (1996) found that, by morphologically or chemically introducing regular microstructured patterns on the substrate, the cell–substrate interaction can be manipulated. In addition, other

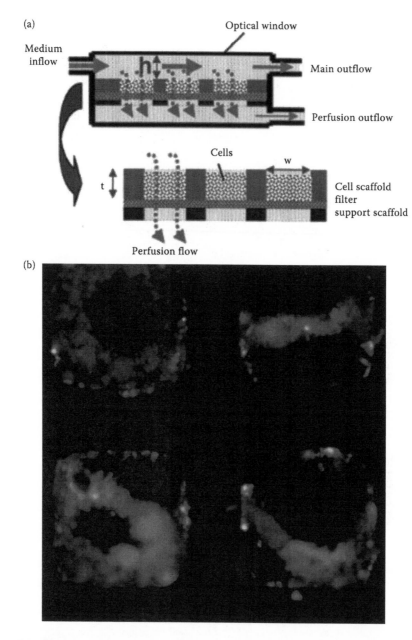

FIGURE 6.2 Schematic of a scaffold fabricated by deep reaction ion etching of silicon incorporated in a bioreactor housing (a) and top view of viable primary hepatocytes, visualized by calcein AM-ethidium homodimer staining, after 13 days of culture in the bioreactor (b). Scaffold dimensions are: diameter (w) = 300 µm, height (t) = 235 µm, reactor headspace height (h) = 805 µm. ((b) from Powers, M.J., et al. 2002a. *Biotechnol. Bioeng.* 78(3): 258; Powers, M.J., et al. 2002b. *Tissue Eng.* 8(3): 506. With permission.)

guided cell growth and cell surface interactions on microstructured patterns have been intensively studied (Hoch et al., 1996; Mrksich, 1998; Detrait et al., 1998; Voldman et al., 1999; Brizzolara, 2000; McFarland et al., 2000; Kotov et al., 2003). A key advantage of micromachining techniques is that it is possible to design and vary wide aspect ratios (height over width) in a controlled manner. Also, they can be effectively used to create microenvironments free of hydrogel mass transport limitations, offering opportunities to control the morphology of the substrate and guide the cell growth by controlling structure parameters and the chemical homogeneity of the structures.

A nonsilicon material widely used in the fabrication of microfluidics and MEMS parts that is offering unique advantages in scaffold fabrication is SU-8 (MicroChem, Newton, MA), an epoxy-based negative photoresist. Using standard photolithography techniques, SU-8 can be used to pattern high aspect ratios (>20) structures with high optical transparence, straight side walls, and excellent thermal stability (Seidemann et al., 2002; Chuang et al., 2003; Voskerician et al., 2003). The first few studies that attempted to integrate US-8 structures were unsuccessful due to improper processing that left cytotoxic residues. This problem has been systematically studied and proper fabrication processes established (Wang et al., 2009, 2010; Vernerkar et al., 2009; Wu et al., 2006). Numerous studies have used SU-8 structures to study cell fates. For example, Chin et al. (2004) used SU-8 platform to study the proliferation dynamics of a heterogeneous adult rat neural stem cell population. Wu et al. (2006) integrated SH-SY5Y human neuroblastoma cells into SU-8 microwell structures and found that in comparison to SU-8 flat surfaces, the cell in microwells developed significantly higher resting membrane potentials. Most recently, Wang et al. (2009) used similar microwell structures interconnected with channels and investigated voltage-gated calcium channel function and concluded that cells with SU-8 patterned microwells were indeed different from cells on planer SU-8 surfaces, suggesting that the high aspect ratio microstructures were not merely "folded" 2D structures. Neural network-like formation by SH-SY5Y cells in the Wang et al. (2009) study is shown in Figure 6.3.

While silicon-based MEMS are well established commercially, the introduction of nonsilicon and, in particular, polymer microfluidic systems is recent (Malek, 2006). Success in entering the general market is heavily dependent on availability of large-volume industrial manufacturing to provide replication. Typical replication methods are hot embossing (Heckele et al., 1998) and injection modeling (Piotter et al., 2004), which provide low-cost mass production of components with large aspect ratios, structural details in submicron range and precisions lower than 2 μm in final polymer products. A microsystem technology technique from the microfluidics field that is finding its way in scaffold fabrication is laser abelation (Pfleging et al., 2007). Laser abelation or micromachining can process a variety of materials, glass, silicon, metals, inorganic, plastics, etc., both conductive and nonconductive, without a vacuum chamber or access to a clean room (Pfleging et al., 2007; Fewster et al., 1994). The technique is of interest in scaffold fabrication for several reasons. First, laser beam spreads so little that it gives out extremely low divergence, providing high precision and repeatability that fit into many special applications. Laser techniques offer unique temporal and spatial characteristics as they generate

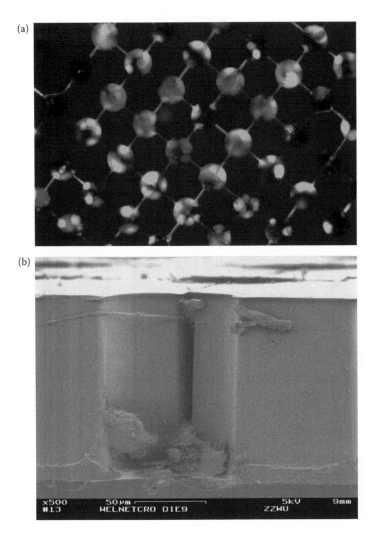

FIGURE 6.3 Image of SH-SY5Y neuroblastoma cells in network micropatterns, fabricated by photolithography from SU-8 photoresist material, on day 8 into differentiation visualized by calcein staining (a) and SEM image of showing a neurite from adjacent well and cell aggregate at the bottom of the well (b). (Adapted from Wang, L. et al. 2009. *Sens. Actuat. B Chem.* 140: 349–355.)

ultrashort pulses of monochromatic (single-wavelength) and coherent (same-phase) light, which is directed under precise control high-density photon energy onto a selected local region of substrate material, resulting in minimal heat-affected zone. Second, the noncontact process of direct laser irradiation not only introduces no tool wear as opposed to traditional micromachining, but also minimizes chance of damage to workpiece material due to process shock or handling. Laser micromachining provides an easy one-step alternative to the conventional chemical etching process without involving any solvent chemicals. Third, laser processing systems, incorporated with advanced computer control with user-friendly programming interfaces, permit easy retooling to fabricate flexible feature size and shape. Figure 6.4 shows the relation between microwell depth and laser pulse number to demonstrate the preciseness of the process and a microwell cross-section of a typical structure in glass. Guan and Kisaalita (2010a, 2010b) have used laser-abelated microstructures with fibroblast to study cell–surface interactions with applications in biofouling.

There are techniques for producing microtissues similar to those that are produced in scaffolds that do not rely on scaffolding. For example, the hanging-drop method has been applied to the production of multicellular tumor spheroids with a wide number of cell types (Sanchez-Bustamante et al., 2005; Kelm et al., 2003).

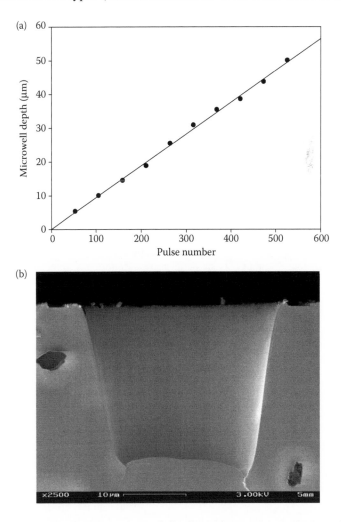

FIGURE 6.4 Relation between microwell depth and laser pulse number at a fluence of 17.55 J/cm² with a repetition rate of 100 s⁻¹ (a) and cross-section view of a fabricated microwell at the same fluence repetition rate, but with a total of 212 (b). (Adapted from Guan, Y. and Kisaalita, W.S. 2010a. *J. Biomed. Mater. Res. Part A* (waiting for a decision).)

Another well-known example is microgravity bioreactors. Approximation of micro-gravitational conditions at ground level has been achieved in specialized cell culture systems using a rotating-wall vessel (RWV) bioreactor developed at NASA—Johnston Space Center. Successful growth and differentiation of 3D cellular aggregates in RWV has been observed in colon cancer cell cultures (Jessup et al., 1993), ovarian tumor cells (Becker et al., 1993), skeletal tissue (Freed et al., 1997), salivary gland cell culture (Goodwin et al., 1993; Lewis et al., 1993), and neuroblastoma cells (unpublished results from the author's laboratory). Unfortunately, both the hanging-drop microgravity tissue production approaches are not easy to scale down (i.e., high-density plates) and are not compatible with HTS requirements. Another example is the use of special media or surface conditions to promote the formation of spheroids. More details on spheroids are provided in Chapter 12.

6.3   NANO AND SCAFFOLD-COMBINED STRUCTURES

Several fabrication techniques have been used to combine nanostructures within a microspace (or pore) large enough to accommodate a multicellular aggregate. These techniques include (1) combining thermally induced phase separation with the porogen-leaching process (Liu and Ma, 2009; Cheng and Kisaalita, 2010), (2) combining thermally induced phase separation with laser micromachining (Wang et al., 2009), (3) combining rapid prototyping and electrospinning techniques (Martins et al., 2009), and (4) codeposition of water-soluble (leachable) and nonwater-soluble fibers (Ekaputra et al., 2008). Other approaches have involved sandwiching cells between two nanostructured materials such as electrospun membranes (Srouji et al., 2008) and lithographically textured polymer membranes (Seunarine et al., 2008). Figure 6.5 shows samples and/or schematics of these structures.

A literature search under this subheading was conducted to identify studies that compared cellular function outcomes separately on smooth (2D), nano-, micro-, and nano–micro combination structures that had potential for HTS applications. Unfortunately, a handful of studies (Cheng and Kisaalita, 2010; Martins et al., 2009; Hartman et al., 2009; Seunarine et al., 2008) that came close to meeting this criterion were found. The findings from these studies are summarized in Table 6.5. It is interesting to note that in all cases, the combined scaffold meets or exceeds the performance of a single level in terms of biological outcomes examined. It is anticipated that more of these types of studies will be published as the popularity of 3D cultures increases, especially in research laboratories.

6.4   TEMPORAL FACTOR

If one imagines holding all other microenvironmental factors (MEFs) at fixed values, it is easy to appreciate the importance of time in culture. To achieve a desired pheno-type both *in vitro* and *in vivo* takes time. However, the time it takes is almost always affected by other factors. The differentiation of MSCs shown in Figure 6.6 is an excellent illustrative example. MSCs are adult progenitor cells with capacity to form cartilage, bone, tendon, ligament, marrow stroma, and other connective tissue as shown. The differentiation of MSCs along multiple lineages to produce fibroblast (tendon and

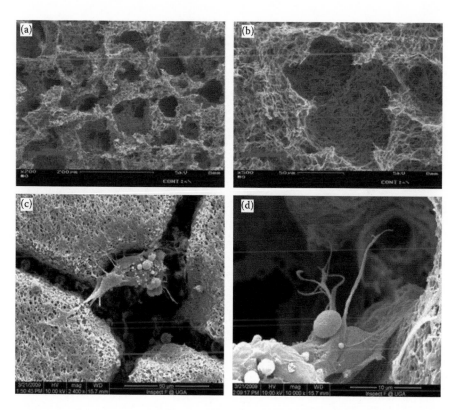

FIGURE 6.5 Examples of nano and scaffold-combined structures. Thermally induced phase separation was combined with porogen leaching to produce the two-level (nanofibrous and micropores) PLLA structure shown in (a) and (b) for more detailed pore structure. [(b) adapted from Cheng, K. and Kisaalita, W.S. 2010. *Biotechnol. Prog.* (DOI: 101002/btpr.391).] Thermally induced phase separation was combined with laser micromachining to produce a microwell PLLA structure with well-to-well connecting channels as shown in (c) with neural progenitor cells in the and neurite extensions along the channel (arrows in d) (Wang and Kisaalita, 2010). Twenty percent poly(ε-caprolactone): collagen (PCL/COL, 80:20, by mass) in 1,1,1,3,3,3-fluoro 2-propanol (HFIP) and either 20% w/v poly(ethylene oxide) or 10% gelatin in HFIP were co-deposited via electrospinning ontoa rotating mandrel (e). Poly(ethyleneoxide) and gelatin were leached, leaving behind PCL/COL as the main fiber shown in (f). ((f) adapted from Ekaputra, A.K. et al. 2008. *Biomacromolecules* 9: 2097–2103.) A 3D plotting technique was use to rapid prototype 30:70 (wt%) blends of starch and poly-caprolactone heated to 140°C through a heated nozzle at a microfiber strand setting of 1 mm. Hierarchical fibrous structures were achieved by integrating electrospun nanofiber meshes, produced from 17% w/v polycaprolactone in 7:3 chloroform: dimethylfomamide, every two consecutive layers of plotted microfibers. SEM micrographs show microfibers without (g) and with (h) nanofiber meshes. ((h) adapted from Martins et al. 2009. *J. Tissue Eng. Regener. Med.* 3:37–42.) Bars represent in μm 100 (a), 50 (b and c), 10 (d), 5(f), and 2000 (g and h).

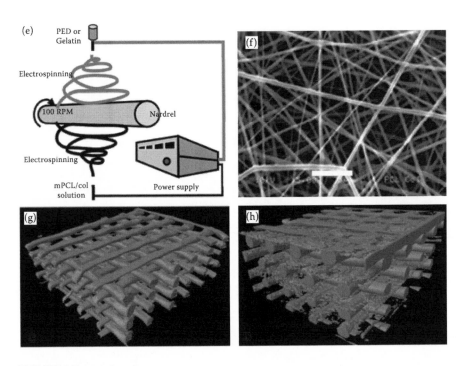

FIGURE 6.5 continued

ligament), chondrocyte (cartilage), osteocyte (bone), and other tissue cells is dependent on all the MEFs (Caplan, 1994) in addition to genetic potential. Therefore it is difficult to conduct experiments to independently show temporal effects.

An alternative way of appreciating the importance of the temporal factor is the time-dependent growth factors to control the resulting phenotype. As indicated in Appendix D, Table D1, not only the concentration of factors is important, but the timing of the addition contributes to the type of phenotype that resulted in cardiac differentiation from human embryonic stem cells. In a more recent study, Carpenedo et al. (2009) engineered a microsphere-mediated delivery of morphogenic factors to embryonic stem cells and induced them to differentiate via aggregation into multicellular spheroids referred to as embryoid bodies (EBs) (Doetschman et al., 1985). Typically, different culture methods and media compositions are manipulated in efforts to control embryonic stem cell differentiation within EBs (Schuldiner et al., 2000; Hopfl et al., 2004), resulting in heterogeneous spatially disorganized differentiation within EBs. On the contrary, during development *in vivo*, morphogens are secreted locally and presented to embryonic cells in a spatially and temporally controlled manner to direct appropriate differentiation and tissue formation (Brennan et al., 2001; Chiang et al., 1996). To mimic the *in vivo* temporal morphogen delivery, Carpenedo et al. (2009) used degradable PLGA [poly(lactic-co-glycolic acid)] microspheres to release retinoic acid directly within EBs. Although not measured directly, presumably, the temporal availability of morphogens to EBs in morphogen-containing media (soluble delivery) and EBs with morphogen-loaded PLGA were different. Carpenedo and coworkers successfully induced the formation of cystic

TABLE 6.5

Summary of Effects of Combining Nano- and High-Aspect Microstructuring on Cell Fate

Scaffold material, fabrication technique, and reference	Cell type and cellular outcome(s) investigated	Outcome level at the 2D or nanoscale	Outcome level at the microscale	Outcome level from the combination
Poly(L-lactic acid) (PLLA) fabricated by combining thermally induced phase separation with porogen leaching (Cheng and Kisaalita, 2010)	Human foreskin fibroblasts, SCRC-1041 (ATCC); phosphorylation of focal adhesion kinase at tyrosine 397—staining was done of day 2 cultures	Staining was punctate and less defined	Staining was between 2D (well-defined streaky pattern) and the less defined staining observed with nanostructures	Staining was the same as observed with nanostructure, suggesting that the combined structure fully captured the adhesion observed on nanostructures in addition to other desirable features like cell aggregate formation
Microfibrous and nanofibrous membranes were produced by electrospinning of calf skin collagen/2,2,2-trifluoroethanol (TFE: 16% w/v) and calf skin collagen/1,1,1,3,3,3-hexafluoro-2-propanol (HFIP: 16% w/v) solutions, respectively, onto a metal plate covered with nonstick aluminum foil (Hartman et al., 2009)	Bone metastatic cell line (C4-2B); attachment, growth, morphology, apoptosis, and actin staining with phalloidin	Nanofibrous—fiber diameter in the range of 111–536 nm	N/A	Microfibrous (micro-/nano-combination—fiber diameter in the range of 800–2500 nm) scaffold represented a better approximation of the in vivo tumor microenvironment with respect to colony formation and apoptotic response to antineoplastic agents when compared to both 2D culture plates and the nanofibrous scaffold
Starch-polycaprolactone micro-structure produced by rapid prototyping and polycarprolactone nano-structure produced by electrospinning (Martins et al., 2009)	Human osteosarcoma-derived cells (Saos-2 cell line) (ECACC); alkaline phosphatase (ALP) activity	N/A	Fabricated by depositing microfibers 30/70% (wt%) blend of starch and polycaprolactone heated to 140°C (see Figure 6.5)	Fabricated as in microscale, except that every two consecutive micro-layers, electrospun nanofiber meshes were integrated; a significant increase in cell proliferation and osteoblastic activity in comparison to the microscale-only structure

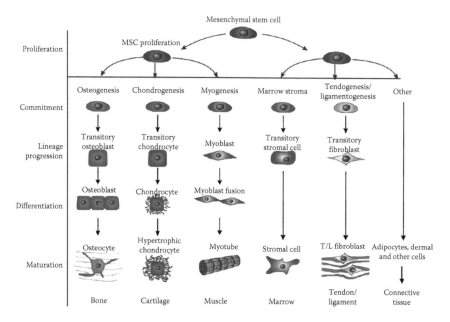

FIGURE 6.6 Schematic of the differentiation of MSCs to highly differentiated multiple phenotypes. (From Caplan, A.I. and Bruder, S.P. 2001. *Trends Mol. Med.* 7(6): 260. With permission.)

spheroids uniquely resembling the phenotype and structure of early streak mouse embryo (E6.75), with an exterior of FOXA2+ (endoderm marker) visceral endoderm enveloping an apiblast-like layer of OCT4+ (epiblast marker) cells. Figure 6.7 shows the SEM ultrastructure of EB differentiated with retinoic acid (morphogen)-containing PLGA microspheres. Although the temporal effects examples presented above have been drawn from the stem cell literature, the effects are by no means restricted to stem cell differentiation.

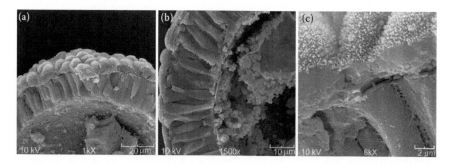

FIGURE 6.7 SEM micrographs of day 10 EBs with retinoic acid microsphere morphogen delivery, revealing the distinct squamous endoderm and pseudostratified epiblastlayer (a and b) as well as a dense coat of microvilli on the surface of the endoderm cells (c), similar to that observed on early streak stage mouse embryos . On the contrary, day 10 EBs with unloaded and without microspheres did not display these features. (From Carpenedo, R.L. et al. 2009. *Biomaterials* 30: 2513. With permission.)

6.5 CONCLUDING REMARKS

The main goal of the chapters of this section is to synthesize the current knowledge and begin to provide meaningful design principles. From the foregoing, the body of work on surface and scaffold technology is extensive; however, it is difficult to translate the knowledge into design principles. This is mainly because of the many cells types, materials, feature geometry and aspect ratios, and cell fate parameters measured (Martínez et al., 2009). As a simple first step toward this goal, an approach for organizing the current literature and the future research is offered here focusing on key cell fates of proliferation, apoptosis, and differentiation. The approach involves establishing a relationship among three dimensionless numbers: the differentiation ratio (D_R), the proliferation ratio (P_R), and the material feature diameter (D_f)-suspended cell diameter (D_s) ratio (D_f/D_s). A plot with a few data points is presented in Figure 6.6 for illustrative purposes.

D_f is the dimension that characterized the material feature such as the electrospun nanofiber diameter (Vasita and Katti, 2006) at the nanoscale or the equivalent diameter of a pore in a polystyrene scaffold (Cheng et al., 2008) or the well diameter of a micropatterned SU-8 structure (Wang et al., 2009). The rationale for the choice of the ratio of the structure feature characteristic dimension to the cells in suspension dimension was based on two observations. First, cell adhesion and spreading have been observed highest on 15 nm tubes and declined significantly with increasing pore sizes (Park et al., 2007). Cells on 30 nm or less tubes showed extensive focal contacts to which actin stress fibers were anchored and with fewer adhesion contacts for higher diameter tubes (e.g., 100 nm), cells were highly migratory. Second, at characteristic dimensions, smaller than the cell size, cells are forced to grow on the surface as opposed to higher dimensions, where they can get into the structure and depending on the geometry, form true 3D microtissues. Taken together, a dimensionless parameter that captures the extremes is likely to be helpful in providing ranges in which particular mechanisms of adhesion or migration predominate while enabling the comparison of data from different materials and their architectures.

As described in the previous chapter, the practice to promote cell adhesion is to coat material surfaces with adhesive proteins such as laminin, fibronectin, poly-L-lysine, and others. These coatings bind to different adhesion receptors initiating communication that may be coating dependent. Using the ratio of the structured material cell fate parameter to that of the "flat material" cell is hypothesized to minimize the coating effect, when comparing results from different systems. To test the idea, data from six recent papers (Table 6.6) have been reanalyzed and plotted as shown in Figure 6.8. The results are promising for several reasons. First, with the exception of a spike from Oh et al. (2009) data, an exponential trend for both differentiation and proliferation is visible. Normally, differentiation and proliferation tend to be inversely related. The anomaly presented in Figure 6.8 is probably due to the fact that, in all cases, the media used for differentiation and proliferation were different. Typically, a growth factor is removed and/or a differentiating agent is added. The inverse relationship would hold for cells cultured under similar conditions. In cases with the same differentiation and proliferation media, apoptosis will probably be predictable by lack of structure enhancement of both proliferation and

TABLE 6.6

Summary of Surface Nano- and Microstructuring Effects on Proliferation and Differentiation

Feature Type	Material and Feature D_f	Cell Type and Cell D_s (μm)	2D Proliferation Measure	3D Proliferation Measure	Proliferation Ratio (PR)	2D Differentiation Measure	3D Differentiation Measure	Differentiation Ratio (DR)	Reference
Nanotube surface	TiO_2 (nm)	rMSC (μm)	$OD_{(420-480\ nm)}$	$OD_{420-480\ nm}$	(3D/2D)	Mineralization— $OD_{405\ nm}$	Mineralization— $OD_{405\ nm}$	(3D/2D)	Park et al. (2007)
	15	30	0.20	0.28	1.4	0.05	0.11	2.2	
	20	30	0.20	0.22	1.1	0.05	—	—	
	30	30	0.20	0.16	0.8	0.05	—	—	
	50	30	0.20	0.10	0.5	0.05	0.04	0.8	
	70	30	0.20	0.08	0.4	0.05	0.02	0.4	
	100	30	0.20	0.05	0.25	0.05	0.02	0.4	
Honeycomb surface	PCL and cap (μm) coated with poly-L-lysine	mNSCC (μm)	—	—	—	Mature neuron— Nestin⁻ and MAP2⁺	Mature neuron—% Nestin⁻ and MAP2	3D/2D	Tsuruma et al. (2008)
	3	20	—	—	—	80.0	13.3	0.17	
	5	20	—	—	—	80.0	10.0	0.13	
	8	20	—	—	—	80.0	50.0	0.63	
	10	20	—	—	—	80.0	60.0	0.75	
	15	20	—	—	—	80.0	70.0	0.88	
Nanotube surface	TiO_2 (nm)	hMSC (μm)	Cells/cm²	cells/cm²	3D/2D	Cell elongation (length/width)	Cell elongation (length/width)	3D/2D	Oh et al. (2009)
	30	30	4423	9807	2.22	1.92	2.88	1.50	
	50	30	4423	8077	1.83	1.92	4.04	2.10	
	70	30	4423	6346	1.43	1.92	5.96	3.10	
	100	30	4423	5769	1.30	1.92	9.88	5.15	

	PEUUR (µm) coated with fibronectin	rBMSC (µm)	Cells/cm²	cells/cm²	3D/2D	Relative Tenomodulin expression	Relative Tenomodulin expression	3D/2D	
Electrospun fibers	0.28	20	134,783	139,130	1.03	1.00	1.03	1.03	Bashur et al. (2009)
	0.82	20	134,783	120,000	0.89	1.00	0.31	0.31	
	2.30	20	134,783	104,348	0.77	1.00	0.26	0.26	
	Collagen (µm)	rbSMC (µm)	[³H]thymidine incorp. ($\times 10^{-5}$ dpm/mg protein)	[³H]thymidine incorp. ($\times 10^{-5}$ dpm/mg protein)	3D/2D				
Honeycomb surface						—	—	—	Suzuki et al. (2009)
	200–300	30	5.24	0.48	0.09		—	—	
	300–500	30	5.25	2.38	0.45		—	—	
		mSMC (µm)							
	≤200	30	13.10	1.14	0.09		—	—	
	200–300	30	13.10	8.19	0.63		—	—	
	PES coated with laminin (µm)	rNSC	Rel fluorescence intensity	Rel fluorescence intensity	3D/2D				
Electrospun fibers	0.283	20	786.9	603.3	0.77	—	—	—	Christopherson et al. (2009)
	0.749	20	786.9	511.5	0.65	—	—	—	
	1.452	20	786.9	472.1	0.60	—	—	—	

Abbreviations: D_f = characteristic diameter of the material feature such as fibers, pillars, tubes, spheres, wells, pores, etc.; D_c = diameter of the cell type in suspension; rMSC = rat mesenchymal stem cells; hMSC = human mesenchymal stem cells; PCL = poly(ε-caprolactone); Cap = amphiphilic polymer—a copolymer of dodecylacrylamide and ω-carboxyhexylacrylamide; mNSC = mice neural stem cells; rNSC = rat neural stem cells; PEUUR = poly(ester urethane) urea; rBMSC = rat bone marrow stromal cells; PES = poly(ethersulfone).

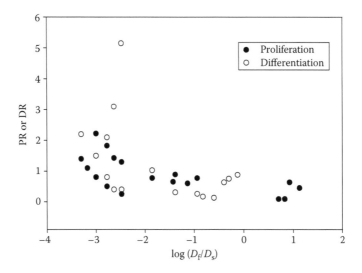

FIGURE 6.8   Relationship between structure architecture and cell fate.

differentiation cell fates. In a similar manner, changes in oxygen gradients may be a factor especially with large microtissues or cellular aggregates. The relationship between oxygen concentration and microtissue size is explored in detail in Box 6.1. The limit of *in vitro* 3D tissue size has probably been set by oxygen diffusion limitations. The scaffold pore or microtissue size recommendations in the literature of, for example, 100 μm (Jones, 2009), 95.9–150.5 μm (O'Brien et al., 2005), and 250 μm (Martin and Vermette, 2005) are consistent with sizes predicted in Box 6.1.

Several limitations in the utility of Figure 6.8 need to be mentioned. As described in the next chapter (Chapter 7), the material's Young's modulus is an important physical MEF. Altering the material characteristic dimension may result in altering the material's modulus. The true independent variable on the *X*-axis may be the material's modulus. Additionally, the interpretation of the results may be further complicated by the time in culture when the cell fate parameter was measured; time in culture is a key MEF as illustrated above. To generate the data presented in Figure 6.8, D_f values of 30 and 20 μm for fibroblasts and nerve cells, respectively, were assumed in the absence of measured values. Homogeneous cell populations display near normal size distribution (Halter et al., 2009). The use of the average cell size in the characteristic dimension ratio is reasonable. However, although this kind of information is not difficult to measure, it is seldom available. Given the few literature data used in Figure 6.8, the credibility of the approach proposed as a first step in establishing design principles is still in question. The author is in the process of expanding the plot with more literature results with measured cell population sizes. More results will provide better understanding. For example, other data may confirm or corroborate the spike shown by Oh et al. (2009).

Before further experimentation to generate more data is attempted, it is helpful to have a plausible mechanistic explanation of the results. The proliferation component

BOX 6.1   OXYGEN CONCENTRATION PROFILE IN CELL AGGREGATES OR 3D MICROTISSUES

When cells form aggregates within porous scaffolds, concentration gradients might exist for any soluble culture-medium component that is consumed or produced by cells – from basic nutrients to effecter molecules. The potential impact of these concentration profiles on cellular or tissue physiology can be profound. Because average local concentrations affect local cell behavior, cells in the middle may behave differently from cells at the surface. If the local concentration in the center is below the minimum required level, cells may not survive. Under the boundary condition, maximal cell aggregate size exist. Correspondingly, there is an upper boundary for the scaffold pore size design.

The magnitude of concentration gradients for cell aggregates in the absence of flow can be readily estimated by rules of thumb derived from a simple mathematical analysis of the competing effects of reaction and diffusion (Figure B6.1.1). If the surface concentration and the minimal concentration $(C = C_L)$ needed for the center of the tissue $(x = L)$ are known, substituting $C = C_L$ at $x = L$ into the concentration profile equation yields

$$DC_L = DC_0 - \frac{QL^2}{2} \tag{6.1}$$

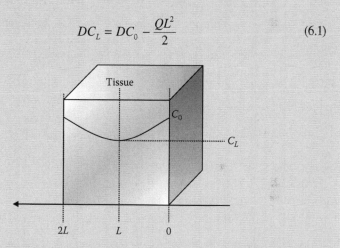

FIGURE B6.1.1   When a substance can diffuse into a slab of tissue and be consumed by cells; the concentration profile for this substance can be estimated by balancing the diffusive transport into the tissue with tissue consumption. In the situation shown, the surface concentration is designated C_0 and the diffusion distance is designated L. The local concentration (C), the distance from the tissue surface (x) and the volumetric consumption rate by the tissue (Q) are related by the differential balance: $D(d^2C/dx^2) = Q$, where the diffusion coefficient (D) is a constant that has been determined for the particular substance. Using a zero-order consumption rate, the solution to the equation for the conditions shown is $C/C_0 = 1 - \Phi^2[(x/L) - (x/L)^2/2]$. Φ^2 is a dimensionless parameter, known as the Thiele modulus: $\Phi^2 = $ (reaction rate)/(diffusion rate) $= L^2Q/(C_0D)$.

continued

Solving the above equation as a function of L yields

$$C_L = C_0 - \frac{QL^2}{2D}$$

(6.2)

or

$$L = \left(\frac{2(C_L - C_0)D}{Q} \right)^{1/2}.$$

(6.3)

Based on the above L-equation, the maximal size of cell aggregate can be calculated if the surface concentration and the lowest concentration ($C = C_L$) at the center of the tissue ($x = L$) are known. Correspondingly, this calculation specifies the maximum pore size of scaffolds that houses the cell aggregates. Plotting the function of C_L versus L, a range of the tissue sizes can be derived based on the reasonable concentrations of particular molecules, which would correspond to the reasonable range of pore sizes for scaffolds.

Taking oxygen diffusion as an example, the center oxygen concentration decreases when the size of the formed tissue increases. Based on the above equations and the data in Table B6.1.1 adapted from Griffith and Swartz (2006), the concentration profile of oxygen at the center of the pore as a function of the pore radius is depicted in Figure B6.1.2. These plots can provide information for the range of pore size (in terms of pore radius) for scaffold design. As a matter of fact, the concentration of oxygen should be higher than zero for cell survival. Thus, the pore size predicted above may be larger than

---

TABLE B6.1.1
Data for Oxygen Concentration Profiles

Cell Properties	Surface Concentration (mol/cm³)	Consumption Rate (mol/cell/s)	Thiele Modulus (Φ^2)	Tissue Thickness ($2L$, in cm²)[a]
Fibroblast (2×10^6 cells/cm³)	1.5×10^{-7}	7×10^{-17}	1.0	0.3
Cartilage (5×10^7 cells/cm³)	1.5×10^{-7}	2×10^{-17}	7.5	0.3
Liver (2×10^8 cells/cm³)	1.5×10^{-7}	4×10^{-16}	6	0.03

Source: Adapted from Griffith, L.G. and Swartz, M.A. 2006. *Nat. Rev. Mol. Cell Biol.* 7: 211–224.

[a] L = diffusion distance.

the actual values. However, the solution listed above is only based on oxygen diffusion; other molecules' diffusions need separate consideration.

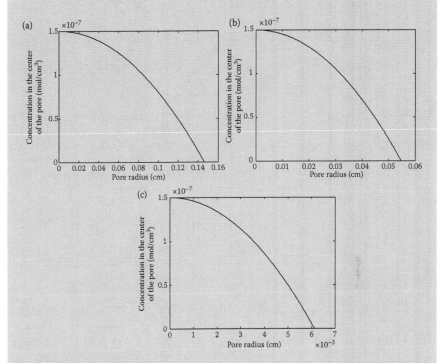

FIGURE B6.1.2 The concentration profile of oxygen at the center of the formed cellular aggregates: (a) for fibroblast cells, (b) for cartilage cells, and (c) for liver cells as a function of scaffold pore radius.

of the curve is consistent with anoikis, a term used to describe a type of apoptosis that results from failure of cells to adhere to the extracellular matrix (Frisch and Ruoslahti, 1997; Matter and Ruoslahti, 2001; Chiarugi, 2008a, 2008b; Chiarugi and Giannoni, 2008). As shown in Figure 6.9, using surfaces nanostructured with nanotubes, smaller structures are hypothesized to be able to support the formation of more focal adhesions in comparison to larger structures. The fewer the focal adhesions, the higher the likelihood of cell loss of anchorage. Loss of anchorage has been shown to result in anoikis in many cell types. However, as pointed out by Chiarugi and Giannoni (2008), cell adhesion dependence is probably made possible by many signal transduction pathways and the final outcome is the sum of all these inputs. For example, Bcl-2 protein expression and protection from apoptosis under serum-free conditions correlated with Bcl-2 transcription (Matter and Ruoslahti, 2001). The Bcl-2 family is made up of proapoptotic and antiapoptotic proteins. The family is classified into three different subfamilies depending on the homology and function of each protein. The first family with antiapoptotic activities is made up of Bcl-2,

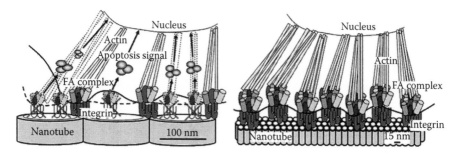

FIGURE 6.9 Hypothetical model showing how spacing at the nanoscale level can influence the number of focal adhesion sites. The smaller nanoscale affords more focal adhesions, resulting in the assembly of actin filaments proving signaling for proliferation (and possibly differentiation, depending on the media used). At the larger nanoscale, focal adhesions can only be sustained at structure interfaces, resulting in fewer sites and partial cell–matrix interaction, which is hypothesized to result in apoptosis (anoikis). (From Park, J. et al. 2007. *Nano Lett.* 7(6): 1689. With permission.)

Bcl-X_L, and Bcl-w (Simpson et al., 2008). More research is needed to define the molecular mediators involved in nanofibers or other nanostructures dependent cellular proliferation, differentiation, or other outcomes.

REFERENCES

Baac, H., Lee, J.H., Seo, J.M., Park, T.Y., Chung, H., Lee, S.D., and Kim, S.J. 2004. Submicron-scale topographical control of cell growth using holographic surface relief grating. *Mater. Sci. Eng. C* 24: 209–212.

Barnes, C.P., Sell, S.A., Boland, E.D., Simpson, D.G., and Bowlin, G.L. 2007. Nanofiber technology: Designing the next generation of tissue engineering scaffolds. *Adv. Drug Deliv. Rev.* 59: 1413–1433.

Bashur, C.A., Shaffer, R.D., Danhlgren, L.A., Guelcher, S.A., and Goldstein, A. 2009. Effect of fiber diameter and alignment of electrospun polyurethane meshes on mesenchymal progenitor cells. *Tissue Eng. Part A* 15(9): 2435–2445.

Becker, J.L., Prewett, T.L., Spaulding, G.F., and Goodwin, T.J. 1993. Three dimensional growth and differentiation of ovarian tumor cell line in high aspect rotating-wall vessel: Morphologic and embryologic considerations. *J. Cell Biochem.* 51: 283–289.

Brennan, J., Lu, C.C., Norris, D.P., Rodriguez, T.A., Beddington, R.S., and Robertson, E.J. 2001. Nodal signaling in epiblast patterns the early mouse embryo. *Nature* 411: 965–969.

Britland, S. and McCaig, C. 1996. Embryonic *Xenopus neurites* integrate and respond to simultaneous electrical and adhesive guidance cues. *Exp. Cell Res.* 226(1): 31–38.

Britland, S., Morgan, H., Wojiak-Stodart, B., Riehle, M., Curtis, A., and Wilkinson, C. 1996. Synergistic and hierarchical adhesive and topographic guidance of BHK cells. *Exp. Cell Res.* 228(2): 313–325.

Brizzolara, R.A. 2000. Patterning multiple antibodies on polystyrene. *Biosens. Bioelectron.* 15: 63–68.

Brunette, D.M. 1986a. Fibroblasts on micromachined substrata orient hierarchically to groves of different dimensions. *Exp. Cell Res.* 164: 11–26.

Brunette, D.M. 1986b. Spreading and orientation of epithelial cells on grooved substrata. *Exp. Cell Res.* 167(1): 203–217.

Brunette, D.M., Kenner, G.S., and Gould, T.R. 1983. Grooved titanium surfaces orient growth and migration of cells from human gingival explants. *J. Dent. Res.* 62(10): 1045–1048.

Campbell, C.E. and von Recum, A.F. 1989. Microtopography and soft tissue response. *J. Invest. Surg.* 2(1): 51–74.

Caplan, A.I. 1994. The mesengenic process. *Clin. Plast. Surg.* 21: 641–650.

Carpenedo, R.L., Bratt-Leal, A.M., Marklein, R.A., Seaman, S.A., Bowen, N.J., McDonald, J.F., and McDervitt, T.C. 2009. Homogeneous and organized differentiation within embryoid bodies induced by microsphere-mediated delivery of small molecules. *Biomaterials* 30: 2513–2515.

Charest, J.L., Bryant, L.E., Garcia, A.J., and King, W.P. 2004. Hot embossing for micropatterned cell substrates. *Biomaterials* 25(19): 4767–4775.

Chehroudi, B., Gould, T.R., and Brunette, D.M. 1988. Effects of a grooved epoxy substratum on epithelial cell behavior *in vitro* and *in vivo. J. Biomed. Mater. Res.* 22(6): 459–473.

Chehroudi, B., Gould, T.R., and Brunette, D.M. 1990. Titanium-coated micromachined grooves of different dimensions affect epithelial and connective-tissue cells differently *in vivo. J. Biomed. Mater. Res.* 24(9): 1203–1219.

Chen, J.L., Chu, B., and Hsiao, B.S. 2006. Mineralization of hydroxyapatite in electrospun nanofibrous poly(L-lactic acid)scaffolds. *J. Biomed. Mater. Res. Part A* 79A: 307–317.

Cheng, K. and Kisaalita, W.S. 2010. Exploring cellular adhesion and differentiation in micro-/ nano-hybrid polymer scaffold. *Biotechnol. Prog.* (accepted, DOI: 101002/btpr.391).

Cheng, K., Lai, Y., and Kisaalita, W.S. 2008. Three-dimensional polymer scaffolds for high throughput cell-based assay systems. *Biomaterials* 29: 2802–2812.

Chesmel, K.D. and Black, J. 1995. Cellular responses to chemical and morphologic aspects of biomaterial surfaces. I. A novel *in vitro* model system. *J. Biomed. Mater. Res.* 29(9): 1089–1099.

Cheung, H.-Y., Lau, K.-T., Lu, T.-P., and Hui, D. 2007. A critical review on polymer-based bioengineered materials for scaffold development. *Composites, Part B* 38: 291–300.

Chiang, C., Ying, L.T.Y, Lee, E., Young, K.E., Corden, J.L., Westphal, H., and Beachy, P.A. 1996. Cyclopia and defective axial patterning in mice lacking Sonic hedgehog gene function. *Nature* 383(6599): 407–413.

Chiarugi, P. 2008a. From anchorage dependent proliferation to survival: Lessons from redox signaling. *IUBMB Life* 60(5): 301–307.

Chiarugi, P. 2008b. Src redox regulation: There is more that meets the eye. *Mol. Cells* 26(4): 329–337.

Chiarugi, P. and Giannoni, E. 2008. Anoikis: A necessary death program for anchorage-dependent cells. *Biochem. Pharmacol.* 76(11): 1352–1364.

Chin, V., Taupin, P., Sanga, S., Scheel, J., Gage, F.H., and Bhatia, S.N. 2004. Microfabricated platform for studying stem cell fates. *Biotechnol. Bioeng.* 88(3): 399–415.

Chou, L., Firth, J.D., Uitto, V.J., and Brunette, D.M., 1995. Substratum surface topography alters cell shape and regulates fibronectin mRNA level, mRNA stability, secretion and assembly in human fibroblasts. *J. Cell Sci.* 108(Pt 4): 1563–1573.

Chou, L., Firth, J.D., Uitto, V.J., and Brunette, D.M., 1998. Effects of titanium substratum and grooved surface topography on metalloproteinase-2 expression in human fibroblasts. *J. Biomed. Mater. Res.* 39(3): 437–445.

Christopherson, G.T., Song, H., and Mao, H.-Q. 2009. The influence of fiber diameter of electrospun substrates on neural stem cell differentiation and proliferation. *Biomaterials* 30: 556–564.

Clark, P., Britland, S., and Connolly, P., 1993. Growth cone guidance and neuron morphology on micropatterned laminin surfaces. *J. Cell Sci.* 105(Pt 1): 203–212.

Clark, P., Connolly, P., Curtis, A.S., Dow, J.A., and Wilkinson, C.D. 1987. Topographical control of cell behavior. I. Simple step cues. *Development* 99(3): 439–448.

Clark, P., Connolly, P., Curtis, A.S., Dow, J.A., and Wilkinson, C.D. 1990. Topographical control of cell behavior: II. Multiple grooved substrata. *Development* 108(4): 635–644.

Clark, P., Connolly, P., Curtis, A.S., Dow, J.A., and Wilkinson, C.D. 1991. Cell guidance by ultrafine topography in vitro. *J. Cell Sci.* 99(Pt 1): 73–77.

Cooper, A., Munden, H.R., and Brown, G.L. 1976. The growth of mouse neuroblastoma cells in controlled orientations on thin films of silicon monoxide. *Exp. Cell Res.* 103(2): 435–439.

Curtis, A.S., Casey, B., Gallagher, J.O., Pasqui, D., Wood, M.A., and Wilkinson, C.D. 2001. Substratum nanotopography and the adhesion of biological cells. Are symmetry or regularity of nanotopography important? *Biophys. Chem.* 94(3): 275–283.

Curtis, A.S., Gadegaard, N., Dalby, M.J., Riehle, M.O., Wilkinson, C.D., and Aitchison, G. 2004. Cells react to nanoscale order and symmetry in their surroundings. *IEEE Trans. Nanobiosci.* 3(1): 61–65.

Dalby, M.J., Gadegaard, N., Riehle, M.O., Wilkinson, C.D., and Curtis, A.S. 2004. Investigating filopodia sensing using arrays of defined nano-pits down to 35 nm diameter in size. *Int. J. Biochem. Cell Biol.* 36(10): 2005–2015.

Deng, Y., Zhao, K., Zhang, X., Hu, P., and Chen, G.Q., 2002. Study on the three-dimensional proliferation of rabbit articular cartilage-derived chondrocytes on polyhydroxyalkanoate scaffolds. *Biomaterials* 23: 4049–4056.

den Braber, E.T., de Ruijter, J.E., Ginsel, L.A., von Recum, A.F., and Jansen, J.A. 1996a. Quantitative analysis of fibroblast morphology on microgrooved surfaces with various groove and ridge dimensions. *Biomaterials* 17(21): 2037–2044.

den Braber, E.T., de Ruijter, J.E., Ginsel, L.A., von Recum, A.F., and Jansen, J.A. 1998. Orientation of ECM protein deposition, fibroblast cytoskeleton, and attachment complex components on silicone microgrooved surfaces. *J. Biomed. Mater. Res.* 40(2): 291–300.

den Braber, E.T., de Ruijter, J.E., Smits, H.T., Ginsel, L.A., von Recum, A.F., and Jansen, J.A. 1996b. Quantitative analysis of cell proliferation and orientation on substrata with uniform parallel surface micro-grooves. *Biomaterials* 17(11): 1093–1099.

Detrait, E., Lhoest, J.-B., Knoops, B., Bertrand, P., and Van den Bosch de Aguilar, P., 1998. Orientation of cell adhesion and growth on patterned heterogeneous polystyrene surface. *J. Neurosci. Methods* 84: 193–204.

Doetschman, T.C., Eistetter, H., Katz, M., Schmidt, W., and Kemler, R. 1985. The *in vitro* development of blastocyte-derived embryonic stem cell line: Formation of visceral yolk sac, blood islands and myocardium. *J. Embryol. Exp. Morphol.* 87: 27–45.

Dowell, N., Turner, A.M.P., Hong, H., Rajan, S., Craighead, H.G., Turner, J.N., and Shain, W., 2000. Role of microfabricated structures on attachment and growth of neurons and astrocytes. *Mol. Biol. Cell* 11: 2831.

Drury, J.L. and Mooney, D.J. 2003. Hydrogels for tissue engineering: Scaffold design variables and applications. *Biomaterials* 24: 4337–4351.

Dunn, G.A. and Brown, A.F. 1986. Alignment of fibroblasts on grooved surfaces described by a simple geometric transformation. *J. Cell Sci.* 83: 313–340.

Eisenbarth, E., Meyle, J., Nachtigall, W., and Breme, J., 1996. Influence of the surface structure of titanium materials on the adhesion of fibroblasts. *Biomaterials* 17(14): 1399–1403.

Ekaputra, A.K., Prestwich, G.D., Cool, S.M., and Hutmacher, D.W., 2008. Combining electrospun scaffolds with electrosprayed hydrogel leads to three-dimensional cellularization of hybrid constructs. *Biomacromolecules* 9: 2097–2103.

Elsdale, T. and Bard, J. 1972. Collagen substrate for the study of cell behavior. *J. Cell Biol.* 41: 298–311.

Fewster, S.D., Coombs, R., Kitson, J., and Zhou, S., 1994. Precise ultrafine surface texturing of implant materials to improve cellular adhesion and biocompatibility. *Nanobiology* 3: 201–214.

Flemming, R.G., Murphy, C.J., Abrams, G.A., Goodman, S.L., and Nealey, P.F., 1999. Effects of synthetic micro- and nano-structured surfaces. *Biomaterials* 20: 573–588.

Freed, L.E., Langer, R., and Martin, I., 1997. Tissue engineering of cartilage in space. *Med. Sci.* 94: 13885–13890.

Freyman, T.M., Yannas, I.V., and Gibson, L.J. 2001. Cellular materials as porous scaffolds for tissue engineering. *Prog. Mater. Sci.* 46: 273–282.

Friedl, P. and Brocker, E.B. 2000. The biology of cell locomotion within three-dimensional extracellular matrix. *Cell. Mol. Life Sci.* 57: 41–64.

Friess, W. 1998. Collagen—biomaterial for drug delivery. *Eur. J. Pharm. Biopharm.* 45: 113–136.

Fujimoto, K., Takahashi, T., Miyaki, M., and Kawaguchi, H., 1997. Cell activation by the micropatterned surface with settling particles. *J. Biomater. Sci. Polym. Ed.* 8(11): 879–891.

Gallagher, J.O., McGhee, K.F., Wilkinson, C.D.W., and Riehle, M.O. 2002. Interaction of animal cells with ordered nanotopography. *IEE Trans. Nanobiosci.* 1: 24–28.

Gelain, F., Horii, A., and Zhang, S. 2007. Designer self-assembling peptide scaffolds for 3-D tissue culture and regenerative medicine. *Macromol. Biosci.* 7: 544–551.

Goodman, S.L., Sims, P.A., and Albrecht, R.M. 1996. Three-dimensional extracellular matrix textured biomaterials. *Biomaterials* 17(21): 2087–2095.

Goodwin, T.J., Prewett, T.L., Wolf, D.A., and Spaulding, G.F. 1993. Reduced shear stress: A major component in the ability of mammalian tissue to form three-dimensional assemblies in simulated microgravity. *J. Cell. Biochem.* 51: 301–311.

Green, A.M., Jansen, J.A., van der Waerden, J.P., and von Recum, A.F. 1994. Fibroblast response to microtextured silicone surfaces: Texture orientation into or out of the surface. *J. Biomed. Mater. Res.* 28(5): 647–653.

Griffith, L.G. and Swartz, M.A. 2006. Capturing complex 3D tissue physiology in vitro. *Nat. Rev. Mol. Cell Biol.* 7: 211–224.

Grinnell, F. 2000. Fibroblast-collagen matrix contraction: Growth-factor signaling and mechanical loading. *Trends Cell. Biol.* 10: 363–365.

Guan, Y. and Kisaalita, W.S. 2010a. Cell adhesion and locomotion on microwell-structured glass substrate. *J. Biomed. Mater. Res. Part A* (waiting for a decision).

Guan, Y. and Kisaalita, W.S. 2010b. The effect of substrate microstructuring on human fibroblast viability, proliferation and cytoskeleton organization. *Biomaterials* (waiting for a decision).

Guenard, V., Valentini, R.F., and Aebischer, P. 1991. Influence of surface texture of polymeric sheets on peripheral nerve regeneration in a two-compartment guidance system. *Biomaterials* 12(2): 259–263.

Habermann, B. 2004. The BAR-domain family of proteins: A case of bending and binding? *EMBO Rep.* 5: 250–255.

Halter, M., Elliot, J.T., Hubbard, J.B., Tona, A., and Plant, A.L. 2009. Cell volume distributions reveal cell growth rates and division times. *J. Theor. Biol.* 257: 124–130.

Hartman, O., Zhang, C., Adams, E.L., Farach-Carson, M.C., Petrelli, N.J., Chase, B.C., and Rabolt, J.F. 2009. Microfabricated electrospun collagen membranes for 3-D cancer models and drug screening applications. *Biomacromolecules* 10: 2019–2032.

Heckele, M., Bacher, W., and Muller, K.D. 1998. Hot embossing—the molding technique for plastic microstructures. *Microsystem Technologies* 4(3): 122–124.

Hoch, H.C., Jelinski, L.W., and Craighead, G.C. (eds) 1996. *Nanofabrication and Biosystems: Integrating Materials Science, Engineering, and Biology.* New York: Cambridge University Press.

Hoch, H.C., Staples, R.C., Whitehead, B., Comeau, J., and Wolf, E.D. 1987. Signaling for growth orientation and cell differentiation by surface topography in uromyces. *Science* 235(4796): 1659–1662.

Hockberger, P.E., Lom, B., Soekarno, C.H., and Healy, K. 1996. Cellular engineering: Control of cell–substrate interactions. In *Nanofabrication and Biosystems: Integrating Materials Science, Engineering, and Biology*, H.C. Hoch, L.W. Jelinski, and G.C. Craighead (eds). New York: Cambridge University Press.

Hopfl, G., Gassman, M., and Desbaillets, L. 2004. Differentiating embryonic stem cells into embryoid bodies. *Methods Mol. Biol.* 254: 79–98.

Hu, J., Liu, X.H., and Ma, P.X. 2008. Induction of osteoblasts differentiation phenotype on poly(L-lactic) nanofibrous matrix. *Biomaterials* 29(28): 3815–3821.

Hunt, J.A., Williams, R.L., Tavakoli, S.M., and Riches, S.T. 1995. Laser surface modification of polymers to improve biocompatibility. *J. Mater. Sci. Mater. Med.* 6: 813–817.

Hutmacher, D.W., Ng, K.W., Kaps, C., Sittinger, M., and Klaring, S. 2003. Elastic cartilage engineering using novel scaffold architectures in combination with a biomimetic cell carrier. *Biomaterials* 24(24): 4445–4458.

Jayaraman, K., Kotaki, M., Zhang, Y., Mo, X., and Ramakrishna, S. 2004. Recent advances in polymer nanofibers. *J. Nanosci. Nanotechnol.* 4(1–2): 52–65.

Jessup, J.M., Goodwin, T.J., and Spaulding, G.F. 1993. Prospects for use of microgravity-based bioreactors to study three-dimensional host–tumor interactions in human neoplasia. *J. Cell. Biochem.* 51: 290–300.

Jones, A.C., Arns, C.H., Hutmacher, D.W., Milthorpe, B.K., Sheppard, A.P., and Knachstedt, M.A. 2009. The correlation of pore morphology, interonnectivity and physical properties of 3D ceramic scaffolds with bone ingrowth. *Biomaterials* 30: 1440–1451.

Kadler, K.E., Holmes, D.F., Trotter, J.A., and Chapman, J.A. 1996. Collagen fibril formation. *Biochem. J.* 316: 1–11.

Kam, L., Shain, W., Turner, J.N., and Bizios, R., 1999. Correlation of astroglial cell function on micro-patterned surfaces with specific geometric parameters. *Biomaterials* 20: 2343–2350.

Kelm, J.M., Timmins, N.E., Brown, C.J., Fussenegger, M., and Nielsen, L.K., 2003. Methods for generation of homogeneous multicellular tumor spheroids applicable to a wide variety of cell types. *Biotechnol. Bioeng.* 83(2): 173–180.

Kononen, M., Hormia, M., Kivilahti, J., Hautaniemi, J., and Thesleff, I., 1992. Effect of surface processing on the attachment, orientation, and proliferation of human gingival fibroblasts on titanium. *J. Biomed. Mater. Res.* 26(10): 1325–1341.

Kotov, N.A., Liu, Y., Wang, S., Cummings, C., Eghtedari, M., Vargas, G., Motamedi, M., Nichols, J., and Cortiella, J., 2003. Inverted colloidal crystal as three-dimensional cell scaffolds. *Langmuir* 20(19): 7887–7892.

Krewson, C.E., Chung, S.W., Dai, W., and Saltzman, W.M., 1994. Cell aggregation and neurite growth in gels of extracellular matrix molecules. *Biotechnol. Bioeng.* 43: 555–562.

Lampin, M., Warocquier, C., Legris, C., Degrange, M., and Sigot-Luizard, M.F. 1997. Correlation between substratum roughness and wettability, cell adhesion, and cell migration. *J. Biomed. Mater. Res.* 36(1): 99–108.

Lewis, M.L., Moriarity, D.M., and Campbell, P.S., 1993. Use of microgravity bioreactors for development of an *in vitro* rat salivary gland cell culture model. *J. Cell. Biochem.* 51: 265–273.

Liu, X. and Ma, P.X. 2009. Phase separation, pore structure, and properties of nanofibrous gelatin scaffolds. *Biomaterials* 30: 4094–4103.

Loesberg, W.A., te Riet, J., van Delft, F.C., Schon, P., Figdor, C.G., Speller, S., van Loon, J.J., Walboomers, X.F., and Jansen, J.A. 2007. The threshold at which substrate nanogroove dimensions may influence fibroblast alignment and adhesion. *Biomaterials* 28(27): 3944–3951.

Lozinsky, V.I. and Plieva, F.M. 1998. Poly(vinyl alcohol) cryogels employed as matrices for cell immobilization. 3. Overview of recent research and developments. *Enzyme Microb. Technol.* 23: 227–242.

Madou, M.J. 2002. *Fundamentals of Microfabrication—The Science of Miniaturization*, 2nd edition, Boca Raton, FL: CRC Press.

Malek, C.G.K. 2006. Laser processing for bio-microfluidics applications (part I). *Anal. Bioanal. Chem.* 385(8): 1351–1361.

Martin, J.Y., Schwartz, Z., Hummert, T.W., Schraub, D.M., Simpson, J., Lankford, J., Jr., Dean, D.D., Cochran, D.L., and Boyan, B.D. 1995. Effect of titanium surface roughness on proliferation, differentiation, and protein synthesis of human osteoblast-like cells (MG63). *J. Biomed. Mater. Res.* 29(3): 389–401.

Martin, Y. and Vermette, P. 2005 Bioreactor s for tissue mass culture: Design, characterization and recent advances. *Biomaterials* 26: 7481–7503.

Martínez, E., Engel, E., Planell, J.A., and Samitier, J. 2009. Effects of artificial micro- and nano-structured surfaces on cell behavior. *Ann. Anat.* 191(1): 126–135.

Martins, A., Chung, S., Pedro, A.J., Sousa, R.A., Marques, A.P., Reis, R.L., and Neves, N.M. 2009. Hierarchical starch-based fibrous scaffold for bone tissue engineering. *J. Tissue Eng. Regener. Med.* 3: 37–42.

Mathews, J.A., Wnek, G.E., Simpson, D.G., and Bowling, G.L. 2002. Electrospinning of collagen nanofibers. *Biomacromolecules* 3: 232–238.

Matsuzaka, K., Walboomers, X.F., Yoshinari, M., Inoue, T., and Jansen, J.A. 2003. The attachment and growth behavior of osteoblast-like cells on microtextured surfaces. *Biomaterials* 24(16): 2711–2719.

Matter, M.L. and Ruoslahti, E. 2001. A signaling pathway from the $\alpha_5\beta_1$ and $\alpha_v\beta_3$ integrins that elevates *bcl-2* transcription. *J. Biol. Chem.* 276(30): 27757–27763.

McFarland, C.D., Thomas, C.H., DeFilippis, C., Steele, J.G., and Healy, K.E. 2000. Protein adsorption and cell attachment to patterned surfaces. *J. Biomed. Mater. Res.* 49: 200–210.

McKenzie, J.L., Waid, M.C., Shi, R., and Webster, T.J. 2004. Decreased functions of astrocytes on carbon nanofibers materials. *Biomaterials* 25: 1309–1317.

Meyle, J., Gultig, K., Brich, M., Hammerle, H., and Nisch, W. 1994. Contact guidance of fibroblasts on biomaterial surfaces. *J. Mater. Sci. Mater. Med.* 5: 463–466.

Meyle, J., Gultig, K., and Nisch, W. 1995. Variation in contact guidance by human cells on a microstructured surface. *J. Biomed. Mater. Res.* 29(1): 81–88.

Meyle, J., Gultig, K., Wolburg, H., and von Recum, A.F. 1993a. Fibroblast anchorage to microtextured surfaces. *J. Biomed. Mater. Res.* 27(12): 1553–1557.

Meyle, J., Wolburg, H., and von Recum, A.F. 1993b. Surface micromorphology and cellular interactions. *J. Biomater. Appl.* 7(4): 362–374.

Miller, C., Shanks, H., Witt, A., Rutkowski, G., and Mallapragada, S. 2001. Oriented Schwann cell growth on micropatterned biodegradable polymer substrates. *Biomaterials* 22(11): 1263–1269.

Mrksich, M. 1998. Tailored substrates for studies of attached cell culture. *Cell. Mol. Life Sci.* 54: 653–662.

Nisbet, D.R., Forsythe, J.S., Shen, W., Finkelstein, D.I., and Horne, M.K. 2009. A review of cellular response on electrospun nanofibers for tissue engineering. *J. Biomater. Appl.* 24: 7–29.

Oakley, C. and Brunette, D.M. 1993. The sequence of alignment of microtubules, focal contacts and actin filaments in fibroblasts spreading on smooth and grooved titanium substrata. *J. Cell Sci.* 106(Pt 1): 343–354.

Oakley, C. and Brunette, D.M. 1995. Response of single, pairs, and clusters of epithelial cells to substratum topography. *Biochem. Cell. Biol.* 73(7–8): 473–489.

O'Brien, F.J., Harley, B.A., Yannas, I.V., and Gidson, L.J. 2005. The effect of pore size on cell adhesion in collagen-GAG scaffolds. *Biomaterials* 26: 433–441.

Oh, S., Brammer, K.S., Li, Y.S.J., Teng, D., Engler, A., Chien, S., and Jin, S. 2009. Stem cell fate dictated solely by altered nanotube dimension. *Proc. Natl. Acad. Sci. USA* 106(7): 2130–2135.

Ohara, P.T. and Buck, R.C. 1979. Contact guidance *in vitro.* A light, transmission, and scanning electron microscopic study. *Exp. Cell Res.* 121(2): 235–249.

Park, J., Bauer, S., von der Mark, K., and Schmuki, P. 2007. Nanosize and vitality: TiO_2 nanotube diameter directs cell fate. *Nanoletters* 7(6): 1686–1691.

Patel, A.J., Lazdunski, M., and Honore, E. 2001. Lipid and mechano-gated 2P domain K^+ channels. *Curr. Opin. Cell Biol.* 13: 422–428.

Petronis, S., Gretzer, C., Kasemo, B., and Gold, J. 2003. Model porous surfaces for systematic studies of material–cell interactions. *J. Biomed. Mater. Res.* 66A: 707–721.

Pfleging, W., Bruns, M., Welle, A., and Wilson, S. 2007. Laser-assisted modification of polystyrene surfaces for cell culture applications. *Appl. Surf. Sci.* 253: 9177–9184.

Piotter, V., Holstein, N., Plewa, K., Ruprecht, R., and Hausselt, J. 2004. Replication of micro components by different variants of injection molding. *Microsystem Technologies– Micro- and Nanosystems–Information Storage and Processing Systems* 10(6–7): 547–551.

Powers, M.J., Domansky, K., Kaazempur-Mofrad, M.R., Kalezi, A., Capitano, A., Upadhyaya, A., Kurzawski, P., Wack, K.E., Beer Stolz, D., Kamm, R., and Griffith, L.G. 2002a. A microfabricated array bioreactor for perfused 3D liver culture. *Biotechnol. Bioeng.* 78(3): 257–269.

Powers, M.J., Janigian, D.M., Wack, K.E., Baker, C.S., Beer Stolz, D., and Griffith, L.G. 2002b. Functional behavior of primary rat liver cells in a three-dimensional perfused microarray bioreactor. *Tissue Eng.* 8(3): 499–513.

Rajnicek, A., Britland, S., and McCaig, C. 1997. Contact guidance of CNS neurites on grooved quartz: Influence of groove dimensions, neuronal age and cell type. *J. Cell Sci.* 110 (Pt 23): 2905–2913.

Rajnicek, A. and McCaig, C. 1997. Guidance of CNS growth cones by substratum grooves and ridges: Effects of inhibitors of the cytoskeleton, calcium channels and signal transduction pathways. *J. Cell Sci.* 110(Pt 23): 2915–2924.

Recknor, J.B., Recknor, J.C., Sakaguchi, D.S., and Mallapragada, S.K. 2004. Oriented astroglial cell growth on micropatterned polystyrene substrates. *Biomaterials* 25(14): 2753–2767.

Rich, A. and Harris, A.K. 1981. Anomalous preferences of cultured macrophages for hydrophobic and roughened substrata. *J. Cell Sci.* 50: 1–7.

Rosdy, M., Grisoni, B., and Clauss, L.C. 1991. Proliferation of normal human keratinocytes on silicone substrates. *Biomaterials* 12(5): 511–517.

Rovensky Yu, A. and Samoilov, V.I. 1994. Morphogenetic response of cultured normal and transformed fibroblasts, and epitheliocytes, to a cylindrical substratum surface. Possible role for the actin filament bundle pattern. *J. Cell Sci.* 107(Pt 5): 1255–1263.

Sanchez-Bustamante, C.D., Kelm, J.M., Mitta, B., and Fussenegger, M. 2005. Heterologous protein production capacity of mammalian cells cultivated as monolayers and microtissues. *Biotechnol. Bioeng.* 93(1): 169–180.

Schmidt, J.A. and von Recum, A.F. 1991. Texturing of polymer surfaces at the cellular level. *Biomaterials* 12(4): 385–389.

Schmidt, J.A. and von Recum, A.F. 1992. Macrophage response to microtextured silicone. *Biomaterials* 13(15): 1059–1069.

Schuldiner, M., Yanuka, O., Itskovitz-Eldor, J., Melton, D.A., and Benvenisty, N. 2000. Effects of eight growth factors on the differentiation of cells derived from human embryonic stem cells. *Proc. Natl. Acad. Sci. USA* 97: 11307–11312.

Senoo, H., Imai, K., Sato, M., Kojima, N., Miura, M., and Hata, R. 1996. Three-dimensional structure of extracellular matrix reversibly regulates morphology, proliferation and collagen metabolism of perisinusoidal stellate cells (vitamin A-storing cells). *Cell Biol. Int.* 20(7): 501–512.

Seunarine, K., Meredith, D.O., Riehle, M.O., Wilkinson, C.D.W., and Gadegaard, N. 2008. Biodegradable polymers tubes with lithographically controlled 3D micro- and nano-topography. *Microelectron. Eng.* 85: 1350–1354.

Sherwood, J.K., Riley, S.L., Palazzolo, R., Brown, S.C., Monkhouse, D.C., Coates, M., Griffith, L.G., Landeen, L.K., and Ratcliffe, A. 2002. A three-dimensional osteochondral composite scaffold for articular cartilage repair. *Biomaterials* 23: 4739–4751.

Simpson, C.D., Anyiwe, K., and Schmmer, A.D. 2008. Anoikis resistance and tumor metastasis. *Cancer Lett.* 271: 177–185.

Srouji, S., Kizhner, T., Suss-Tobi, E., Livne, E., and Zussman, E. 2008. 3-D nanofibrous electrospun multilayered construct in an alternative ECM mimicking scaffold. *J. Mater. Sci. Mater. Med.* 19: 1249–1255.

Suzuki, T., Ishii, I., Kotani, A., Masuda, M., Hiata, K., Ueda, M., Ogata, T., Sakai, T., Ariyoshi, N., and Kitada, M. 2009. Growth inhibition and differentiation of cultured smooth muscle cells depend on cellular crossbridge across the tubular lumen of type I collagen matrix honeycombs. *Microvasc. Res.* 77: 143–149.

Teixeira, A.I., McKie, G.A., Foley, J.D., Bertics, P.J., Nealey, P.F., and Murphy, C.J. 2006. The effect of environmental factors on the response of human corneal epithelial cells to nanoscale substrate topography. *Biomaterials* 27(21): 3945–3954.

Thapa, A., Miller D.C., Webster, T.J., and Haberstroh, K.M. 2003b. Nano-structured polymers enhance bladder smooth muscle cell function. *Biomaterials* 24: 2915–2926.

Thapa, A., Webster, T.J., and Haberstroh, K.M. 2003a. Polymers with nano-dimensional surface features enhance bladder smooth muscle cell adhesion. *J. Biomed. Mater. Res. A* 67(4): 1374–1383.

Tsuruma, A., Tanaka, M., Yamamoto, S., and Shimomura, M. 2008. Control of neural stem cell differentiation on honeycomb films. *Colloids Surf. A: Physiochem. Eng. Aspects* 313/314: 536–540.

Turner, A.M.P., Dowell, N., Turner, S.W.P., Kam, L., Isaacson, M., Turner, J.N., Craighead, H.G., and Shain, W., 2000. Attachment of astroglial cells to microfabricated pillar arrays of different geometries. *J. Biomed. Mater. Res.* 51: 430–441.

Turner, S., Kam, L., Isaacson, M., Craighead, H.G., Shain, W., and Turner, J. 1997. Cell attachment on silicon nanostructures. *J. Vac. Sci. Technol. B* 15: 1325–1341.

van Kooten, T.G., Whitesides, J.F., and von Recum, A. 1998. Influence of silicone (PDMS) surface texture on human skin fibroblast proliferation as determined by cell cycle analysis. *J. Biomed. Mater. Res.* 43(1): 1–14.

Vogel, V. and Sheetz, M. 2006. Local force geometry sensing regulates cell functions. *Nat. Rev. Mol. Cell Biol.* 7: 265–275.

Voldman, J., Gray, M.L., and Schmidt, M.A. 1999. Microfabrication in biology and medicine. *Annu. Rev. Biomed. Eng.* 1: 401–425.

Wang, L. and Kisaalita, W.S. 2010. Characterization of micropatterned and nano-fibrous scaffolds for neural network activity readout for high throughput screening. *J. Biomed. Mater. Res. Part B* (accepted).

Wang, L., Mumaw, J., Stice, S., and Kisaalita, W.S. 2010. 3D neural-network-based HTS platform for neural degeneration disease targets. *Biomaterials* (submitted).

Wang, L., Wu, Z.-Z., Xu, B., Zhao, Y., and Kisaalita, W.S. 2009. SU-8 microstructure for quasi-three-dimensional cell-based biosensing. *Sens. Actuat. B Chem.* 140: 349–355.

Webb, A., Clark, P., Skepper, J., Compston, A., and Wood, A. 1995. Guidance of oligodendro-cytes and their progenitors by substratum topography. *J. Cell Sci.* 108(Pt 8): 2747–2760.

Weiss, P. 1958. Cell contact. *Int. Rev. Cytol.* 7: 391–424.

Wilkinson, P.C., Shields, J.M., and Haston, W.S. 1982. Contact guidance of human neutrophil leukocytes. *Exp. Cell Res.* 140(1): 55–62.

Wojciak-Stothard, B., Curtis, A., Monaghan, W., MacDonald, K., and Wilkinson, C. 1996. Guidance and activation of murine macrophages by nanometric scale topography. *Exp. Cell Res.* 223(2): 426–435.

Wojciak-Stothard, B., Curtis, A.S., Monaghan, W., McGrath, M., Sommer, I., and Wilkinson, C.D. 1995a. Role of the cytoskeleton in the reaction of fibroblasts to multiple grooved substrata. *Cell. Motil. Cytoskeleton* 31(2): 147–158.

Wojciak-Stothard, B., Denyer, M., Mishra, M., and Brown, R.A. 1997. Adhesion, orientation, and movement of cells cultured on ultrathin fibronectin fibers. *In Vitro Cell. Dev. Biol. Anim.* 33(2): 110–117.

Wojciak-Stothard, B., Madeja, Z., Korohoda, W., Curtis, A., and Wilkinson, C. 1995b. Activation of macrophage-like cells by multiple grooved substrata. Topographical control of cell behavior. *Cell. Biol. Int.* 19(6): 485–490.

Woo, K.M., Chen, V.J., and Ma, P.X. 2003. Nano-fibrous scaffolding architecture selectively enhances protein adsorption contributing to cell attachment. *J. Biomed. Mater. Res. Part A* 67A(2): 531–537.

Wood, A. 1988. Contact guidance on microfabricated substrata: The response of teleost fin mesenchymal cells to repeating topographical patterns. *J. Cell Sci.* 90(Pt 4): 667–681.

Woerly, S., Plant, G.W., and Harvey, A.R. 1996. Cultured rat neuronal and glial cells entrapped within hydrogel polymer matrices: A potential tool for neural tissue replacement. *Neurosci. Lett.* 205: 197–201.

Wu, Z.-Z., Zhao, Y., and Kisaalita, W.S. 2006. Interfacing SH-SY5Y human neuroblastoma cells with SU-8 microstructures. *Colloids Surf. B: Biointerfaces* 52: 14–21.

Yang, Y., Hoe, U., Wang, X., Horii, A., Yokoi, H., and Zhang, S. 2009. Designer self-assembling peptide nanomaterials. *Nano Today* 4: 193–210.

Yang, F., Murugan, R., Ramakrishna, S., Wang, X., Ma, Y.-X., and Wang, S. 2004. Fabrication of a nanostructured porous PLLA scaffolds intended for nerve tissue engineering. *Biomaterials* 25: 1891–1900.

Yim, E.K.F. and Leong, K.W. 2005. Significance of synthetic nanostructures in dictating cellular response. *Nanomed. Nanotechnol. Biol. Med.* 1: 10–21.

Zhang, S., Gelain, F., and Zhao, X. 2005. Designer self-assembling peptide nanofibers scaffolds for 3D tissue cell cultures. *Semin. Cancer Biol.* 15: 413–420.

Zhao, Y.P., Ye, D.-X., Wang, G.-C., and Lu, T.-M. 2003. Designing nanostructures by glancing angle deposition. *SPIE Proc.* 5219: 59.

Zimmerberg, J. and McLaughlin, S. 2004. Membrane curvature: How BAR domains bend bilayers. *Curr. Biol.* 14: R250–R252.

7 Material Physical Property and Force Microenvironmental Factors

7.1 BASICS

7.1.1 YOUNG'S MODULUS, STIFFNESS, AND RIGIDITY

Imagine a material in equilibrium and under a simple load (subjected to a force, F) as shown in Figure 7.1a. If an imaginary section is introduced at XX, normal to the plane of F, the total force carried must be equal to F and the distribution of F as internal forces of cohesion is called stress (σ). If the force is uniformly distributed over the section, then

$$\sigma = \frac{F}{A},\qquad(7.1)$$

where A is the cross-sectional area of the material. In Figure 7.1a, the forces are tensile, but can be compressive as well. Strain (ε) is a measure of deformation produced in the material under F, defined as elongation (x) divided by original materials length (l):

$$\varepsilon = \frac{x}{l}.\qquad(7.2)$$

Strain is proportional to the stress that causes it (Hooke's law), which is obeyed within certain limits of stress by some materials (e.g., glass, timber, concrete, etc.) and not obeyed by others (e.g., soft materials like hydrogels). Within the limits for which Hooke's law is obeyed, the ratio of the stress to the strain produced is called the modulus of elasticity or Young's modulus (E) (Figure 7.1b):

$$E = \frac{\sigma}{\varepsilon},\qquad(7.3)$$

E is usually the same in both tension and compression and is a constant for a given material. An average value of E can be established for a given range of stress for materials that do not obey Hooke's law. Where a number of forces are acting together on a material, the resultant strain will be the sum of the individual strains caused by each force acting separately (principle of superposition). The commonly used word in the literature to describe the physical characteristic of materials is "stiffness."

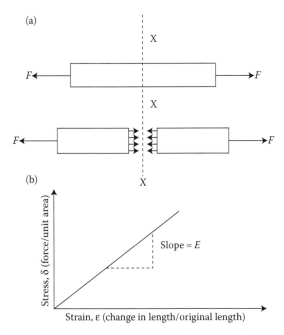

FIGURE 7.1   Schematics of a material under a tensile force, F (a) and the stress–strain curve (b).

Stiffness is the property of a solid body to resist deformation, used interchangeably with "rigidity." A material with a high E exhibits high stiffness or rigidity and vice versa. Other less technical words used to describe stiffness are "pliability" and "compliance."

7.1.2   SHEAR MODULUS OR MODULUS OF RIGIDITY

Imagine a material in equilibrium and under a shear load (F) as shown in Figure 7.2. As in compressive or tensile loading, shear stress (τ) is given by

$$\tau = \frac{F}{A} \tag{7.4}$$

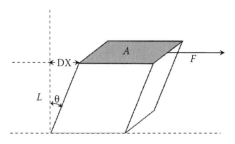

FIGURE 7.2   Schematic of material under shear force (F).

and shear strain (γ) is given by

$$\gamma = \frac{\Delta x}{l} = \tan\theta, \tag{7.5}$$

where Δx is the transverse displacement and l is the original length. The shear modulus or sometimes called modulus of rigidity (G) characterized the deformation of a solid under a force parallel to one of its surface and is given by

$$G = \frac{\tau}{\gamma}. \tag{7.6}$$

Young's modulus and shear modulus are related by the equation

$$E = 2(1 + \nu)G, \tag{7.7}$$

where ν is Poisson's ratio, which is a measure of the Poisson effect. When a material is stretched or compressed in one direction, it contracts or expands in the direction perpendicular to the direction of the tensile or compressive force, a phenomenon called the Poisson effect. Numerically, $\nu = -(\varepsilon_{transverse}/\varepsilon_{axial})$, and most materials have Poisson's ratio ranging between 0 and 0.5. A Poisson's ratio of 0.490–0.499, for incompressible materials, has been assumed for hydrogels and soft biological tissue by numerous investigators (e.g., Palmeri et al., 2008; Ghosh et al., 2007; Manduca et al., 2001).

7.1.3 MATERIAL PHYSICAL PROPERTIES CHARACTERIZATION

The most common methods used for measuring physical properties of rigid and soft materials are direct tensile and compressive tests, dynamic mechanical analysis, atomic force microscopy (AFM), microrheology, and magnetic/ultrasonic resonance elastography (MRE/URE) or resonance. In uniaxial tensile testing, samples are clamped and stretched at constant expansion rates and from these results the Young's modulus is calculated according to Equation 7.3. In uniaxial compression, cylindrical samples between two smooth impermeable plates are compressed and the results are used to calculate Young's modulus as described above. Dynamic measurement analysis requires the use of a sinusoidal shear load on the sample. The applied stress (σ^*) and resultant shear (γ^*) are measured with shear stress and strain sensors, respectively. According to Anseth et al. (1996), the complex shear modulus (G^*) is defined as

$$G^* = G' + iG'' = \frac{\sigma^*}{\gamma^*}, \tag{7.8}$$

G' (elastic or storage modulus) is defined as the real part of G^* and characterizes the relative degree of a material to recover. The imaginary component of G^* (G'', viscous or loss modulus) characterizes the relative degree of material to flow or the "viscous response." For hydrogels, most tensile tests are conducted at a constant extension rate with varying loads until the sample fails. The stress (τ) is plotted against $\lambda - \lambda^{-2}$, where λ is the extension ratio (defined as the ratio between the final length and the initial lengths of the material). The slope of the plot is equal to G (shear modulus).

In AFM measurements, the sample is compressed by the indenting tip as shown in Figure 7.3. The loading force can be calculated from the deflection and the spring constant of the cantilever. The Young's modulus of the material can be calculated from force-indentation curves fitted to the Hertz model, extended by Sneddon (1965), which describes the elastic deformation of two bodies in contact under load, through the relationship between applied force and indentation depth (Lekka et al., 1999). Assuming that the shape of the AFM tip is approximated by a paraboloid, the force (F) as a function of indentation is

$$F = \frac{4}{3}\sqrt{RE'}\Delta z^{3/2}, \tag{7.9}$$

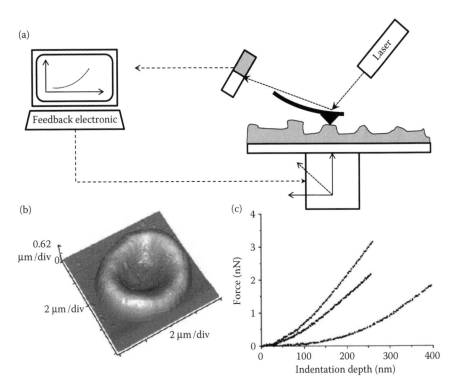

FIGURE 7.3 Schematic diagram of the AFM setup and sample measurement. (a) The sharp tip of the cantilever is used to probe the sample surface, typically mounted on a piezoelectric scanner with x–y motion capability as well as the z-direction for indentation. The back of the cantilever reflects the laser light onto the position-sensitive photodiode that measures the cantilever deflection. Based on the spring constant of the cantilever, the loading force can be calculated. (b) The topographical image of a red blood cell, and (c) the relationships between the force and indentation depth from the same cell at different locations. (From Dulińska, I., et al. 2006. *J. Biochem. Biophys. Methods* 66: 5. With permission.)

where Δz is the indentation depth, R is the tip radius of curvature, and E' is the reduced Young' modulus of the tip-sample system defined as follows:

$$\frac{1}{E'} = \frac{1 - v_{tip}^2}{E_{tip}} + \frac{1 - v_{sample}^2}{E_{sample}}, \tag{7.10}$$

where E_{tip} and E_{sample} are Young's moduli of the tip and sample, respectively; v_{tip} and v_{sample} are Poisson's ratios of the tip and sample, respectively. In the case of hydrogels or biological samples, the Poisson's ratio (v_{sample}) ranges between 0.490 and 0.499 (Manduca et al., 2001) and E_{sample} is much smaller than E_{tip}, simplifying the equation to

$$E' = \frac{E_{sample}}{1 - v_{sample}^2}. \tag{7.11}$$

The field in which the deformation and flow of materials are studied conventionally is known as rheology. In most rheological measurements, devices known as rheometers are used for macroscopic responses to macroscopic stress, shear, or flow. For such measurements, the samples have to be several millimeters in size and at least 1 cm³ in volume (Kimura, 2009). Owing to their inhomogeneity, macrorheological measurement techniques are not applicable to soft matter; instead they are suited to homogeneous systems. Also, the frequency range of the measurements by rheometers is high in tens of Hz, not compatible with the viscoelastic characteristic of soft matter. Microrheologic techniques, on the contrary, enable the study of rheological properties with micrometer- or submicrometer-sized probe particles. Either the fluctuation of the probe particles is detected and the mean square displacement is calculated from the trajectory of the particles or the scattered light intensity or particles are manipulated by time-dependent external forces (electric, magnetic, optical). Typical macrorheological methods are listed in Table 7.1 together with their applicable frequency and moduli range (Waigh, 2005). The mechanical properties of the soft matter (e.g., the viscoelastic modulus) can be calculated from the mean squares displacement (Pai et al., 2008; Solomon and Lu, 2001; Mason, 2000) or the applied force and induced displacement (Kimura, 2009; Valegol and Lanni, 2001).

TABLE 7.1

Microrheological Methods and Frequency and Modulus Range

Method	Frequency Range	Modulus (Pa)
Video tracking	10^{-2}–10^{1}	1^{-4}–1
Laser tacking	10^{-2}–10^{4}	10^{-2}–10^{2}
Diffusive wave spectroscopy (transmit)	10^{1}–10^{6}	10^{-3}–10^{3}
Magnetic tweezers	10^{-2}–10^{3}	10^{-3}–10^{5}

Source: Adapted from Waigh, T.A. 2005. *Rep. Prog. Phys.* 68(3): 685–724.

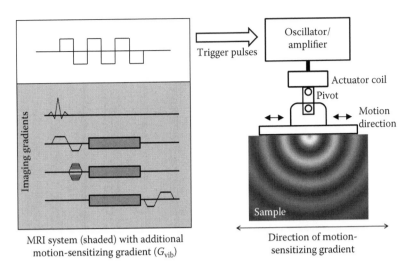

MRI system (shaded) with additional
motion-sensitizing gradient (G_{vib})

Direction of motion-
sensitizing gradient

FIGURE 7.4 Schematic diagram of the MRE system setup; composed of a magnetic resonance imaging (MRI) system, an oscillator, and an electromechanical actuator. Typically, MRI operates with imaging gradients (G_{slices}, G_{phase}, and G_{read}) used to encode the spatial positions of the MR signal. The radiofrequency (RF) pulses provided by the imager synchronize the oscillator that operates the electromechanical actuator coupled to the surface of the sample. In the presence of the motion-sensitizing gradient (G_{vib}), the cyclic motion of the spins results in a measurable phase shift in the received MR signal, which is used to calculate the displacement in each volume element. These data enable the visualization of the propagating shear waves within the sample and a reconstruction of the corresponding viscoelastic properties. (Adapted from Brandl, F., Sommer, F., and Goepferich, A. 2007. *Biomaterials* 28: 134–146.)

The MRE approach is illustrated in Figure 7.4. The system is typically composed of three main components: MRI system with a motion-sensitizing gradient, an oscillator, and an electromechanical actuator. Shear waves are generated within the sample using the electromechanical actuator in close contact with the sample. The visualized wave images are used to reconstruct the viscoelastic properties of the sample at each location (Manduca et al., 2001). The development of the techniques was inspired by the need to extend palpation (a physician's physical examination that assesses the stiffness of tissue for detection of tumors and other applications restricted to parts of the body that are accessible by the hand) to parts of the body that are not easily accessible by hand. The technique is therefore nondestructive, noninvasive, and allows the visualization of changes in elastic properties in a spatial manner. It has been extensively applied to the determination of physical properties of hydrogels (Tables 7.2 and 7.3) and tissue explants *ex vivo* (Hoyt et al., 2008; Venkatesh et al., 2008). For detailed treatment of the principles of the approach, see Madelin et al. (2004) and Manduca et al. (2001).

Another approach belonging to the elastography category is the ultrasonic resonance measurements (Gao et al., 1996; Palmeri et al., 2008). The speed with which shear waves are propagated in tissue has been used to quantify the tissue shear modulus. Waves are generated within the tissue using focused, impulsive, acoustic

TABLE 7.2
Modulus of Elasticity for Different Cells/Tissues[a]

Tissue Type	Animal Source	Testing Methodology	Modulus of Elasticity (kPa)	Reference
Achilles' tendon	Rat	Tension	310,000	Almeida-Silveira et al. (2000)
Articular cartilage	Bovine	Compression	350	Freed et al. (1997)
Skeletal muscle	Rat	Tension	100	Limder-Ganz and Gefen (2004)
Carotid artery	Mouse	Perfusion	90	Guo et al. (2006)
Spinal cord	Human	Tension	89	Bilston and Thibault (1996)
Thyroid cancer[b]	Human	Compression	45	Lyshchik et al. (2005)
Spinal cord	Rat	Tension	27	Fiford and Bilston (2005)
Cardiac muscle	Mouse	Tension	20–150	Wu et al. (2000)
Skeletal muscle	Mouse	AFM indentation	12	Engler et al. (2004a, 2004b)
Thyroid	Human	Compression	9	Lyshchik et al. (2005)
Lung	Guinea pig	Tension	5–6	Yuan et al. (2000)
Breast tumor	Human	Compression	4	Miyaji et al. (1997)
Kidney	Swine	Microrheology	2.5	Nasseri et al. (2002)
Premalignant breast[c]	Human	AFM indentation	2.2	Wellman et al. (1999)
Fibrotic liver	Human	Compression	1.6	Yeh et al. (2002)
Liver	Human	Compression	0.64	Yeh et al. (2002)
Lymph containing metastases	Human	Vibrational resonance	0.33	Miyaji et al. (1997)
Brain	Swine	AFM indentation	0.26–0.49	Miller et al. (2000)
Mammary gland	Human	Compression	0.16	Paszek et al. (2005)
Lymph node	Human	Vibrational resonance	0.12	Miyaji et al. (1997)
Fat	Human	AFM indentation	0.017	Wellman et al. (1999)

Source: Adapted from Levental, I., Georges, P.C., and Janmey, P.A. 2007. *Soft Matter.* 3: 299–306.

[a] When multiple values were available, the value at the lowest strain rate and lowest prestrain was used as an approximation of the "resting" modulus of elasticity of the tissue.

[b] Thyroid papillary adenocarcinoma.

[c] Mammary ductal carcinoma *in situ.*

radiation force excitation and the resulting response in terms of displacement is tracked ultrasonically in time. Algorithms have been developed to qualify the shear modulus from time-displacement data (e.g., Palmeri et al., 2008). The target application of this kind of technique is for noninvasive quantification of shear modulus to diagnose diseases like liver fibrosis and steatosis.

The interest in measuring physical properties at the single-cell level is predicated on a well-accepted notion that cell functions are essentially determined by their

TABLE 7.3

Elastic Moduli for Various Substrates

Substrate	Testing Method	Elastic Moduli (kPa)	Reference
Polystyrene tissue culture plates	Instron 5566 electromechanical tensile tester	2–4,000,000	Chen et al. (2009)
PLGA (fibrous)	Kawabata electromechanical tester	323,000	Li et al. (2002)
Polyelectrolyte multilayer films, assembly pH 6.5	AFM indentation	142,000	Chen et al. (2009)
Borosilicate glass	AFM indentation	$67,670 \pm 990$ (air) $69,830 \pm 820$ (water)	Constantinides et al. (2008)
Soya protein (Zein) (porous)	MARUI electromechanical tester	28,200–86,000 (depending on porosity)	Gong et al. (2006)
Nanofibrous poly(L-lactic acid) (PLLA) fabricated by a combination liquid–liquid phase separation	Instron electromechanical compressive tester	9650 ± 430	Wanga and Kisaalita (2010)
Electrospun collagen membranes	AFM indentation	~7000	Hartman et al. (2009)
Polystyrene (fibrous)	Instron 5566 electromechanical tensile tester	3680	Baker et al. (2006)
PLGA (porous)	Instron Model 5561 electromechanical tester	3000	Zhang et al. (2006)
Porous polystyrene fabricated by the leachable porogen technique	Instron electromechanical compressive tester	77	Lai et al. (2010)
Polypropylene	AFM indentation	2560 ± 210 (air) 2700 ± 150 (water)	Constantinides et al. (2008)
Polyelectrolyte multilayer films, assembly pH 4.0	AFM indentation	1700	Chen et al. (2009)
Poly-L-lactic acid co-ε-caprolactone (porous)	Electromechanical tensile tester	1250 (irradiated) 1170 (dried) 930 (wetted)	Vaquette et al. (2008)
Chitosan (porous sponge)	Instron 8511 electromechanical tensile tester	800	Kim et al. (2001)
PCL (porous)	Zwick BZ2.5/TN1S universal electormechanical tester	500	Olah et al. (2006)

TABLE 7.3   (continued)
Elastic Moduli for Various Substrates

Substrate	Testing Method	Elastic Moduli (kPa)	Reference
Polyelectrolyte multilayer films (PEMs)[a], assembly pH 2.0	AFM indentation	200	Chen et al. (2009)
Alginate (sponge)	Not specified	120–380	Shapiro and Cohen (1997)
Alginate (fibrous)	Instron 5564 electromechanical tester	24.3	Partap et al. (2007)
Alginate (unmodified)	Instron 5564 electromechanical tester	9.8	Partap et al. (2007)
PEG-fibronectin (0–2% PEG-diacrylate (DA) crosslinker)	MTS Synergie 100 with 10 N load cell electromechanical tester	0.448–5.408	Peyton et al. (2006)
Poly(acrylamide) hydrogels, 0.03–0.26% bis-acrylamide	Sheets deformed with a tensile force of 0.103 N	0.015–0.070	Pelham and Wang (1997)
Hyaluronan–gelatin hydrogels	Oscillatory shear measurement	0.011–3.5	Vanderhooft et al. (2009)

[a] Alternating layers of poly(acrylic acid) and poly(allylamine hydrochloride) adjusted to the same pH resulting in ionically crosslinked PEMs.

Unless otherwise stated, data are presented as means ± one standard deviation.

structure. The structural organization of cells is characterized by certain mechanical properties. For cells to function, they must be able to sense and exert force. How they do this is briefly outlined below.

7.1.4   Contractile Force Generation in Cells

Three types of filaments can be found in all eukaryotic cells: microtubules, intermediate filaments, and microfilaments (or actin filaments) that form the internal scaffolding of cells called the cytoskeleton. Microtubules are long protein polymers of α- and β-tubulin dimers along which organelles are moved. They have a diameter of 25 nm and their lengths may vary from 200 nm to 25 mm. Intermediate filaments have an average diameter of 10 nm, between that of microfilaments and microtubules. Microfilaments are the thinnest of all the filaments and are linear polymers of the protein, actin. They may contain as few as 10–20 to more than 1000 actin subunits. Another protein, tropomyosin, is a regular constituent of microfilaments; it serves as a fibrous reinforcement along actin chains and in some cases participates in the regulation of microfilament sliding. Like microtubules, microfilaments exhibit polarity; they possess a plus end and a minus end with respect to rate of assembly. Contractile forces in cells are generated by crossbridging interactions of microfilaments and myosin (Discher et al., 2005). Myosins occur in two forms (I and II). Myosin I is involved in several types of membrane-associated movement in nonmuscle cells. Myosin II is responsible for microfilament sliding that takes place in muscle cells as well as nonmuscle cell movements. Figure 7.5 presents a schematic of a simplified version of crossbridging, which hydrolyzes ATP to power microfilament sliding that results in contractile forces. For adherent cells, some of these forces are transmitted to the substrate (Tan et al., 2003; Balaban et al., 2001). Adherent cells interrogate the substrate by pulling via the actin–myosin cytoskeleton machinery, sense resistance and convert the signal into a biochemical signal (mechanosensing and mechanotransduction), and responds (mechanoresponse) to the resistance through the cytoskeleton (Wells, 2008; Saha et al., 2008; Engler et al., 2006; Discher et al., 2005). The cytoskeleton is an essential component in cellular motility. As cells move in the ECM, *in vitro* inside a scaffold or on a 2D surface, they experience

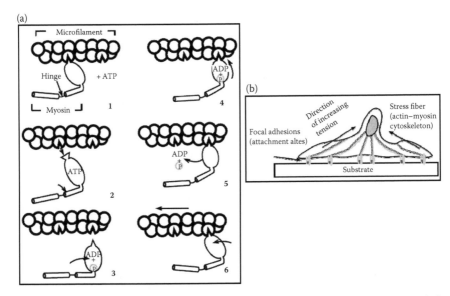

FIGURE 7.5   (a) Simplified myosin crossbridge cycle: (1) Beginning of the cycle—myosin is tightly linked to an actin subunit and bent at an angle of 45°; the ATP-binding site on the myosin is active. (2) The myosin–actin binding site changes on binding ATP such that it no longer fits the actin subunit, thus releasing the crossbridge. The bound ATP is hydrolyzed with the products of the reaction (ADP and P) remaining bound. (3) Another conformation change results from ATP breakdown shifting the crossbridge to the 90° position and reactivating the actin-binding site at the tip. (4) The crossbridge attaches to another actin subunit of the filament. (5) Products of ATP breakdown are released. (6) The release of the products results in the shift from the 90° to the 45° position, pulling the microfilament a distance and the cycle is ready for a repeat. (Adapted from Wolfe, S.L. 1993. *Molecular and Cellular Biology.* Belmont, CA: Wadsworth Publishing Company.) (b) Schematic of a cell attached to a substrate via integrin-containing adhesions that are linked to the actin–myosin cytoskeleton.

external forces, in response to their "pulling." The magnitudes of the forces involved in different cell types and measurement methodologies have recently been reviewed by Ananthakrishnan and Ehrlicher (2007). The subject of force sensing is further explored in detail in the following section.

7.1.5 FORCE AND GEOMETRY SENSING

There are a number of proteins through which the force signal can be transmitted in and out of the cell in a bidirectional manner including exposure of cryptic peptide sequence, gating of a mechanosensitive ion channel, and strengthening of strained receptor–ligand interactions, known as catch bonds (Thomas, 2008).

As outlined in Chapter 5, many cell adhesion molecules consist of tandem-repeat sequences. This is also true of many other ECM proteins. Subjecting these molecules to forces may result in changing the molecular folding in such a manner that specific molecular-recognition sites are hidden or exposed. For example, the cell-adhesion protein fibronectin consists of more than 50 modular repeats, which links the ECM to contractile cytoskeleton through its tripeptide sequence, RGD. When folded, many RDG sites are accessible in the loop region, but when stretched, many cryptic binding sites buried in the folds are exposed (Little et al., 2008; Smith et al., 2007; Vogel, 2006; Pankov and Yamada, 2002). In a similar fashion, force-induced conformation changes can result in activating enzymes and/or their substrates. For example, recent studies have shown that *Src* substrate is "made ready" for phosphorylation by mechanical unfolding (Harrison, 2003).

The gating of mechanosensitive channels can occur in one of two ways, either by membrane tension or by force-bearing filaments connected to the intra- and/or extracellular channel protein domains—the tethered model (Kung, 2005; Sukharev and Corey, 2004; Goodman and Schwarz, 2003). An example, recognized more than two decades ago (Guharay and Sachs, 1984) that is present in most eukaryotic cells (Pendersen and Nilius, 2007), is the stretch-activated cation channel, which displays a range of permeabilities, suggesting a heterogeneous composition. Another example, initially proposed to be gated by the tethered model is the mechanosensitive calcium channel, which is now known to be also gated by tension (Maroto et al., 2005).

When a receptor–ligand bond regulated by tensile mechanical force becomes longer lived at higher forces, it is called a catch bond (Marshall et al., 2003). Two well-known types of adhesive proteins, shown to form catch bonds, are selectins (leukocyte cell adhesive proteins—see Chapter 5) (Evans et al., 2004) and bacterial protein called FimH (Thomas et al., 2002). Catch bonds only bind strongly when needed to resist force. More recently, more receptors have been shown to form catch bonds, for example, myosin has been shown to form catch bonds with actin (Guo and Guilford, 2006). In all these cases, catch bonds can be weakened or eliminated by activating mutations in the interdomain region distal to the binding site (Thomas, 2009), suggesting that mechanical force regulates interdomain conformation and subsequently binding. The concept of geometry sensing was introduced in the previous chapter and will not be repeated here. Suffice to mention that sensing is not limited to force but is also extended to geometry (Vogel and Sheetz, 2006).

7.2   STIFFNESS-DEPENDENT RESPONSES

7.2.1   BIOLOGICAL AND NONBIOLOGICAL MATERIALS' STIFFNESS

The modulus of elasticity of biological materials and in particular cells/tissues range from approximately 100 Pa for the soft tissues such as the brain to approximately hundreds of MPa for stiff tissues such as cartilage, as shown in Table 7.2. However, the rigidity of materials that cells interact with in traditional tissue culture vessels is orders of magnitude higher, as shown in Table 7.3. Although 3D scaffold materials exhibit lower E values, they are still orders of magnitude higher in some cases. The exception is hydrogel-based materials as shown in Table 7.3. However, making polymer-based materials in the porous form, in these cases creating 3D scaffolds, resulted in lower E. On the contrary, similar treatments with hydrogels resulted in increased E values. For example, while polystyrene decreased by 100 orders of magnitude, alginate increased by approximately 10 orders of magnitude (Table 7.3).

Another interesting observation is that diseased cells/tissues exhibit higher stiffness in comparison to their normal counterparts. For example, in a recent study, Venkatesh et al. (2008) used MR elastography and examined 44 liver tumors including 14 metastatic lesions, 12 hepatocellular carcinomas, 9 hemangiomas, 5 cholangiocarcinomas, and 3 cases of focal nodular hyperplasia. The shear stiffness of the tissues is shown in Figure 7.6. A cutoff of 5 kPa differentiated normal tissue and benign tumors from malignant tumors. At the single-cell level, Dulińska et al. (2006) used AFM and measured the stiffness of pathological red blood cells (erythrocytes) from patients with hereditary spherocytosis, thalassemia, and glucose 6-phosphate-dehydrogenase (G6PD) deficiency. Before measurement, the erythrocytes were immobilized on a poly-L-lysine-coated glass surface with 0.5% glutaraldehyde. In agreement with Venkatesh et al. (2008), the Young's modulus of diseased red

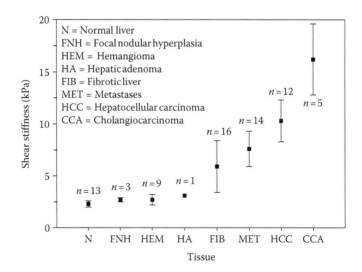

FIGURE 7.6   Shear stiffness of normal, benign, and malignant liver tissues determined by MRE. (Adapted from Venkatesh, S.K. et al. 2008. *AJR* 190: 1534–1540.)

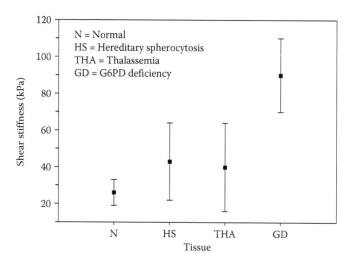

FIGURE 7.7   Young's moduli for different types of anemias determined by AFM. (Adapted from Dulińska, I. et al. *J. Biochem. Biophys. Methods* 66: 1–11.)

blood cells was higher than in normal cells (Figure 7.7). Similar results have been reported for prostate (Hoyt et al., 2008), breast (Murayama et al., 2008), and other tumors (Paszek et al., 2005).

The exact cause of high stiffness in malignant tumors is not well understood. What is well documented is that tumors exhibit altered integrins (Oertl et al., 2006; Guo and Giancotti, 2004; Nadav and Katz, 2001). Why tumors are stiffer than normal tissue has been addressed by Paszek et al. (2005, p. 241) and they reported that, "tumors are rigid because they have stiff stroma [connective tissue] and elevated Rho [small GTR-binding protein responsible for termination of RNA synthesis]-dependent cytoskeletal tension that drives focal adhesions, disrupts adherent junctions, perturbs tissue polarity, enhances growth, and hinders lumen formation." These observations are consistent with *in vitro* stiffness-dependent findings reported in detail below, speaking of the "abnormal nature" of 2D cultures on substrates with nonphysiological stiffness.

For a long time, it was not possible to accurately establish stiffness-dependent responses as it was not possible to separate the chemical from stiffness cues; changing material stiffness would result in also changing the adhesive ligand density (Walton et al., 2007). For example, increasing the collagen concentration results in the fabrication of more stiff hydrogels, while at the same time increasing the collagen fiber-binding sites. The need to separate these cues has led to the development of new synthetic or semisynthetic hydrogels offering improved control over independence of biochemical and stiffness cues (Vanderhooft et al., 2009; Chen et al., 2009; Engler et al., 2006; Thompson et al., 2005, 2006; Peyton et al., 2006; Semler et al., 2005; Yeung et al., 2005; Engler et al., 2004a, 2004b; Pelham and Wang, 1997). These materials have enabled the experimentations that have yielded the results reported in the following subsections.

7.2.2 STIFFNESS-DEPENDENT MORPHOLOGY AND ADHESION

Not all cell types appear to be sensitive to surface stiffness. However, all stiffness-sensitive cells studied so far spread more and adhere better to stiffer surfaces. Actually some cannot grow on very soft surfaces with stiffness equal to or below 0.05 kPa. Levental et al. (2007) have summarized the effect of stiffness on cell morphology. Examples include increased neuron branching (Flanagan et al., 2002) in the ~0.1 kPa shear modulus (G') range; epithelial cell *in vivo*-like organization including polarity (Paszek et al., 2005) in the G' range of 0.3 kPa; and myoblast of muscle displaying actomyosin striation (Engler et al., 2004a) in the G' range of 4 kPa. It is interesting to note that in all these cases, the optimal shear modulus echoes tissues from which those cells were derived (Levental et al., 2007). Other studies have also made similar observations (Ghosh et al., 2007).

A key feature of stiffness is its effect on cell spreading. Solon et al. (2007) conducted an elegant study that demonstrated this relationship with fibroblast on fibronectin-laminated polyacrylamide gels (Figure 7.8). Numerous studies have confirmed this trend in other cells including bovine aorta endothelial cells (Yeung et al., 2005), astrocytes (Georges et al., 2006), and rat kidney (Pelham and Wang, 1997), mammary epithelial (Paszek et al., 2005), and smooth muscle (Engler et al., 2004a, 2004b) cells. Integrins seem to play key roles; for example, the expression of α_5-integrin in fibroblasts correlated with cell surface area in response to gel stiffness (Yeung et al., 2005). The stiffness dependence of fibroblasts and endothelial cells was no longer evident when cells became confluent, or in the case of fibroblasts even when two cells make contact (Yeung et al., 2005), for which no explanation has been provided to date. Interestingly, neutrophils (white blood cells) have no preference for a given stiffness and spread equally well on soft and stiff surfaces (Yeung et al., 2005). Also, platelets were shown to be insensitive to stiffness (Georges et al., 2006). It is worth noting that neutrophils and platelets under normal

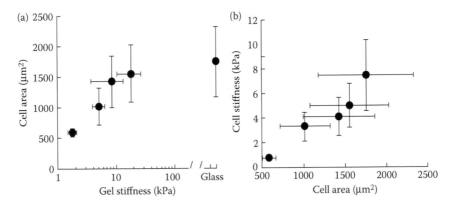

FIGURE 7.8 Relationships between cell and substrate (gel) stiffness and cell spread area. (a) Projected cell area as a function of gel stiffness and (b) cell stiffness as a function of projected cell area. Each point on the graphs in (a) and (b) is the mean ± SD of 12–40 cells. (Adapted from Solon, J. et al. *Biophys. J.* 93: 4453–4461.)

physiological conditions are nonadherent, suggesting the importance of ECM adhesive ligand and receptor interaction in understanding stiffness-dependent effects.

Cell–ECM interconnections are mediated through adhesion complexes, which are specialized subcellular sites that contain specific adhesion receptors, cytoskeletal elements, and a wide range of interconnecting proteins (Geiger et al., 2001). It is not surprising that stiffness affects the size of adhesion complexes as shown by Paszek et al. (2005); a more than sevenfold increase in large adhesions per cell was observed between fibroblasts cultured on soft (0.4 kPa) and stiff (5.0 kPa) fibronectin-coated polyacrylamide gels. Also, α_5 integrins, FAKp397, vinculin, and actin stress fibers were upregulated. Both the traction force and the cell spreading (area) increased with substrate shear modulus. Matter and Ruoslahti (2001) used Chinese hamster ovary (CHO) cells and showed that integrin-mediated adhesion was necessary for survival and loss of adhesion resulted in apoptosis. Furthermore, Bcl-2 (a protein associated with protection from apoptosis) transcription was elevated in cells that attached to fibronectin through $\alpha_v\beta_1$ or to vitronectin through $\alpha_v\beta_3$ but not elevated through cell attaching through the $\alpha_v\beta_1$ integrin. These results further underscore the importance of not only stiffness but the ligand type and its interaction with integrins.

7.2.3 STIFFNESS-DEPENDENT MIGRATION

Cells exert less tension on softer surfaces but craw faster. Cells move in response to external signals in their microenvironment. As indicated in Chapter 5, these signals may be chemical (diffusible or nondiffusible) detected by receptor proteins located on the cell membrane, and transmitted by them via signaling cascades to the cell interior. Other signals handled in a similar manner may be physical (force and/or surface stiffness). For most cells on the move, the process can be divided into three stages, involving constant restructuring of the actin cytoskeleton (Ananthakrishnan and Ehrlicher, 2007) as shown in Figure 7.9. In the first stage, the cell "grows" an actin network at its leading edge. In the second stage, the cell adheres and deadheres to the substrate at the leading and the rear body end, respectively. In the third stage, contractile forces, generated largely by the action of the actomyosin network, pull the cell forward. Depending on the nature of the signal (chemoattractant or chemorepellent), the cell polymerized in the region closest to the signal to move toward or away from the signal. This signal tracking is beautifully demonstrated by NIH 3T3 fibroblasts that were cultured on 8% acrylamide gels with varying stiffness, achieved by bis-acrylamise (0.06–0.03%). As shown in Figure 7.10, a cell moved from the soft side toward the gradient and another cell moved from the stiff side toward the gradient (Lo et al., 2000). In the same study, cells exerted approximately 6 and 11 kdyn/cm^2 forces on 140 and 300 kdyn/cm^2 (Young's modulus) gels, respectively. They also crawled faster on the softer gel (0.44 versus 0.26 µm/min). In a rather similar later study, with smooth muscle cells, cells accumulated toward the stiff end of a soft-to-stiff gradient gel (Peyton and Putnam, 2005; Zaari et al., 2004). Similar behavior has been demonstrated with epithelial cells (Pelham and Wang, 1997). In smooth muscle cells, the crawling speed was highest at intermediate stiffness levels in a manner similar to adhesive ligand concentration effect (Goodman et al., 1989). Mathematical models of

(1) Protrusion of the leading edge

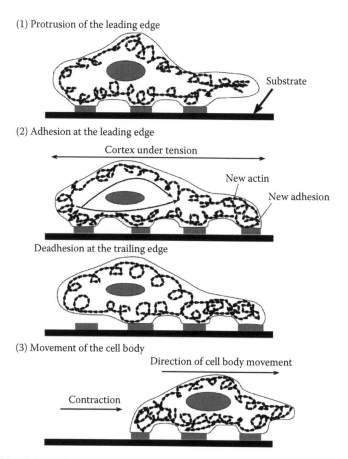

(2) Adhesion at the leading edge

Cortex under tension

New actin

New adhesion

Deadhesion at the trailing edge

(3) Movement of the cell body

Direction of cell body movement

Contraction

FIGURE 7.9   Schematic of the three stages of cell movement. After committing to a direction of motion, the cell protrudes in the direction by actin polymerization at the leading edge. Protrusion is followed by adhesion and deadhesion to the surfaces the cell is moving and leaving, respectively. The cell then used contractile forces in the cell body and the rear to pull the whole cell body forward. (Adapted from Alberts, B., et al. 2002. *Molecular Biology of the Cell*, 4th edition. New York: Garland Sciences.)

ligand concentration effects have suggested the effect to be a shift in balance between ligand-mediated traction and ligand-mediated anchorage (Zaman et al., 2005).

As the above-mentioned studies reveal, most of the stiffness-locomotion studies have been conducted with 2D gels. However, fibroblast migration in poly(ethylene glycol) (PEG) hydrogels containing integrin-binding peptides Ac-GCGYGRDGSPG (the adhesion domain is underlined) was studied (Raeber et al., 2007). The gels were rendered MMP-degradable by crosslinking the PEG macromers with the bifunctional MMP-sensitive peptide sequence Ac-GCRD-GPQG ↓ IWGQ-DRCG. Gels that contained RGDSP and substrates for MMPs supported 3D fibroblast migration, which was almost completely inhibited with GM6001 (broad MMP inhibitor). Exposure to TNF-α (used as a model cytokine) led to a marked increase

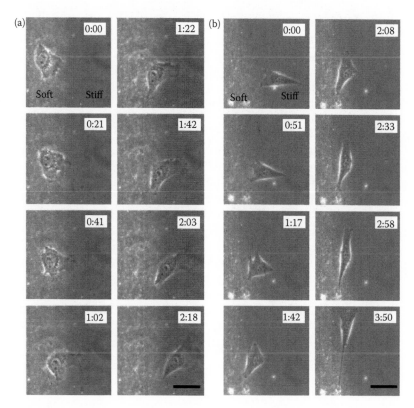

FIGURE 7.10 Fibroblast (NIT 3T3) cell movement on gel substrates with stiffness gradient. Fluorescence beads were incorporated in the gel to visualize changes in stiffness. By illuminating the gels simultaneously with phase and fluorescence optics, cell movement in relation to stiffness were observable. (a) Illustration of a cell that moved from the soft side toward the gradient. The cell turned approximately 90° into the stiff side of the substrate. The increase in spread area as the cell passed the boundary is observable. (b) Illustration of a cell that moved from the stiff side toward the gradient, changing its direction as it entered the gradient and moving along the boundary. Bar = 40 μm. (From Lo, C.-M. et al. *Biophys. J* 79: 147, 2000. With permission.)

in migration and higher expression of $\alpha_v\beta_3$ integrin receptor. But blocking of the receptor with antibodies did not affect migration. Increase in proMMP-9 was observed. In fibroblasts, TNF-α has been demonstrated to indirectly regulate the proteolytic activity of MMP-9 in conjunction with TGF-β (Han et al., 2002), leading to the speculation that increase in migration might have been due to TNF-α-induced MMP-9 expression. The 2D stiffness studies need to be replicated in a 3D microenvironment such as the one described above for results that will be physiologically more relevant.

7.2.4 STIFFNESS-DEPENDENT GROWTH AND DIFFERENTIATION

A number of studies have shown that other microenvironmental factors being equal, surfaces or scaffolds optimally modulate cells with cell-source tissue-like stiffness.

The ability to fabricate hydrogels exhibiting different stiffness without changing the surface or scaffolds chemistry has enabled investigations into material stiffness effects. In studies conducted with different media, generally increased growth has been associated with increasing stiffness. For example, the colony sizes of mucoepidermoid carcinoma (MEC) cells grown in basement membrane/collagen gels on increasing stiffness (0.170–1.2 kPa) increased from approximately 50 to 180 µm (Paszek et al., 2005). Chondrocytes exhibited increased growth and proliferation markers on hard gels with G' values of approximately 10 kPa (Genes et al., 2004; Subramanian and Lin, 2005).

Although the stiffness optima for different kinds of adherent cells vary widely, it is generally true that cell proliferation and differentiation increase with the stiffness of the matrix. However, when Wang et al. (2000) cultured normal and H-R*as*-transformed NIH 3T3 fibroblasts on collagen-coated polyacrylamide gels with stiffness ranging between 4.7 and 14 kPa, the spreading area of normal (control) cells increased with increasing stiffness, but the transformed cells were unresponsive to changes in substrate stiffness, suggesting the importance of R*as* in mediating stiffness effects.

Optimal stiffness with respect to differentiation has been reported. For example, myoblasts exhibited actin–myosin striation on collagen-coated PA gels of 12 kPa (Figure 7.11) (Engler et al., 2004a). Many other studies have reported similar finding, including stem cell differentiation. Sample results are summarized in Table 7.4. By plotting the optimal stiffness against the stiffness of the tissue of the cell origin (Figure 7.12), it is evident that cells differentiate optimally on surfaces with

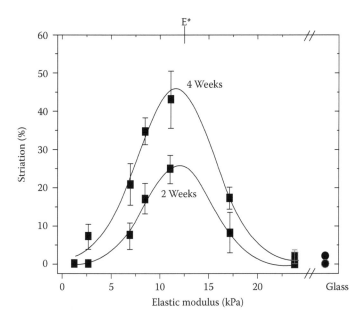

FIGURE 7.11 Relationship between myoblasts differentiation and collagen-coated polyacrylamide gels stiffness. The striation of myoblast tubes to myotubes depended on the elastic modulus of the substrate. The optimum gel modulus was found to be 12 kPa (Gaussian curve fitting) for both 2- and 4-week cultures. The circle represents differentiation on glass. Error bars are ±SEM. (From Engler, A.J., et al. 2004a. *J. Cell Biol.* 166: 877–887. With permission.)

TABLE 7.4

Stiffness-Dependent Cell Growth and Differentiation Responses

Surface/Scaffold Material	Cell Type	Function and/or Differentiation Marker	Optimal Stiffness, kPa (Reference)	Cell-Source Tissue Stiffness Range, kPa (Reference)
Collagen-coated PA gels	Skeletal myoblasts (C2C12)	Striation	12.0 (Engler et al., 2004a)	5.5–29.3 (Papazoglou et al., 2006)
Synthetic hydrogels based on different concentrations of acrylamide, bis-acrylamide, and poly(acrylamide) with peptide modifications to enable cell adhesion	Neural stem cells isolated from hippocampi of adult female Fisher 344 rats	β-tubulin III, a neuronal marker	~0.5 (Saha et al., 2008)	0.1–1.0 (Engler et al., 2006)
Collagen-coated polyacrylamide (PA) gels	Human mesenchymal stem cells	P-NFH	~0.25 (Engler et al., 2006)	0.1–1.0 (Engler et al., 2006)
Collagen-coated polyacrylamide (PA) gels	Human mesenchymal stem cells	MyoD	~9.0 (Engler et al., 2006)	8.0–17.0 (Engler et al., 2006)
Collagen-coated polyacrylamide (PA) gels	Human mesenchymal stem cells	CBFα1	~32.0 (Engler et al., 2006)	25–40 (Engler et al., 2006)
Collagen-coated polyacrylamide (PA) gels	Human osteoblasts (hFOB, ATCC)	CBFα1	~31 (Engler et al., 2006)	25–40 (Engler et al., 2006
Collagen-coated polyacrylamide (PA) gels	Murine myoblast (C2C12)	MyoD	~11.0 (Engler et al., 2006)	8.0–17.0 (Engler et al., 2006)
Basement membrane and collagen gels (1 mg/mL)	MEC cells	Acinus formation	~0.170	0.167 (Paszek et al., 2005)

tissue-like stiffness. It is interesting to note that in most of these cases, differentiation recapitulates a morphology or function that is not attainable in standard 2D tissue culture plates, such as muscle cell striation (Engler et al., 2004a), liver cell albumin secretion (Chen et al., 2009), and mammary acinus formation (Paszek et al., 2005).

7.2.5 Substrate Stiffness-Dependent Cell's Internal Stiffness

Fibroblasts have been shown to synchronize their internal stiffness with the stiffness of the surface they adhere on, within a physiologically relevant stiffness range. By

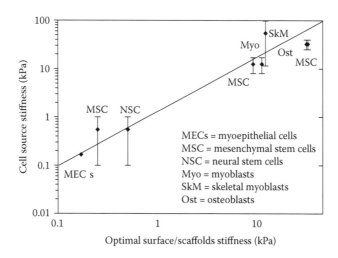

FIGURE 7.12   Parity plot of optimal substrate stiffness versus cell-source tissue stiffness.

varying the bis-acrylamide (0.03–3%) in 7.5 acrylamide, Solon et al. (2007) success-fully fabricated hydrogels with a 0.2–10 kPa stiffness range. Crosslinked gels were coated with fibronectin and fixed and taped to Petri dishes. Their stiffness was measured with AFM. NIH 3T3 fibroblasts were cultured on these 2D gels and fixed after 24 h in culture with 4% paraformaldehyde. The stiffness of the cells was also measured by AFM. For a range of surface stiffness (1–5 kPa), fibroblasts harmo-nized their average stiffness with that of the gel surface without the formation of stress fibers. Beyond the 5 kPa stiffness, cells remained softer than the surface prob-ably because the stiffness harmonizing mechanism was not functional. Stress fibers became visible at higher stiffness where cells were able to fully spread. As shown in Figure 7.13, cell stiffness increased from 6 to 8 kPa for 20 kPa gels as well as for cells on glass. At the higher surface stiffness, cells exhibited bigger stress fibers, which is characteristic of cells cultured on glass and was also observed in this study. Overall, these results are significant in that they conclusively suggested that fibro-blasts adapt to the stiffness of the microenvironment in which they reside and they fail to adapt at high stiffness (>10 kPa) outside the physiological range, characterized by the appearance of stress fibers.

A search of the literature revealed no similar study conducted with other cells types. But the validity of the results for other cells can be indirectly inferred as size for the different cells is observed to increase with increasing surface stiffness, a relationship that was demonstrated in the Solon et al. (2007) study. Such findings have been reported with normal rat kidney (NRK) (Pelham and Wang, 1997), mam-mary epithelial (Paszek et al., 2005), smooth muscle (Engler et al., 2004a, 2004b), and neuronal and glial (Georges et al., 2006) cells. It is possible that the use of substrates with stiffness outside the physiological range (i.e., saturation range) puts cells in a position to behave like "transformed," which could in some cases explain the limitation of 2D cultures in standard culture plates recapitulating physiologically complex relevant features, like acin formation in liver cells.

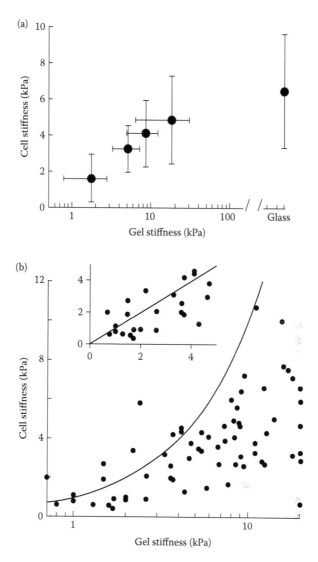

FIGURE 7.13   Influence of substrate stiffness on cell stiffness. (a) Cell stiffness as a function of the stiffness of the gel on which the cell is cultured (mean ± SD). (b) Mean individual cell stiffness measurement versus mean stiffness of adjacent gel. The line shows gel stiffness and the inset is an enlargement of the stiffness range up to 5 kPa. (From Solon, J. et al. 2007. *Biophys. J.* 93: 4458. With permission.)

7.3   FORCE-DEPENDENT RESPONSES

The studies of mechanical load-induced changes in the structure, composition, and function of living tissues, alternatively referred to as mechanobiology, have predominantly been carried out with cells from force-bearing tissue. A representative sample of such studies is presented in Table 7.5. As shown, the cells used include, fibroblasts,

TABLE 7.5

Force-Dependent Responses of Cells

Type of Load/Force/Duration	Cell Type	Response	Reference
Cyclic stretch, 1.5 Hz, 20%, 1–4 days	Fetal rat cardiac fibroblasts	Time-dependent response: in the absence of fetal calf serum (FCS), procollagen synthesis decreased as stretch time increased from 24 to 48 h. In the presence of FCS, both procollagen synthesis and synthesis rate increased as stretch time increased from 24 to 48 h	Butt and Bishop (1997)
Static uniaxial strain, 10% and 20% magnitude for 24 h	Rat cardiac fibroblasts	Magnitude different effect: 10% strain stimulated collagen III and fibronectin mRNA levels, while 20% strain caused collagen III and fibronectin mRNA level decrease. Both 10% and 20% increased TGF-β1 activity (10% had higher activity than 20%)	Lee et al. (1999)
Static equibiaxial tensile or compressive, −6%, −4%, −3%, 0%, 3%, 6%, for 24 h	Rat cardiac fibroblasts	Pure tensile and compressive area strains induced divergent responses in ECM mRNA levels: decreased level at −3%, −6%, and 6% strain and increased level at 3% strain. TGF-β1 activity was dependent on the magnitude of applied area strain (increase as strain increased), regardless of type	Lee et al. (1999)
Cyclic stretch, 20%, 1.5 Hz, 6–72 h	Human pulmonary fibroblasts	Substrates-specific effect: increase type I collagen expression on laminin, elastin-coated substrates, but not on fibronectin substrates. The cell aligned perpendicular to the force vector. Time-dependent manner: on elastin matrix, procollagen mRNA level increase first as duration increased, after 12 h, then decreased as duration increased	Breen (2000)
Cyclic 20% elongation, 1.5 Hz, 24 or 48 h	Adventitial pulmonary artery fibroblast	Growth factor-dependent response: load had no effect on cell replication or procollagen production when only with 1% FCS, but increased procollagen production when with 10% FCS. Load also enhanced the stimulatory effect of PDGF on procollagen production	Bishop et al. (1998)
Cyclic uniaxial stretch, 0.5 Hz, 4 h duration with magnitude of 4% and 8%	Human tendon fibroblasts	Stretch magnitude-dependent response: as the magnitude of stretch increased, cell proliferation, collagen I and TGF-β1 gene expression, amount of TGF-β1 protein in medium and collagen I protein also increased	Yang et al. (2004)

Loading condition	Cell/tissue type	Response	Reference
Cyclic longitudinal stretch, 1 Hz, 5% magnitude with 15 or 30 min duration, and 6, 12, and 24 h recover time	Human tendon fibroblasts	Recover time-dependent response: as recover time after stretch increased, cell proliferation increased	Barkhausen et al. (2003)
Cyclic equibiaxial tensional and compressive forces, 0.5 Hz, 10% for 24 h	Human periodontal ligament fibroblasts	Differential response to differential type of mechanical loading: mRNA of collagen 1, secretion of collagen 1 and fibronectin protein increased with tension load but decreased with compression load. Total protein secreted into the medium, MMP-2 (matrix metalloproteinase-2) and TIMP-2 (tissue inhibitor of metalloproteinase-2). RNA levels increased with both tension and compression load	He et al. (2004)
Cyclic uniaxial compression, 0.1 Hz, 3%, for 48 h	Bovine calves cartilage disk explant	Increased protein synthesis, enhanced effect of IGF-1-stimulated protein synthesis	Bonassar et al. (2000)
Cyclic uniaxial compression, 0.1 Hz, 2%, for 2, 4, 16, 24 and 48 h duration	Bovine calves cartilage disk explants	Time-dependent response: as compression time increased, protein synthesis and IGF-1 uptake increased. The increase saturated after 24 h	Bonassar et al. (2000)
Cyclic stretch, 0.25 Hz, 24%, for 3 h	Bovine chondrocytes	Increased collagen II and aggrecan mRNA when applied with load. Response involves integrins in signal transduction	Holmvall et al. (1995)
Cyclic uniaxial stretch, 1 Hz, magnitude of 1000 μstrain, 48 h of 1000 μstrain led to an increase of proliferation and early osteoblast activities related to matrix production (CICP). In contrast, activities that are characteristic of the differentiated osteoblast (AP activity) and relevant for matrix mineralization (OC release) were decreased	Human osteoblasts Kaspar et al. (2000)	Cyclic strain at a physiologic magnitude	
Cyclic hydrostatic pressure, 5 MPa, 0.5, 0.25, 0.05, and 0.0167 Hz, for 1.5 or 20 h	Bovine articular cartilage explants or cell culture (chondrocytes)	Differential response for differential load frequency and duration: for 1.5 h duration, increased proteoglycan synthesis in tissue explants by 0.5 Hz only and inhibition of proteoglycan synthesis in chondrocytes by pressure with all frequencies. For 20 h duration, synthesis was inhibited by low frequency but stimulated by high frequency	Parkkinen et al. (1993)

continued

TABLE 7.5 (continued)
Force-Dependent Responses of Cells

Type of Load/Force/Duration	Cell Type	Response	Reference
Cyclic strain, 0–10%, 1 Hz, 0–24 h	Bovine aortic endothelial cells	Force- and time-dependent manner: increase in MMP-2 activity and expression by increase in force or time duration, through p38- and ERK-dependent pathways	von Offenberg et al. (2004)
Cyclic loading, 0.1 Hz or static loading, 4%, for 4 h	Bovine aortic endothelial cells	Decrease in cell density and DNA synthesis, in both cyclic and static loading. Static more severe	Woodell et al. (2003)
Cyclic stretch, 20%, 1 Hz, for up to 24 h	Rat vascular smooth muscle cells	Time- and elongation-dependent manner: secretion of TGF-β proteins increased with increased time or elongation of stretch. Stretch also stimulated mRNA expression of TGF-β, fibronectin, type I and type IV collagen. The stimulation can be inhibited by neutralizing antibody against TGF-β and tyrosine kinase inhibitors	Joki et al. (2000)
Cyclic stretch, 10%, 1 Hz, 0–72 h	Rat primary smooth muscle cells	Time-dependent increase in L-arginine transport. Increased collagen I synthesis after 72 h stretch	Durante et al. (2000)
Cyclic biaxial stretch, 4%, 1 Hz, 0–48 h	Human primary smooth muscle cells	Time-dependent increase in proteoglycans: versican, biglycan, and perlican mRNA (increase with stretch duration) and decrease in decorin mRNA	Lee et al. (2001)
Uniaxial strain, cyclic 10%, 1 Hz; or static 5%, for 24, 48, or 72 h	Human primary smooth muscle cells	Differential response for different types of strain: increase in MMP-2 mRNA and protein, and MMP-9 secretion in static strain; decrease in MMP-2 mRNA and protein in cyclic strain	Asanuma et al. (2003)
Cyclic stretch, rate of 10% elongation, 0.5 Hz, for 24 h	Rabbit vascular smooth muscle cells	Stretched SMCs had increased collagen and total protein synthesis and elevated secretion of TGF-β. The increase is mediated via an autocrine–paracrine mechanism of Ang II	Li et al. (1998)
Rectangular electrical pulses 2 ms, 5 V/cm, 1 Hz, applied continuously for 5 days	Neonatal rat ventricular myocytes seeded onto ultrafoam collagen sponges	Electrical stimulation induced cell alignment and coupling, increased amplitude of synchronous contraction by a factor of 7	Radisic et al. (2004)

Source: Adapted from Wang, J.H.-C. and Thampty, B.P. 2006. *Biomechan. Model Mechanobiol.*, 5: 5. With permission.

chondrocytes, endothelial, and muscle cells. The loading is typically cyclic and cellular responses are time dependent. Stimulation of TGF-β expression, increase in proliferation, and increase in production of ECM components such as collagen and matrix metalloproteinases are ubiquitous. The TGF-β expression is one excellent demonstration of the interaction between two MEFs (physical and soluble factors). TGF-βs are a family of at least five 25 kDa dimeric proteins, but only three (TGF-β1, -β2, and -β3) have been found in mammals (Roberts and Sporn, 1993). They are synthesized in an inactive form and activation can occur by matrix metalloproteinases, other proteases and glucosidases, or by acidic conditioning (Grimaud et al., 2002). Mauck et al. (2003) applied dynamic deformation loading together with either TGF-β1 or IGF-1, and increased the aggregate modulus of engineered articular cartilage greater than either stimulus applied alone. Kashiwagi et al. (2004) transected and sutured Achilles tendons of 90 rats and treated them with TGF-β1 doses of 10 and 100 ng. First a dose-dependent increase in the expression of procollagen I and II mRNAs was observed. Second, the failure load and stiffness of the healing tendons were increased by the treatment at 2 and 4 weeks in comparison to control (only exposed to phosphate-buffered saline). Since the animals were not immobilized, it is reasonable to assume that the healing tendons in both the control and treatment were equally exposed to the same kind of normal usage loading. Early healing as expressed by the failure load and tension strongly suggested the synergistic effect of TGF-β1 and mechanical load.

As with TGF-β1, a single application of one of several growth factors has been shown to stimulate the repair of ruptured tendons. Aspenberg (2007) has assembled results for PDGF, IGF-1, VEGF, bone morphogenetic proteins (BMPs), growth differentiation factors (GDF-5, -6, and -7), and thrombocyte concentrate (PRP). In all cases, the response was dependent on the mechanical microenvironment. Cytokines are a family mainly made up of the interleukins (ILs), the tumor necrosis factors (TNFs), the interferons, and the colony stimulating factors (CTFs). These small proteins are mainly known to regulate inflammatory and immune responses. But their roles in pathways linking mechanical forces and cellular outcomes are of increasing interest. For example, IL-1 has been shown to decrease ECM protein synthesis in cartilage. Cartilage explants compressed in the presence of IL-1 receptor antagonist increase proteoglycan synthesis while decreased biosynthesis was observed in the absence of the compression (Murata et al., 2003).

7.4   CONCLUDING REMARKS

The results of the stiffness-mediated effects presented support the notion that many cell types change their morphology and gene expression profile when cultured on surfaces that are chemically equivalent, but different in terms of stiffness. Also, within a physiologically relevant stiffness range, fibroblast and probably other cells match their internal stiffness to that of the surface they are adhering to. The emerging design principle from this collective knowledge is that, whenever possible, surfaces or scaffolds whose stiffness is within the physiologically relevant range should be used. It should be noted that physiologically relevant stiffness ranges are different for different cells types. Hydrogels seem to be the most ideal materials with which physiological stiffness can be matched; however, they have numerous disadvantages, including less optimal optical and mechanical properties to allow optical readout signals in and out of the sample and to withstand the mechanical forces imposed by fluid handling equipment, respectively. A possible way out in HTS applications is to use stiff scaffolds with spatial dimensions that promote the formation of cellular aggregates where the cells deposit their own physiologically relevant EMC and their internal stiffness depends more on cell–cell contact as opposed to cell–stiff scaffold contact. The questions that need answering are how these cellular aggregates match their mechanical properties to those of their tissue of origin.

Given what we know now, how does one begin to formulate design principles that integrate the three MEFs outlined in this and the previous two chapters? This question represents a translational research grand challenge that can be stated differently—how are these multiple cues integrated by the cells? A better understanding of information integration at the cellular level is a prerequisite to meet the grand challenge. The preliminary data required to begin answering the grand challenge question can be found in our current understanding of the interaction between growth factor receptor binding (or GPCRs)- and integrin binding-mediated signaling that can lead to the cellular responses of proliferation, apoptosis, and differentiation.

Why integrins and GPCRs? The common molecular entity players in the three broad categories of the MEFs examined in this and the previous two chapters are integrins; they are the main mechanoreceptors that link the cytoskeleton to the EMC. As outlined in Chapter 5, the largest of their three domains (ECM domain) binds to substrates, serving as both adhesive receptor and mechanotransducers (Katsumi et al., 2004). Through the receptor function, signals are transmitted across the membrane in response to binding to ECM ligands, resulting in the regulation of some if not all the MEF-mediated outcomes discussed. Integrins are able to do this because they bind binding proteins like paxillin, caveolin, and focal adhesion kinase through which they are able to recruit kinases to activate pathways that ultimately lead to phosphorylation (Iqbal and Zaidi, 2005). In addition to stimulation via integrins, cells are simultaneously subjected to long-range or soluble factors (e.g., cytokines) and fluid shear stresses via sensors that include GPCRs and stretch-sensitive ion channels (Discher et al., 2009).

Integrin-mediated adhesion has long been implicated in the activation of signal transduction pathways including the mitogen-activated protein kinase (MAPK) pathway. The MAPKs are a family of serine/threonine kinases that play an essential role in signal transduction by regulating gene expression in the nucleus in response to changes in the cellular environment. They include the extracellular signal-regulated protein kinases (ERK), c-Jun N-terminal kinases (JNK), p38s and ERK5. MAPKs require dual phosphorylation in two conserved threonine (Thr) and tyrosine (Tyr) residues within their activation loop for their full activation (Canagarajah et al., 1997). These phosphorylation sites are separated by only one amino acid; thereby defining a tripeptide motif Thr-X-Tyr and this amino acid X differentiates between the different groups of MAPKs. Upon activation, MAPKs phosphorylate and control the activity of key cytoplasmic molecules and nuclear proteins, which in turn can regulate gene expression (Turjanski et al., 2007).

Before fully discussing signaling interactions, consider a second emerging design principle. The "picture" of how the different regulators that are associated with the

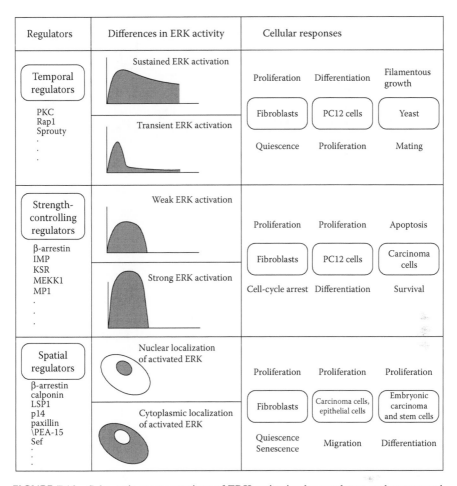

FIGURE 7.14 Schematic representations of ERK activation by regulators and corresponding cellular responses. (From Ebisuya, M., Kondoh, K. and Nishida, E. 2005. *J. Cell Sci.* 118: 3000. With permission.)

three different MEFs presented in this and the previous two chapters activate ERK, resulting in different cellular outcomes, has recently been "painted" by Ebisuya et al. (2005) by assembling results from different signaling studies in different cell types, as presented in Figure 7.14 (abbreviations used are explained in Box 7.1). The themes of ERK activation that seem to determine the nature of the cellular outcome includes sustained ERK, transient ERK, weak ERK, strong ERK, cytoplasmic-, and nuclear-localization. From a design principle point of view, depending on the cell type, it should be possible to create a microenvironment "understood" or characterized by the type of ERK activation with which to produce a desired cellular outcome. How to reliably control the microenvironment to achieve a desired ERK activation seems to be the missing link. Also, quantitative descriptions of sustained versus transient and weak versus strong need defining. It should be pointed out that almost all the

BOX 7.1   KEY TO MAPK PATHWAY
ABBREVIATIONS USED IN THIS CHAPTER*

B-Raf: A member of the Raf family of serine–threonine protein kinases of the mitogen-activated protein kinase kinase kinase (MAPKKK) type that play a key role in growth factor signaling; others family members include c-Raf-1 (MAPKKK) and A-Raf (MAPKKK)

Crk/C3G: Crk is an adaptor protein; C3G is a guanine nucleotide exchange factor for the Ras guanosine triphosphatase

ERK: Extracellular signal-regulated kinase

FAK/Src: FAK stands for focal adhesion kinase and Src is a family of non-receptor protein tyrosine kinases; c-Src is a member of the family

Fyn: Src-family kinases that associate with T-cell antigen receptor and phosphorylate a wide variety of intracellular signaling molecules

Grb2: Growth factor receptor-bound protein 2

MEK: MAPK/ERK kinase

P130CAS: Stands for 130 kDa Crk-associated substrate, also known as $p130^{CAS}$, an adaptor protein that is tyrosine phosphorylated and is involved in focal adhesion organization

PAK: Stands for p21-activated kinase; serine–threonine kinase activated by Rho-family GTPases

PKA: Stands for protein kinase A; also known as adenosine 3′,5′-monophosphate (cAMP)-dependent protein kinase; tetrameric serine–threonine protein kinase activated by binding of cAMP to the regulatory subunits

Rac: Small guanosine triphosphatase; member of the Rho subfamily of the Ras superfamily, which has been implicated in regulation of the cytoskeleton and cell motility

Raf: A family of serine–threonine protein kinases of the MAPKKK type that play a key role in growth factor signaling; see B-Raf

Rap1: Small guanosine triphosphatase; member of the Ras subfamily of the Ras superfamily, which has been implicated in cell growth and the regulation of gene expression

Ras: Small guanosine triphosphatase; the founding member of a superfamily of GTPases, which has been implicated in cell growth and the regulation of gene expression

RasGAP: Stands for Ras-GTPase activating protein

Shc: An adaptor protein involved in oncogenesis that contains both an Src homology 2 (SH2) domain and a phosphotyrosine-binding (PTB) domain

Yes: Members of the Src-family tyrosine kinases that are activated during the transition from G_2 Phase to M Phase of the cell cycle. It is highly homologous to proto-oncogene protein PP60(C-SRC)

* Obtained from http://stke.sciencemag.org/cgi/glossarylookup

studies that Ebisuya et al. (2005) relied on for the "picture" were conducted with 2D cultures. However, it should be noted that in a recent study, the 3D matrix has been shown to induce sustained activation of ERK via the Srk/Ras/Raf signaling pathway (Damianove et al., 2008). The study was conducted with NIH 3T3 fibroblasts cultured in cell-derived 3D matrices (Cukierman, 2002; Cukierman et al., 2001). Unfortunately, the cell density was maintained too low in comparison to typical microtissue cell populations. The cells were devoid of cell–cell interactions typical of most cell aggregates in 3D scaffolds, raising questions as to how the finding fits into "real" 3D cultures. Nevertheless, it is one of the first studies to report a 2D/3D comparison in terms of ERK activation.

A model of integrin signaling to ERK is shown in Figure 7.15 (Walker and Assoian, 2005; Schwartz and Ginsberg, 2002; Barberis et al., 2000). In this model, integrins stimulate the activation of ERK activity through multiple transduction pathways including FAK/Src, caveolin/Fyn, PAK, or Rac. These molecules regulate multiple points in the Ras-Raf-MEK-ERK pathway as shown. The cooperation or interaction with growth factor tyrosine kinases (Shc and Ras) is shown as an example

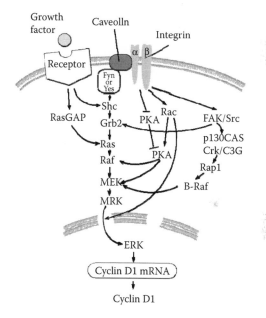

FIGURE 7.15   Integrin-dependent and MAPK pathways interaction regulating ERK activity. Growth factor receptors initially trigger recruitment of the Grb2/Sos complex to the plasma membrane, leading to sequential activation of Ras, Raf, MEK, and ERK, which translocates to the nucleus to phosphorylate Elk-1 and other transcription factors. Integrins stimulate the activation of ERK through multiple signal transduction pathways. For example, ERK can be activated through FAK/Src, caveolin, Fyn, PAK, or Rac and these molecules regulate multiple points in the Ras/Raf/MEK/ERK pathway. In addition to regulating the activation of ERK through PAK, Rac stimulates ERK translocation to the nucleus. (Adapted from Ebisuya, M., Kondoh, K., and Nishida, E. 2005. *J. Cell Sci.* 118: 2997–3002; Walker, J.L. and Assoian, R.K. 2005. *Cancer Metastasis Rev.* 24: 383–393; Schwartz, M.A. and Ginsberg, M.H. 2002. *Nat. Cell Biol.* 4: E65–E68.)

of how the long range (growth factor) and short range (ECM proteins like fibronectin binding to integrin) are integrated by the cell to come to an outcome like proliferation. The stimulation of ERK translocation to the nucleus by Rac is shown.

Cell anchorage to fibronectin, a process primarily mediated by integrins, resulted in 2–3-fold greater activation of MAPK by peptide mitogens as compared with suspended cells and the effects of anchorage on Raf-1 or Raf-B activation were much greater (8–20-fold) (Lin et al., 1997). Similar activation of MEK1 is also brought about by integrin-dependent PAKs. ERK can be activated by introducing active upstream components like Raf or MEK1 in suspended cells but still the signal is poorly relayed to the nucleus (Alpin et al., 2001). Accumulation of ERK in the cell and subsequent phosphorylation of Elk-1 (a transcription factor) are adhesion dependent. Even though activation of ERK occurs in the cytoplasm, it must translocate into the nucleus to exert many of its actions. Even when ERK is efficiently activated, integrins still regulate the cascade, as adhesion to the ECM is required for efficient accumulation of activated ERK in the nucleus (Alpin et al., 2001). The ERK target, Elk-1, and enhanced Elk-1 transactivation potential are more efficient in adherent cells (Alpin et al., 2001). Nuclear localization of ERK and phosphorylation of Elk-1 are also inhibited by the disruption of the actin cytoskeleton, further highlighting the importance of integrin–actin cytoskeleton connection (Alpin et al., 2001). Therefore, in suspended cells, the ERK pathway is only weakly activated and this may affect many of its downstream substrates, which play an important role in transcription. This may help explain the lack of sensitivity to stiffness by the nonadherent cells mentioned above. It is possible that the short-lived ERK1/2 in these cells is due to the absence of Src in their adhesion complexes or of the right integrin–ligand combination.

The general take-home message from the foregoing is that a better understanding of ERK should result in more robust design principles. Assuming that these principles are within reach, how would one know for sure that when these principles are applied, the *in vivo* niche is recapitulated? In most of the studies presented in this book, a single function or morphological feature not observed in 2D cultures has been used to suggest the merit of 3D cultures. But can the field of tissue engineering do better than this? For example, can the field borrow from how health sciences capture complex biological phenomenon through biomarkers? Is it possible to identify easily measurable ubiquitous biomarkers that would be common to cells based on the tissue of their origin? Success or failure in establishing robust design principles depends on answers to these questions, which are explored further in the next chapter.

REFERENCES

Alberts, B., Bary, D., Lewis, J., Raff, M., Roberts, K., and Watson, J.D. 2002. *Molecular Biology of the Cell*, 4th edition. New York: Garland Sciences.

Almeida-Silveira, M.I., Lambertz, D., Perot, C., and Goubel, F. 2000. Changes in stiffness induced by hindlimb suspension in rat Archilles tendon. *Eur. J. Appl. Physiol.* 81(3): 252–257.

Alpin, A.E., Stewart, S.A., Assoian, R.K., and Juliano, R.L. 2001. Integrin-mediated adhesion regulates ERK nuclear translocation and phosphorylation of Elk-1. *J. Cell Biol.* 153: 273–282.

Ananthakrishnan, R. and Ehrlicher, A. 2007. The forces behind cell movement. *Int. J. Biol. Sci.* 3: 303–317.

Anseth, K.S., Bowman, C.N., and Brannon-Peppas, L. 1996. Mechanical properties of hydrogels and their experimental determination. *Biomaterials* 17: 1647–1657.

Asanuma, K., Magid, R., Johnson, C., Nerem, R.M., and Galis, Z.S. 2003. Uniaxial strain upregulates matrix-degrading enzymes produced by human vascular smooth muscle cells. *Am. J. Physiol. Heart Circ. Physiol.* 284(5): H1778–H1784.

Aspernberg, P. 2007. Stimulation of tendon repair: Mechanical loading, GDFs platelets. A mini-review. *Int. Orthop.* 31: 783–789.

Baker, S.C., Atkin, N., Gunning, P.A., Granville, N., Wilson, K., Wilson, D., and Southgate, J. 2006. Characterization of electrospun polystyrene scaffolds for three-dimensional in vitro biological studies. *Biomaterials* 27: 3136–3146.

Balaban, N.Q., Schwarz, U.S., Riveline, D., et al. 2001. Force and focal adhesion assembly: A close relationship studied using elastic micropatterned substrates. *Nat. Cell Biol.* 3(5): 466–472.

Barberis, L., Wary, K.K., Fiucci, G., Liu, F., Hirsch, E., Brancaccio, M., Altruda, F., Tarone, G., and Giancotti, F. 2000. Distinct roles of adaptor proteins Shc and focal adhesion kinase in integrin signaling to ERK. *J. Biol. Chem.* 47(24): 36532–36540.

Barkhausen, T., van Griensven, M., Zeichen, J., and Bosch, U. 2003. Modulation of cell functions of human tendon fibroblasts by different repetitive cyclic mechanical stress patterns. *Exp. Toxicol. Pathol.* 55(2–3): 153–158.

Bilston, L.E. and Thibault, L.E. 1996. The mechanical properties of the human cervical spinal cord in vitro. *Ann. Biomed. Eng.* 24(1): 67–74.

Bishop, J.E., Butt, R., Dawes, K., and Laurent, G. 1998. Mechanical load enhances the stimulatory effect of PDGF on pulmonary artery fibroblast procollagen synthesis. *Chest* 114(1 Suppl): 25S.

Bonassar, L.J., Grodzinsky, A.J., Srinivasan, A., Davila, S.G., and Trippel, S.B. 2000. Mechanical and physicochemical regulation of the action of insulin-like growth factor-I on articular cartilage. *Arch. Biochem. Biophys.* 379(1): 57–63.

Brandl, F., Sommer, F., and Goepferich, A. 2007. Rational design of hydrogels for tissue engineering: Impact of physical factors on cell behavior. *Biomaterials* 28: 134–146.

Breen, E.C. 2000. Mechanical strain increases type I collagen expression in pulmonary fibroblasts *in vitro. J. Appl. Physiol.* 88(1): 203–209.

Butt, R.P. and Bishop, J.E. 1997. Mechanical load enhances the stimulatory effect of serum growth factors on cardiac fibroblast procollagen synthesis. *J. Mol. Cell Cardiol.* 29(4): 1141–1151.

Canagarajah, B.J., Khokhlatchev, A., Cobb, M.H., and Goldsmith, E.J. 1997. Activation mechanisms of the MAP kinase ERK2 by dual phosphorylation. *Cell* 90: 233–869.

Chen, A.A., Khetani, S.R., Lee, S., Bhatia, S.N., and Van Vliet, K.J. 2009. Modulation of hepatocyte phenotype *in vitro* via mechanical tuning of polyelectrolyte multilayers. *Biomaterials* 30: 1113–1120.

Constantinides, G., Kalcioglu, Z.I., McFarland, M., Smith, J.F., and Van Vliet, K.J. 2008. Probing mechanical properties of fully hydrated gels and biological tissues. *J. Biomech.* 41: 3285–3289.

Cukierman, E. 2002. Preparation of extracellular matrices produced by cultured fibroblasts. *Curr. Protocols Cell. Biol.* 10: 9.1–9.14.

Cukierman, E., Pankov, R., Stevens, D.R., and Yamada, K.M. 2001. Taking cell-matrix adhesion to the third dimension. *Science* 294(5547): 1708–1712.

Damianove, R., Stefanove, N., Cukierman, E., Momchilova, A., and Pankov, R. 2008. Three-dimensional matrix induces sustained activation of ERK1/2 via Src/Ras/Raf signaling pathway. *Cell Biol. Int.* 32: 229–234.

Discher, D., Dong, C., Fredberg, J.J., Guilak, F., Ingber, D., Janmey, P., Kamm, R.D., Schmid-Schönbein, G.W., and Weinbaum, S. 2009. Biomechanics: Cell research and applications for the next decade. *Ann. Biomed. Eng.* 37(5): 847–859.

Discher, D.E., Janmey, P., and Wang, Y.-L. 2005. Tissue cells feel and respond to the stiffness of their substrates. *Science* 310: 1139–1143.

Dulin´ska, I., Targosz, M., Strojny, W., Lekka, M., Czuba, P., Balwierz, W., and Szymon´ski, M. 2006. Stiffness of normal and pathological erythrocytes studies by means of atomic force microscopy. *J. Biochem. Biophys. Methods* 66: 1–11.

Durante, W., Liao, L., Reyna, S.V., Peyton, K.J., and Schafer, A.I. 2000. Physiological cyclic stretch directs L-arginine transport and metabolism to collagen synthesis in vascular smooth muscle. *FASEB Journal* 14(12): 1775–1783.

Ebisuya, M., Kondoh, K., and Nishida, E. 2005. The duration, magnitude and compartmentalization of ERK MAP kinase activity: Mechanisms of providing signaling specificity. *J. Cell Sci.* 118: 2997–3002.

Engler, A., Backkova, L., Newnman, C., Hategan, A., Griffin, M., and Discher, D. 2004b. Substrate compliance versus ligand density in cell on gel responses. *Biophys. J.* 86: 617–628.

Engler, A.J., Griffin, M.A., Sen, S., Bonemann, C.G., Sweeney, H.L., and Discher, D.E. 2004a. Myotubes differentiate optimally on substrates with tissue-like stiffness: pathological implications for soft or stiff microenvironments. *J. Cell Biol.* 166(6): 877–887.

Engler, A.J., Sen, S., Sweeney, H.L., and Discher, D.E. 2006. Matrix elasticity directs stem cell lineage specification. *Cell* 126(4): 677–689.

Evans, E., Leung, A., Heinrich, V., and Zhu, C. 2004. Mechanical switching and coupling between two dissociation pathways in a P-selectin adhesive bond. *Proc. Natl. Acad. Sci. USA* 101: 11281–11286.

Filford, R.J. and Bilston, L.E. 2005. The mechanical properties of rat spinal cord in vitro. *J. Biomech.* 38(7): 1509–1515.

Flanagan, L.A., Ju, Y.E., Marg, B., Osterfield, M., and Janmey, P.A. 2002. Neurite branching on deformable substrates. *Neuroreport* 13: 2411–2415.

Freed, L.E., Langer, R., Pellis, N.R., and Vunjak-Novakovic, G. 1997. Tissue engineering of cartilage in space. *Proc. Natl Acad. Sci. USA* 94(25): 13885–13890.

Gao, L., Parker, K.L., Lerner, R.M., and Lavinson, S.F. 1996. Imaging of the elastic properties of tissue—A review. *Ultrasound Med. Biol.* 22(8): 959–977.

Geiger, B., Bershadskey, A., Pankov, R., and Yamada, K.M. 2001. Transmembrane crosstalk between extracellular matrix–cytoskeleton crosstalk. *Nat. Rev. Mol. Cell. Biol.* 2: 793–805.

Genes, N.G., Rowley, J.A., Mooney, D.J., and Bonassar, L.J. 2004. Effect of substrate mechanics on chondrocyte adhesion to modify alginate surfaces. *Arch. Biochem. Biophys.* 422(2): 161–167.

Georges, P.C., Miller, W.J., Meaney, D.F., Sawyer, E.S., and Janmey, P.A. 2006. Matrices with compliance comparable to that of brain tissue select neuronal over glial growth in mixed cortical cultures. *Biophys. J.* 90: 3012–3018.

Ghosh, K., Pan, Z., Guan, E., Ge, S., Liu, Y., Nakamuro, T., Ren, X.-D., Rafailovich, M., and Clark, R.A.F. 2007. Cell adaptation to a physiologically relevant ECM mimics with different viscoelastic properties. *Biomaterials* 28: 671–679.

Gong, S., Wang, H. Sun, Q., Xue, S.-T., and Wang J.-Y. 2006. Mechanical properties and in vitro biocompatibility of porous zein scaffolds. *Biomaterials* 27: 3793–3799.

Goodman, M.B. and Schwarz, E.M. 2003. Transducing touch in *Caenorhabditis elegans*. *Annu. Rev. Physiol.* 65: 429–452.

Goodman, S.L., Risse, S.L., and von der Mark, K. 1989. The E8 subfragment of maminin promotes locomotion of myoblasts over extracellular matrix. *J. Cell Biol.* 109(2): 799–809.

Grimaud, E., Heyman, D., and Redini, F. 2002. Recent advances in TGF-β effects on chondrocytes metabolism. Potential therapeutic roles TGF-β in cartilage disorders. *Cytokines Growth Factor Rev.* 13: 241–257.

Guharay, F. and Sachs, F. 1984. Stretch-activated single ion channel currents in tissue-cultured embryonic chick skeletal muscle. *J. Physiol.* 352: 685–701.

Guo, B. and Guilford, W.H. 2006. Mechanics of actomyocin bonds in different nucleotide states are tuned to muscle contraction. *Proc. Natl. Acad. Sci. USA* 103(26): 9844–9849.

Guo, W. and Giancotti, F.G. 2004. Integrin signaling during tumor progression. *Nat. Rev. Mol. Cell Biol.* 5: 816–826.

Guo, X.M., Ren, X., Levin, E.R., and Kassab, G.S. 2006. Estrogen modulates the mechanical homeostasis of mouse arterial vessels through nitric oxide. *Am. J. Physiol. Heart Circ. Physiol.* 290: H1788–H1797.

Han, Y.P., Nien, Y.D., and Garner, W.L. 2002. Tumor necrosis factor-alpha-introduced proteolytic activation of pro-matrix metalloproteinases-9 by human skin is controlled by downregulating tissue inhibitor of metalloproteinase-1 and mediated by tissue-associated chymotrypsin-like proteinase. *J. Biol. Chem.* 277(30): 27319–27327.

Harrison, S.C. 2003. Variation on Src-like theme. *Cell* 112: 737–740.

Hartman, O., Zhang, C., Adams, E.L., Farach-Carson, M.C., Petrelli, N.J., Chase, B.C., and Rabolt, J.F. 2009. Microfabricated electrospun collagen membranes for 3-D cancer models and drug screening applications. *Biomacromolecules* 10: 2019–2032.

He, Y., Macarak, E.J., Korostoff, J.M., and Howard, P.S. 2004. Compression and tension: Differential effects on matrix accumulation by periodontal ligament fibroblasts *in vitro*. *Connect Tissue Res.* 45(1): 28–39.

Holmvall, K., Camper, L., Johansson, S., Kimura, J.H., and Lundgren-Akerlund, E. 1995. Chondrocyte and chondrosarcoma cell integrins with affinity for collagen type II and their response to mechanical stress. *Exp. Cell Res.* 221(2): 496–503.

Hoyt, K., Castaneda, B., Zhang, M., Nigwekar, P., di Sant'Agnese, P.A., Joseph, J.V., Strang, J., Rubens, D.J., and Parker, K.J. 2008. Tissue elasticity properties a biomarker for prostate cancer. *Cancer Biomarkers* 4: 213–225.

Iqbal, J. and Zaidi, M. 2005. Molecular regulation of mechanotransduction. *Biochem. Biophys. Res. Commun.* 328(3): 751–755.

Joki, N., Kaname, S., Hirakata, M., Hori, Y., Yamaguchi, T., Fujita, T., Katoh, T., and Kurokawa, K. 2000. Tyrosine-kinase dependent TGF-beta and extracellular matrix expression by mechanical stretch in vascular smooth muscle cells. *Hypertens Res.* 23(2): 91–99.

Kashiwagi, K., Mochizuki, Y., Yasunaga, Y., Ishida, O., Deie, M., and Ochi, M. 2004. Effects of transforming growth factor-beta 1 on early stages of healing of the Achilles tendon in a rat model. *Scand. J. Plast. Reconstr. Surg. Hand. Surg.* 38(4): 193–197.

Kaspar, D., Seidl, W., Neidlinger-Wilke, C., Ignatius, A., and Claes, L. 2000. Dynamic cell stretching increases human osteoblast proliferation and CICP synthesis but decreases osteocalcin synthesis and alkaline phosphatase activity. *J. Biomech.* 33(1): 45–51.

Katsumi, A., Orr, A.W., Tzima, E., and Schwartz, M.A. 2004. Integrins in mechanotransduction. *J. Biol. Chem.* 279(13): 12001–12004.

Kim, S.E., Cho, Y.W., Kang, E.J., Kwon, I.C., Lee, E.B., Kim, J.H., Chung, H., and Jeong S.Y. 2001. Three-dimensional porous collagen/chitosan complex sponge for tissue engineering. *Fibers and Polymers* 2(2): 64–70.

Kimura, Y. 2009. Microrheology of soft matter. *J. Phys. Soc. Jpn.* 78(4): 041005-1–041005-8.

Kung, C. 2005. A possible unifying principle for mechanosensation. *Nature* 436: 647–654.

Lai, Y., Cheng, K., and Kisaalita, W.S. 2010. Porous polystyrene scaffolds for three-dimensional cell-based assays. *Nat. Methods* (submitted).

Lee, A.A., Delhaas, T., McCulloch, A.D., and Villarreal, F.J. 1999. Differential responses of adult cardiac fibroblasts to *in vitro* biaxial strain patterns. *J. Mol. Cell Cardiol.* 31(10): 1833–1843.

Lee, R.T., Yamamoto, C., Feng, Y., Potter-Perigo, S., Briggs, W.H., Landschulz, K.T., Turi, T.G., Thompson, J.F., Libby, P., and Wight, T.N. 2001. Mechanical strain induces specific changes in the synthesis and organization of proteoglycans by vascular smooth muscle cells. *J. Biol. Chem.* 276(17): 13847–13851.

Lekka, M., Laidler, P., Gil, D., Lekki, J., Stachura, Z., and Hrynkiewcz, A.Z. 1999. Elasticity of normal cancerous human blabber cells studied by scanning force microscopy. *Eur. Biophys. J.* 28: 312–316.

Levental, I., Georges, P.C., and Janmey, P.A. 2007. Soft biological materials and their impact on cell function. *Soft Matter* 3: 299–306.

Li, Q., Muragaki, Y., Hatamura, I., Ueno, H., and Ooshima, A. 1998. Stretch-induced collagen synthesis in cultured smooth muscle cells from rabbit aortic media and a possible involvement of angiotensin II and transforming growth factor-beta. *J. Vasc. Res.* 35(2): 93–103.

Li, W.-J., Laurencin, C.T., Caterson, E.J., Tuan, R.S., and Ko, F.K. 2002. Electrospun nanofibrous structure: A novel scaffold for tissue engineering. *J. Biomed. Mater. Res.* 60: 613–621.

Lin, T.H., Chen, Q., Howe, A., and Juliano, R.L. 1997. Networks and crosstalk: Integrin signaling spreads. *Nat. Cell* 272(14): 8849–8852.

Linder-Ganz, E. and Gefen, A. 2004. Mechanical compression-induced pressure sores in rat hindlimb: Muscle stiffness, histology, and computational models. *J. Appl. Physiol.* 96(6): 2034–2049.

Little, W.C., Smith, M.L., Ebneter, U., and Vogel, V. 2008. Assay to mechanically tune and optically probe fibrillar fibronectin conformations from fully relaxed to breakage. *Matrix Biol.* 27(5): 451–461.

Lo, C.-M., Wang, H.-B., Dembo, M., and Wang, Y.-L. 2000. Cell movement is guided by rigidity of the substrate. *Biophys. J.* 79: 144–152.

Lyshchik, A., Higashi, T., Asato, R., Tanaka, S., Ito, J., Hiraoka, M., Brill, A.B., Saga, T., and Togashi, K. 2005. Elastic moduli of thyroid tissues under compression. *Ultrasound Imaging* 27(2): 101–110.

Madelin, G., Baril, N., De Certaines, D.J., Franconi, J.-M., and Thiaudière, E. 2004. NMR characterization of mechanical waves. In *Annual Reports on NMR Spectroscopy*, G.A. Webb (ed.), pp. 203–244. New York: Academic Press.

Manduca, A., Oliphant, T.E., Dresner, M.A., Mahowald, J.L., Kruse, S.S., Amromin, E., Felmlee, J.P., Greenleaf, J.F., and Ehman, R.L. 2001. Magnetic resonance elastography: Non-invasive mapping of tissue elasticity. *Med. Image Anal.* 5: 237–254.

Maroto, R., Raso, A., Wood, T.G., Kurosky, A., Martinac, B., and Hamill, O.P. 2005. TRPC1 forms the stretch-activated cation channel in vertebrate cells. *Nat. Cell Biol.* 7: 179–185.

Marshall, B.T., Long, M., Piper, J.W., Yago, T., McEver, R.P., and Zhu, C. 2003. Direct observation of catch bonds involving cell-adhesion molecules. *Nature* 423: 190–193.

Mason, T.G. 2000. Estimating the viscoelastic moduli of complex fluids using the generalized Stokes–Einstein equation. *Rheol. Acta* 39: 371–378.

Matter, M.L. and Ruoslahti, E. 2001. A signaling pathway from the $\alpha_5\beta_1$ and $\alpha5_v\beta_3$ integrins that elevated ncl-2 transcription. *J. Biol. Chem.* 276(30): 27757–27763.

Mauck, R.L., Nicoll, S.B., Seyhan, S.L., Ateshian, G.A., and Hung, C.T. 2003. Synergistic action of growth factors and dynamic loading for articular cartilage tissue engineering. *Tissue Eng.* 9(4): 597–611.

Miller, K., Chinzei, K., Orssengo, G., and Berdnarz, P. 2000. Mechanical properties of brain tissue in-vivo: Experiment and computer simulation. *J. Biomech.* 33(11): 1369–1376.

Miyaji, K., Furuse, A., Nakajima, J. Kohno, T., Ohtsuka, T., Yagyu, K., Oka, T., and Omata, S. 1997. *Cancer* 80(10): 1920–1925.

Murata, M., Bonassar, L.J., Wright, M., Mankin, H.J., and Towle, C.A. 2003. A role for the interleukin-1 receptor in the pathway linking static mechanical compression to decrease proteoglycan synthesis in surface articular cartilage. *Arch. Biochem. Biophys.* 413(2): 229–235.

Murayama, Y., Haruta, M., Katakeyama, Y., Shiina, T., Sakuma, H., Takenoshita, S., Omata, S., and Constantinou, C.E. 2008. Development of new instrument for examination of stiffness in the breast using haptic sensor technology. *Sens. Actuat. A—Phys.* 143(2): 430–438.

Nadav, L. and Katz, B.Z. 2001. The molecular effects of oncogenesis on cell-extracellular matrix adhesion (review). *Int. J. Oncol.* 19(2): 237–247.

Nasseri, S., Bilston, L.E., and Phan-Thien, N. 2002. Viscoelestic properties of pig kidney in shear, experimental results and modeling. *Rheol. Acta* 41(1-2): 180–192.

Oertl, A., Reija, B., Makarevic, J., Weich, E., Hofler, S., Jones, J., Jonas, D., Bratzke, H., Baer, P.C., and Blaheta, R.A. 2006. Altered expression of beta 1 integrins in renal carcinoma cell lines exposed to the differentiation inducer valproic acid. *Int. J. Mol. Med.* 18(2): 347–354.

Olah, L., Filipczak, K., Jaemermann, Z., Czigany, T., Borbas, L., Sosnowski, S., Ulanski, P., and Rosiak, J.M. 2006. Synthesis, structural and mechanical properties of porous polymeric scaffolds for bone tissue regeneration based on neat poly(epison-caprolactone) calcium carbonate. *Polym. Adv. Technol.* 17(11–12): 889–897.

Pai, A., Sundd, P., and Tees, F. 2008. *In situ* microrheological determination of neutrophils stiffening following adhesion in a model capillary. *Ann. Biomed. Eng.* 36(4): 596–603.

Palmeri, M.L., Wang, M.H., Dahl, J.J., Frinkley, K.D., and Nightingale, K.R. 2008. Quantifying hepatic shear modulus *in vivo* using acoustic radiation force. *Ultrasound Med. Biol.* 34(4): 546–558.

Pankov, R. and Yamada, K.M. 2002. Fibronectin at a glance. *J. Cell Sci.* 115: 3861–3863.

Papazoglou, S., Rump, J., Braun, J., and Sack, I. 2006. Shera wave group velocity inversion in MR elastography of human skeletal muscle. *Magn. Reson. Med.* 56: 489–497.

Parkkinen, J.J., Ikonen, J., Lammi, M.J., Laakkonen, J., Tammi, M., and Helminen, H.J. 1993. Effects of cyclic hydrostatic pressure on proteoglycan synthesis in cultured chondrocytes and articular cartilage explants. *Arch. Biochem. Biophys.* 300(1): 458–465.

Partap, S., Hebb, A.K., Rehman, I., and Darr, J.A. 2007. Formation of porous natural-synthetic polymer composites using emulsion templating and supercritical fluid assisted impregnation. *Polym. Bull.* 58: 849–860.

Paszek, M.J., Zahir, N., Johnson, K.R., Lakins, J.N., Rozenberg, G.I., Gefen, A., Reinhart-King, C.A., et al. 2005. Tensional homeostasis and malignant phenotype. *Cancer Cell* 8: 241–254.

Pelham, R.J. and Wang, Y.L. 1997. Cell locomotion and focal adhesion are regulated by substrate flexibility. *Proc. Natl. Acad. Sci. USA* 94(25): 13661–13665.

Penderson, S.A. and Nilius, B. 2007. Transient receptor potential channels in mechanosensing and cell volume regulation. *Methods Enzymol.* 428: 183–207.

Peyton, S.R. and Putnam, A.J. 2005. Extracellular matrix rigidity governs smooth muscle cell motility in biphasic fashion. *J. Cell. Physiol.* 204(1):198–209.

Peyton, S.R., Kim, P.D., Chajar, C.M., Seliktar, D., and Putnam, A.J. 2008. The effects of matrix stiffness and RhoA on the phenotypic plasticity of smooth muscle cells in 3-D biosynthetic hydrogel system. *Biomaterials* 29: 2597–2607.

Peyton, S.R., Raub, C.B., Keshchrumrus, V.P., and Putnam, A.J. 2006. The use of poly(ethylene glycol) hydrogels to investigate the impact of ECM chemistry and mechanics on smooth muscle cells. *Biomaterials* 27(28): 4881–4893.

Radisic, M., Park, H., Shing, H., Consi, T., Schoen, F.J., Langer, R., Freed, L.E., and Vunjak-Novakovic, G. 2004. Functional assembly of engineered myocardium by electrical stimulation of cardiac myocytes cultured on scaffolds. *PNAS* 101(52): 18129–18134.

Raeber, G.P., Lutolf, M.P., and Hubbell, J.A. 2007. Mechanisms of 3-D migration and matrix remodeling of fibroblasts within artificial ECMs. *Acta Biomat.* 3: 615–629.

Roberts, A.B. and Sporn, M.B. 1993. Physiological actions and clinical applications of transforming growth factor-β (TGF-β). *Growth Factors* 8: 1–9.

Saha, K., Keung, A.J., Irwin, E.F., Li, Y., Little, L., and Scheffer, D.V. 2008. Substrate modulus directs neural stem cell behavior. *Biophys. J.* 95: 4426–2238.

Schwartz, M.A. and Ginsberg, M.H. 2002. Networks and crosstalk: Integrin signaling spreads. *Nat. Cell Biol.* 4: E65–E68.

Semler, E.J., Lancin, P.A., Dasguputa, A., and Moghe, V.P. 2005. Engineering hepatocellular morphogenesis and function via ligand-presenting hydrogels with graded mechanical compliance. *Biotechnol. Bioeng.* 89(3): 296–307.

Shapiro, L. and Cohen, S. 1997. Novel alginate sponges for cell culture and transplantation. *Biomaterials* 18: 583–590.

Smith, M.L., Gourdon, D., Little, W.C., Kubow, K.E., Enguiluz, R.A., Luna-Morris, S., and Vogel, V. 2007. Force-induced unfolding of fibronectin in the extracellular matrix of living cells. *PLOS Biol.* 5(10): 2243–2254.

Sneddon, I.N. 1965. The relation between load and penetration in axisymmetric Boussinesq problem for a punch of arbitrary profile. *Int. J. Eng. Sci.* 3: 47–57.

Solomon, M.J. and Lu, Q. 2001. Rheology and dynamics of particles in viscoelastic media. *Curr. Opin. Colloid Interface Sci.* 6(5–6): 430–437.

Solon, J., Levental, I., Sengupta, K., Georges, P., and Janmey, P.A. 2007. Fibroblasts adaptation and stiffness to soft substrates. *Biophys. J.* 93: 4453–4461.

Subramanian, A. and Lin, H.Y. 2005. Crosslinked chitosan: Its physical properties and the effects of matrix stiffness on chondrocytes cell morphology and proliferation. *J. Biomed. Mater. Res. Part A* 75A(3): 742–753.

Sukharev, S. and Corey, D.P. 2004. Mechanosensitive channels: Multiplicity of families and gating paradigms. *Science's STKE* 219: 1–24.

Tan, J.L., Tien, J., Pirone, D.M., Gray, D.S., Bhadriraju, K., and Chen, C.S. 2003. Cells lying on a bed of microneedles: An approach to isolate mechanical force. *Proc. Natl Acad. Sci. USA* 100(4): 1484–1489.

Thomas, W.E. 2008. Catch bonds in adhesions. *Annu. Rev. Biomed. Eng.* 10: 39–57.

Thomas, W.E. 2009. Mechanochemistry of receptor–ligand bonds. *Curr. Opin. Struct. Biol.* 19: 50–55.

Thomas, W.E., Trintchina, E., Forero, M., Vogel, V., and Sokurenko, E. 2002. Bacterial adhesion to target cells enhanced by shear-force. *Cell* 109: 913–923.

Thompson, M.T., Berg, M.C., Tobias, I.S., Licheter, J.A., Rubner, M.F., and Van Vliet, K.J. 2005. Tunning compliance of nanoscale polyelectrolyte multilayers to modulate cell adhesion. *Biomaterials* 26(34): 6835–6845.

Thompson, M.T., Berg, M.C., Tobias, I.S., Licheter, J.A., Rubner, M.F., and Van Vliet, K.J. 2006. Biochemical functionalization of polymeric cell substrata can alter mechanical compliance. *Biomacromolecules* 7(6): 1990–1995.

Turjanski, A.G., Vasqué, J.P., and Gutkind, J.S. 2007. MAP kinases and the control of nuclear events. *Oncogene* 26: 3240–3253.

Valegol, D. and Lanni, F. 2001. Cell traction force on soft biomaterials. I Microrheology of type I collagen gels. *Biophys. J.* 81: 1786–1792.

Vanderhooft, J.L., Alcoutlabi, M., Magda, J.J., and Prestwich, G.D. 2009. Rheological properties of cross-linked hyaluronan-gelatin hydrogels for tissue engineering. *Macromol. Biosci.* 9: 20–28.

Vaquette, C., Frochot, C., Rahouadj, R., Muller, S., and Wang, X. 2008. Mechanical and biological characterization of a porous poly-L-lactic acid-co-epsilon-caprolactone scaffold for tissue engineering. *Soft Mater.* 6(1): 25–33.

Venkatesh, S.K., Yin, M., Glocner, J.F., Takahashi, N., Araoz, P.A., Talwalkar, J.A., and Ehman, R. 2008. MR elestography of liver tumors: Preliminary results. *AJR* 190: 1534–1540.

Vogel, V. 2006. Mechanotransduction involving multimodular proteins: Converting force into biochemical signals. *Annu. Rev. Biophys. Biomol. Struct.* 35: 459–488.

Vogel, V. and Sheetz, M. 2006. Local force and geometry sensing regulate cell functions. *Nat. Rev. Mol. Cell Biol.* 7: 265–275.

von Offenberg Sweeney, N., Cummins, P.M., Birney, Y.A., Cullen, J.P., Redmond, E.M., and Cahill, P.A. 2004. Cyclic strain-mediated regulation of endothelial matrix metallo-proteinase-2 expression and activity. *Cardiovasc. Res.* 63(4): 625–634.

Waigh, T.A. 2005. Microrheology of complex fluids. *Rep. Prog. Phys.* 68(3): 685–724.

Walker, J.L. and Assoian, R.K. 2005. Integrin-dependent signal transduction regulating cyclin D1 expression and G1 phase cell cycle progression. *Cancer Metastasis Rev.* 24: 383–393.

Walton, E.B., Oommen, B., and Van Vliet K.J. 2007. How stiff and thin can engineered extra-cellular matrix be? Modeling molecular forces at the cell-matrix interface. *Conf. Proc. IEEE Eng. Med. Biol. Soc.* 2007: 6419–6421.

Wang, H.B., Dembo, M., and Wang, Y.L. 2000. Substrate flexibility regulates growth and apoptosis of normal but not transformed cells. *Am. J. Physiol. Cell. Physiol.* 279: C1345–C1350.

Wang, J.H.-C. and Thampatty, B.P. 2006. An introductory review of cell mechanobiology. *Biomechan. Model. Mechanobiol.* 5: 1–16.

Wang, L. and Kisaalita, W.S. 2010. Characterization of micropatterned nano-fibrous scaffolds for neural network activity readout for high-throughput screening. *J. Biomed. Mater. Res. Part B – Applied Biomaterials* (accepted).

Wellman, P., Howe, R., Dalton, E., and Kern, K. 1999. *Harvard Biorobotics Laboratory Technical Report*, Harvard University.

Wells, R. 2008. The role of matrix stiffness in regulating cell behavior. *Hepatology* 47(4): 1394–1400.

Wolfe, S.L. 1993. *Molecular and Cellular Biology*. Belmont, CA: Wadsworth Publishing Company.

Woodell, J.E., LaBerge, M., Langan, E.M., 3rd, and Hilderman, R.H. 2003. *In vitro* strain-induced endothelial cell dysfunction determined by DNA synthesis. *Proc. Inst. Mech. Eng. H* 217(1): 13–20.

Wu, Y.M., Cazorla, O., Labeit, D., Labeit, S., and Granzier, H. 2000. Changes in titin and col-lagen underlie diastolic stiffness diversity of cardiac muscle. *J. Mol. Cell. Cardiol.* 32(12): 2151–2161.

Yang, G., Crawford, R.C., and Wang, J.H. 2004. Proliferation and collagen production of human patellar tendon fibroblasts in response to cyclic uniaxial stretching in serum-free conditions. *J. Biomech.* 37(10): 1543–1550.

Yeh, W.C., Li, P.C., Jeng, Y.M., Hsu, H.C., Kuo, P.L., Li, M.L., Yang, P.M., and Lee, P.H. 2002. Elastic modulus measurements of human liver and correlation with pathology. *Ultrasound Med. Biol.* 28(4): 467–474.

Yeung, T., Georges, P.C., Flanagan, L.A., Marg, B., Ortz, M., Funaki, M., Zahir, N., Ming, W., Weaver, V., and Janmey, P.A. 2005. Effects of substrate stiffness on cell morphology, cytoskeletal structure, and adhesion. *Cell Motil. Cytoskeleton* 60: 24–34.

Yuan, H.C., Kononov, S., Cavalcante, F.S.A., Lutchen, K.R., Ingenito, E.P., and Suki, B. 2000. Effects of collagenase and elastase on the mechanical properties of lung tissue strips. *J. Appl. Physiol.* 89(1): 3–14.

Zaari, N., Rajagopalan, P., Kim, S.K., Engler, A.J., and Wong, J.Y. 2004. Photopolymerization in microfluidic gradient generators: Microscale control of substrate compliance to manipulate cell response. *Adv. Mater.* 16(23–24): 2133–2137.

Zaman, M.H., Kamm, R.D., Matsudaira, P., and Lauffenburger, D.A. 2005. Computational model for cell migration in three-dimensional matrices. *Biophys. J.* 89(2): 1389–1397.

Zhang, J., Zhang, H., Wu, L., Ding, J. 2006. Fabrication of three dimensional polymeric scaf-folds with spherical pores. *J. Mater. Sci.* 41: 1725–1731.

8 Proteomics as a Promising Tool in the Search for 3D Biomarkers

8.1 WHY SEARCH FOR THREE-DIMENSIONALITY BIOMARKERS?

In the concluding remarks section of Chapter 3, one of the intended outcomes of the data presented was providing "evidence in support of the hypothesis that ubiquitous biomarkers for three dimensionality exist and can be found." In this chapter, the question of biomarkers is further explored in more detail; it is highly relevant to the establishment of design principles as suggested in the concluding remarks section of Chapter 7. The dilemma with writing this chapter was that the overall "story" of this book is incomplete without a major piece on biomarkers, yet there is hardly any published work on the topic that can provide the foundation for a full chapter. What does one do? While pondering on this question, the author was reminded of a couple of well-written papers on biomarkers for diseases like Alzheimer's and Parkinson's, for which the challenge was similar—despite the vast number of studies, the diagnosis (and for that matter treatment) is unsatisfactory. The authors of such papers tend to focus on the need for biomarkers for diagnosis and the best tools and/or methodology to find them. An excellent example of one of these papers is a review of Parkinson's disease research in the context of neuroproteomics as a promising tool by Pienaar and Daniels (2008). This paper and others like it provided the inspiration for the approach used in writing this chapter. As the chapter title suggests, and to the best of the author's knowledge, proteomics tools offer great promise in the search for three-dimensionality biomarkers. Although Chapter 3 identified specific promising areas to search, the treatment here is kept general for the benefit of those who might have different ideas of where to search.

Before proceeding, it is necessary to first answer the question—why look for three-dimensionality biomarkers. One of the two compelling reasons has been articulated in the previous chapter. Briefly, apart from the concept of "three-dimensional matrix adhesion," originally proposed by Cukierman et al. (2001) as a possible indication or "diagnosis" or marker for a culture state of three dimensionality, the field of tissue engineering has not provided knowledge on the basis of which a consensus for three dimensionality and the associated CPR could be established. As a result, claims of "physiologically more relevant" are readily made for cells cultured on any surface or scaffold that provides loosely defined 3D geometry, either at the nano- or microstructure levels or their combinations, as long as the resulting cell phenotypes are different between the 2D and 3D geometries. Second, the concept of using combinatorial approaches to fabricate libraries of polymers or other material scaffolds (Simon et al.,

2007; Yang et al., 2008b) for tissue engineering or cell-based drug discovery call for a high-throughput assay by which "hit materials" can be quickly identified for further development. The development of such assays or biosensors can potentially be guided by a cell–material interaction outcome (Yang et al., 2008a). An interaction with a material that yields cells that emulate *in vivo* conditions would be most desirable. Three-dimensionality biomarkers would provide the intellectual basis for material discovery platform development. Figure 8.1 portrays an example of a conceptual framework in which a scaffold material discovery can be conducted in HTS (Yang et al., 2008a). As shown in Figure 8.1a and b, poly(desaminotyrosyl-tyrosine ethyl ester carbonate) (pDTEc) and poly(desaminotyrosyl-tryrosine octyl ester carbonate) (pDTOc) share a structurally identical backbone, but different side chains (ethyl of pDTEc and octyl of pDTOc). Because of this, these two polymers exhibit different properties, which resulted in differences in cells cultured on films (2D) of the polymers and their blends. Cells on pDTEc exhibited enhanced cell spreading, adhesion, and proliferation in comparison to pDTOc (Bailey et al., 2006; Ertel and Kohn, 1994). If a balance between proliferation and differentiation is desired, then an optimal polymer blend that meets the need may exist and may easily be found through HTS with a polymer blend combinatorial scaffold library. The library can be generated in a fluid handling instrument represented by a two-syringe pump system in Figure 8.1c, which would yield arrays of porogen-leached scaffolds of varying polymer compositions (Figure 8.1d). In the proof-of-concept study of the system shown in Figure 8.1, Yang et al. (2008a, 2008b) used Fourier transformed infrared (FTIR) spectroscopy to verify

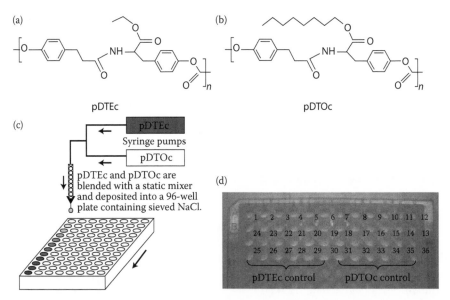

FIGURE 8.1 Conceptual framework for HTS scaffold material discovery. (a) and (b) Chemical structures of pDTEc and pDTOc, respectively. (c) Schematic of combinatorial fabrication of pPDTEc-pDTOc polymer blend scaffold library. (d) Porogen-leached and freeze-dried sample of the scaffold library in a 96-well plate. (From Yang, Y. et al. 2008a. *Adv. Mater.* 20: 2037–2043. With permission.)

scaffold polymer compositions. Extending such a study to the question of how well the polymer blends support the "three dimensionalitiness" of the cells cultured within, calls for biomarkers measurable in HTS readouts.

Taken together, the subfield or field of 3D culture needs validated biomarkers. The "elephant in the room" is whether such a thing exists. Based on material presented in this book, a case can be made for at least to proceed with the search. Relevant work done that can guide the first steps have focused on cell adhesion (Cukierman et al., 2001) and the related signaling pathways (Damianova et al., 2008), further considered in the following two subsections.

8.2 CELLULAR ADHESIONS

Cell–ECM adhesions have been well established—based primarily on *in vitro* studies on 2D surfaces and in particular with fibroblasts and epithelial cells (Adams, 2001; Sastry and Burridge, 2000). Different types of adhesions were initially observed almost three decades ago to be present within a single cell at any one time (Izzard and Lochner, 1980). On 2D surfaces *in vitro*, cell contacts can occur in three different forms: focal complexes (FC), focal adhesion (FA), and fibrillar (FB) adhesions. Figures 8.2 and 8.3 show a confocal-like image and protein composition cartoons, respectively, of the three predominant types of adhesion found in adherent cells plated on flat substrates. As cells move over a surface, they "feel" or sense the environment by extending and retracting filopodia that are endowed with a high concentration of

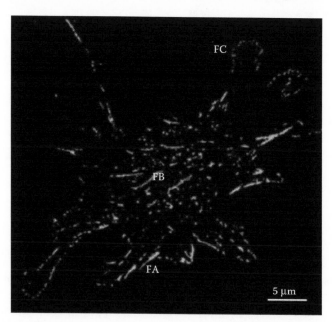

FIGURE 8.2   Image depicting the three types of adhesions predominantly found in adherent cells plated on 2D substrates; focal complex (FC), focal adhesion (FA), and fibrillar adhesion (FB). (Adapted from Worth, D.C. and Parsons, M. 2008. *Int. J. Biochem. Cell Biol.* 40: 2397–2409.)

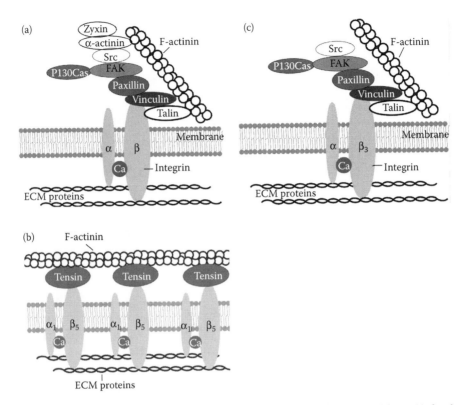

FIGURE 8.3 Cartoons of cell–ECM adhesions including protein compositions: (a) focal adhesion (FA), (b) fibrillar adhesion (FB), and (c) focal complex (FC). (Adapted from Worth, D.C. and Parsons, M. 2008. *Int. J. Biochem. Cell Biol.* 40: 2397–2409.)

integrin receptors. FCs are involved in "feeling" the environment before more stable contacts are established. They are small, transient structures, which are usually seen immediately behind the leading edge of spreading or migrating cells, supporting nascent filopodial growth and lamellipodial actin networks (Worth and Parsons, 2008). On the contrary, FAs, located at the base of the cell are considered more mature FCs. They are larger, as shown in Figure 8.2. FAs contain more actin-binding proteins responsible for providing mechanical stability by strong adhesion to ECM protein (e.g., vitronectin) using integrins such as $\alpha_v\beta_3$. FA enables tractional forces to be transmitted from the cell to the ECM and vice versa (Yamada et al., 2003). The integrins organize intracellular adhesion complexes composed of proteins like focal adhesion kinase (FAK), paxillin, tensin, kinectin, talin, as well as classical signaling molecules such as Src and ERK. FB adhesions are long; highly stable complexes that run parallel to bundles of fibronectin. The dynamic movement of the $\alpha_5\beta_1$ integrin along actin microfilament bundles from the cell periphery toward the cell center is hypothesized to stretch the fibronectin, exposing cryptic sites for assembling the linear fibronectin fibril (Yamada et al., 2003). The $\alpha_5\beta_1$-organized intracellular complex is rich in tensin (Pankov et al., 2000). Tensin is a 220-kDa protein that is concentrated at FAs and contains three distinct actin-binding sites (Lo et al., 1994).

The term "3D-matrix adhesion" was coined by Cukierman et al. (2001), based on the observation that there was another type of adhesion that differed from FA and FB adhesions as known from 2D studies. The difference was primarily in the content of $\alpha_5\beta_1$ and $\alpha_v\beta_3$ integrins, paxillin, tyrosine phosphorylation of FAK, and other cytoskeletal components. A more complete summary of the differences is provided in Table 8.1. In 3D scaffolds, Cukierman et al. (2001) found that paxillin and α_5 integrin colocalized to unusual cell–matrix adhesions, parallel to fibronectin-containing extracellular fibers as opposed to localizing separately to focal and fibrillar adhesions, similar to colocalization defining matrix adhesion *in vivo*. It was speculated that the FA and FB adhesion studies *in vitro* represented exaggerated precursors on *in vivo* 3D-matrix adhesion. When cell preparations are stained with dye-conjugated antibodies against phosphorylated FAK at tyrosine 397 (FAK PY397), "3D-matrix adhesion" is characterized by diffuse, less well-defined streaky patterns (punctate) as opposed to the well-defined long fibrillar patterns found with 2D cell preparations. Figure 8.4 shows the difference between "3D-matrix adhesion" and the classic 2D adhesion. 2D adhesion shows a streaky pattern, while 3D shows a more punctated and less well-defined pattern of FAK PY397 labeling.

In the absence of biomarkers for three dimensionality, several investigators have used FAK PY397 staining to ascertain "3D-matrix adhesion" for cultures on either nano-/microstructured surfaces or scaffolds to infer the physiological relevance of the structure. For example, Schindler et al. (2005) found fibroblasts cultured on a synthetic nanofiber scaffold (UltraWeb™) to exhibit similar punctate patterns as opposed to well-defined streaky patterns and the marketing literature for the product makes claims of "more physiological relevance." Berry et al. (2005) reported that fibroblasts cultured on nanotubes exhibited punctate actin throughout the cell body,

TABLE 8.1
Molecular Composition of Cell Adhesions[a]

Compound	Focal Adhesion	Fibrillar Adhesion	3D-Matrix Adhesion
A$_5$ integrin	Localized only at the periphery of adhesions	Present but different	Present but different
β_1 integrin	Present	Present	Present
β_3 integrin	Present	Deficient	Deficient
Paxillin	Present	Deficient	Present
Tensin	Present	Present	Present
Talin	Present	Present	Present
Vinculin	Present	Deficient	Present
Phosphorylated tyrosine	Present	Deficient	Present
FAK	Present	Deficient	Present
Tyrosine 397-phosphorylated FAK (FAK PY397)	Present	Deficient	Deficient

[a] FC is not added to this table because it is considered an early stage of FA.

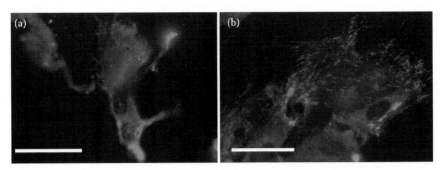

FIGURE 8.4 Immunofluorescence images of fibroblasts (SCRC-1041) stained for FAK PY397 after 2 days in culture on a glass substrate (b) and in a fibrous (nano)-large pores (micro) combined PLLA scaffolds (a). The cell staining on glass was streaky, while that on PLLA scaffolds was more punctate and less well defined. [Adapted from Cheng, K. and Kisaalita, W.S. 2010. *Biotechnol. Prog.* (accepted, DOI: 101002/btpr.391).]

while cells in the microtube controls exhibit dash-shaped adhesions throughout the cell, reflecting their well-spread morphology compared to the small adhesions and smaller sized morphology on nanotubes. In more recent studies, Yim et al. (2007) and Patel et al. (2007) have reported the induction of "3D-matrix adhesion" in addition to promoting neural stem cell differentiation by nanofibrous surfaces. While FAK PY397 staining seems to work well with some cell types, experience in some laboratories has been mixed. In the author's own laboratory (Cheng and Kisaalita 2008, 2010), stem neural progenitor cells were stained in parallel with fibroblasts, both cultured on PLLA scaffolds. The pattern for the progenitor cells was not distinct "3D-matrix adhesion." This probably explains the limited use of the approach despite the demand for ascertaining "three-dimensionalitiness" of nano-/microstructured surfaces and scaffolds applications. Additionally, the microscopy nature of the procedure renders it low throughput and not attractive in HTS applications as envisioned in Figure 8.1, for example.

8.3   SIGNALING PATHWAYS

The relevance of the model of integrin signaling to FAK shown in Figure 7.15 to cellular outcome differences observed has been confirmed in a recent study by Damianova et al. (2008). As pointed out in the concluding remarks section of Chapter 7, this study reported 3D matrix-induced activation of ERK via the Src/Ras/ Raf signaling pathway that was sustained. The studies were conducted with NIH 3T3 fibroblasts cultured in cell-derived 3D matrices. As shown in Figure 7.14, sustained ERK activation in 2D results in proliferation and differentiation in fibroblasts and PC12 cells, respectively, underscoring the importance of cell type as one considers searching for biomarkers. Although the cell density used in the Damianova et al. study was low and not typical of cell aggregates found in 3D cultures, the results suggest a powerful signaling basis in support of the existence of biomarkers. The search for three-dimensionality biomarkers is therefore justifiable. In the following subsection, a promising proteomics approach for discovering these biomarkers is presented in general terms.

8.4 OVERVIEW OF PROTEOMICS TECHNIQUES

The concept and definition of proteomics were introduced in Chapter 3 and will not be repeated here. The general steps in a proteomics strategy suitable for three-dimensionality biomarker study are presented in Figure 8.5. The common or central technologies or procedures employed in the many possible techniques and analysis tools are the following: some form of protein separation procedure, techniques for peptide detection, and protein identification or bioinformatics.

8.4.1 Protein Separation by Two-Dimensional Polyacrylamide Gel Electrophoresis (2DE)

Sample preparation for 2DE typically involves lysing the cells, solubilizing the protein in commercial solutions containing chaotropes, detergents, reducing agents, carrier ampholytes and protease inhibitors (Rabilloud and Chevallet, 2000). In 2DE preparation, proteins are first separated according to their charge (first dimension) using isoelectric focusing (Görg et al., 1985). An isoelectric point is the pH at which the net charge on a protein is zero and, as such, the protein does not migrate if subjected to an electric field. The gels are fabricated in such a way that they exhibit a pH gradient and as the proteins migrate, their movement stops when they reach their isoelectric point. It is possible to use gels with different pH ranges, for example, narrow ranges (5–6, 4–7) and a wide range (3–10) (Simpson, 2003). Proteins are then separated orthogonally (second dimension) according to their molecular weight in gels containing sodium dodecyl sulfate (SDS). SDS imparts a negative charge enabling the separation according to mass. The 2DE approach enables the resolution of variants (isoforms) of the same protein. According to Klose (1975), it is possible

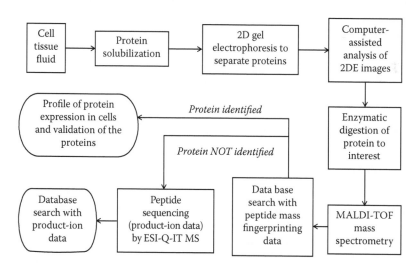

FIGURE 8.5 A flow diagram for a typical proteomics study with two-dimensional gel electrophoresis (2DE), MS, and database searches tailored to three-dimensionality biomarkers study.

to resolve up to 10,000 proteins in a single gel. A variety of chemical stains and fluorescence probes are available to quantitatively visualize the proteins (Tokarski et al., 2006; Smejkal et al., 2004; Mackintosh et al., 2003; Neuhoff et al., 1988; Merril et al., 1981). As traditionally done, 2DE comparative studies as would be the case in 2D/3D/*in vivo* study can be complicated, especially for poorly expressed proteins. In such cases, samples can be labeled with different fluorescent dyes (e.g., Cy3 and Cy5) followed by separation on the same gel. This approach has been shown to improve the dynamic range while reducing technical variability (Wu, 2006).

8.4.2 PEPTIDE DETECTION

After enzymatic digestion of excised protein spots, the mass spectrometry (MS) instrumentations of choice are the matrix-assisted laser desorption/ionization time-of-flight mass spectrometry (MALDI-TOF-MS) and the electrospray ionization-quadruple ion trap mass spectrometry (ESI-Q-IT-MS). MALDI-TOF-MS is sensitive (fmol peptide levels), accurate (better than 50 ppm), and fast (3–5 min per sample). However, it is not suitable for sequencing peptides in proteolytic digestions. One way of solving this problem is to derivatize the peptides prior to analysis (Keough et al., 1999). MALDI-TOF-MS yields peptide-mass fingerprints, while ESI-Q-IT-MS yields product-ion spectra (peptide sequence data). Typically ESI-Q-IT-MS is directly coupled to liquid chromatography, which initially separates the peptide mixtures. Although ESI-Q-IT-MS is slow (1 h per analysis), as shown in Figure 8.4, it provides quality product-ion data that facilitate the identification of proteins that cannot be identified with mass fingerprint data. A popular strategy in many proteomics laboratories is to use a two-tiered protein identification approach where the fast MALDI-TOF-MS is used first and for the proteins that are not conclusively identified on the basis of peptide-mass fingerprint data, a more time-consuming ESI-Q-IT-MS is used to generate the product-ion data (Beranova-Giorgianni, 2003). An alternative ionization method is the surface-enhanced laser desorption/ionization (SELDI) that uses protein array technology, for example, IMAC30 ProteinChip (Escher et al., 2007), used in combination with TOF-MS. Unfortunately SELDI only detects signals but does not identify proteins or peptides and is therefore not suitable for the biomarker study proposed herein. For more detailed additional techniques such as the ICAT (isotope coded affinity tag) and iTRAQ (isobaric tagging for relative and absolute protein quantification), see Pienaar and Daniels (2008).

8.4.3 PROTEIN IDENTIFICATION

Proteins of interest are identified by searching protein-sequence databases with peptide-mass fingerprints, product-ion data, or peptide sequence tags as inputs. Database search programs usually come with the spectrometer software package. The software compares the experimentally generated data with the theoretical data calculated for each protein in the database. The search usually returns a list of closely matching candidates. There are several publicly available databases. An excellent example is the SWISSPROT database (www.ebi.ac.uk/uniprot/ or www.isb-isb.ch) maintained by the Swiss Institute of Bioinformatics and the European Bioinformatics Institute.

8.5 STUDY DESIGN AND METHODS

The key factors that have guided the strategy proposed herein are (1) low-abundance proteins that are not easily detected in conventional 2D gel electrophoresis may be important, (2) membrane proteins with poor solubility in standard protein extraction solutions may be important, and (3) sample availability is not a limiting factor since cells will be either cultured or freshly dissected from animals. The basic conditions that can guide sample preparation are provided in Box 8.1. Desirable protein for potential biomarkers should be those that are differentially expressed as protein expression studies described in Chapter 3. In executing the experiment, precautions or additional steps are needed to address the low-abundance proteins and membrane proteins with poor solubility potential problems.

8.5.1 ADDRESSING LOW-ABUNDANCE AND POOR SOLUBILITY PROTEINS

Reduction of high-abundance proteins that may obscure the less abundant and probably more interesting proteins can be achieved in several ways. Some investigators use affinity separation such as microbeads with covalently conjugated antibodies (Ahmed and Rice, 2005; Ahmed, 2009). This is critical in blood-based proteomics to remove proteins like albumin and fibronectin. It is also critical in cultured cell

---

BOX 8.1 BASIC EXPERIMENTAL CONDITIONS

Cell types sources: Fibroblast, nerve, and muscle; for each source consider three types of cell line, stem cell derived, and primary cells. The proposed choices here have been previously justified (Chapter 4). But a different substitute is acceptable, provided a corresponding *in vivo* counterpart is practically obtainable. The cell in 3D microenvironment should be able to exhibit measurable CPR as defined in Chapter 4. This is particularly important when it comes to validating the three-dimensionality biomarker.

In vivo surrogate: Freshly dissected cells from corresponding sources identified above.

2D control: Coated (e.g., poly-ornithine and laminin) multiwell glass or polystyrene plates.

Low-aspect-ratio (nano-/microstructured) 3D surfaces: Coated with the same biochemical as with 2D plates. If possible, use the same base material as used with 2D control with Young's moduli in physiological range.

High-aspect-ratio microstructured 3D scaffolds and spheroids: Spheroids and scaffolds with Young's moduli in physiological range, where possible, coated with the same biochemical(s) as with 2D plates. If possible, use the same material as used in 2D control.

Cell fate and temporal effects: Use proliferation and differentiation media to study the cell fate effect on biomarkers and with differentiation media study before and after differentiation for cell lines and stem cells; effect of time in culture for primary cells may be a good way of studying temporal effects.

proteomics to remove serum protein components as well as substrate coatings like collagen. Another approach is an alternative to 2DE of liquid chromatography coupled to MS. The separated proteins are directly fed into the ESI-Q-IT-MS. The various separation techniques (affinity, size exclusion, reverse phase, and charge chromatography available in liquid chromatography) make it easier to separate and detect species that are missed in 2DE due to low-abundance levels (Washburn et al., 2002; Wolters et al., 2001).

8.5.2 BIOMARKER VALIDATION

The expression profile of identified proteins should be established followed by validation studies. A combination of experimental approaches is advised as usually no one approach is without fault when it comes to validation. Additionally, a combination of two or more experimental approaches is likely to provide new insights into the meaning of three dimensionality *in vitro*. One approach can utilize small-interfering RNA (siRNA) (Colombo and Moll, 2008). Another approach can utilize small molecule antagonists (Zhou et al., 2006; Gosh et al., 2006; Dayam et al., 2006). A third approach can utilize protein-activating antibodies delivered on microbeads into the microenvironment (Carpenedo et al., 2009). Detailed validation approach decisions as well as potential pitfalls will depend on the suspected nature of the protein biomarkers. For example, let us assume the differentially expressed proteins are several integrins and their expression profiles have been confirmed in several cell types and scaffold geometries, exhibiting a required CPR function. In such a case the application of general validation approaches proposed above for integrins can face potential pitfalls. The addition of siRNA or small molecules into cultures will provide a knockdown, or reduction of function. The siRNA and the small molecule inhibitors may not completely abolish the presence of integrins, and they will allow for continued integrin-mediated cell–cell interactions; however, it is expected that the desired results can be obtained in such a system without complete gene knockout. The addition of these small RNA constructs that are complementary to the integrin messenger RNA (mRNA) will be able to limit translation. The binding of the siRNA to the integrin mRNA will activate a mechanism similar to the cellular defensive mechanism for double-stranded RNA degradation. This process is continual and the products from the first degradation mediate future inactivation of additional transcripts. While this is a very effective system, it is not capable of a complete knockout of the gene of interest. If there is an inability to knockdown integrin function, multiple small molecule inhibitors of integrins are in clinical trials or commercially available as drugs in addition to small molecules being discovered and described in the literature. While there are limited combinations of small molecules available, their use can still provide vital information. If the use of siRNA or the addition of small molecules does not disrupt the ability of the cells to aggregate and does not alter protein production, the further experiments need not be preformed. However, if the aggregates dissociate, the third approach experiments should be performed.

Microbeads range in size between 20 and 75 μm. Uniform incorporation of microbeads in embryoid bodies from human embryonic stem cells has been described (Carpenedo et al., 2009). Although in this case the microspheres were made out of PLGA and the purpose was slow release of morphogens, there is no reason why the same procedures cannot be followed with polystyrene microbeads functionalized with monoclonal antibodies. In further experiments, the expression of identified biomarker(s) could be increased or knocked. If the results from this experiment show enhanced CPR, it will provide clear evidence in support of the integrins as three-dimensionality biomarkers. If the results are negative or ambiguous, it will still be possible that integrins are true biomarkers, but the current experiment is insufficient in proving so. While the use of siRNA will most likely provide the most reliable data, transfections are far from perfect. The ability to use small molecules to perform a similar experiment allows for comparable results to be obtained with a different experimental procedure. However, as stated before, it is possible to miss the validation of the integrins due to cell aggregate dissociation should this occur.

8.6 CONCLUDING REMARKS

To successfully accomplish the task in a timely manner requires the participation of multiple laboratories with skills in culturing the cell types of interest. This is one of the reasons for writing this chapter in the present format—to interest 3D culture investigators in coming together to find 3D biomarkers. However, there is a need for coordinating these efforts, especially as far as MS experiments go. As Robert Serve pointed on in the News Focus section of the September 26, 2008, issue of *Science*, a study conducted in 2007 (p. 1760) "… compared the ability of 87 different labs to use MS to identify correctly 12 different proteins spiked into *E. coli* sample. No lab got them all, and only one correctly identified 10 of the 12." In a coordinated study, it might be helpful to accomplish all the MS experimentation at a single reputable facility.

These experiments could be conducted in such a manner that more data will be generated for inclusion in Figure 6.8, facilitating further the establishment of design principles in the field as explained in Chapter 6. The absence of a few common genes in the transcription and proteomics studies reviewed in Chapter 3 would on the surface raise questions about the existence of ubiquitous biomarkers for three dimensionality. But on the other hand, the few 2D/3D comparative proteomics studies available to date and reviewed in Chapter 3 did not address the important issues of poor solubility and low-abundance proteins.

REFERENCES

Adams, J.C. 2001. Cell matrix contact structures. *Cell. Mol. Life Sci.* 58(3): 371–392.

Ahmed, E. 2009. Sample preparation and fractionation for proteome analysis and cancer biomarker discovery by mass spectrometry. *J. Sep. Sci.* 32: 771–798.

Ahmed, N. and Rice, G.E. 2005. Strategies for revealing lower abundance proteins in two-dimensional protein maps. *J. Chromatogr. B—Anal. Technol. Biomed. Life Sci.* 815(1–2): 39–50.

Bailey, L.O., Becker, M.L., Stephen, J.S., Gallant, N.D., Mahoney, C.M., Washburn, N.R., Rege, A. and Amis, E.J. 2006. Cellular response to phase-separated blends of tyrosine-derived polycarbonates. *J. Biomed. Mater. Res. Part A* 76A(3): 491–502.

Beranova-Giorgianni, S. 2003. Proteome analysis by two-dimensional gel electrophoresis and mass spectrometry: Strengths and limitations. *Trends Anal. Chem.* 22(5): 273–281.

Berry, C.C, Dalby, M.J., McCloy, D., and Affrossman, S. 2005. The fibroblast response to tubes exhibiting internal nanotopography. *Biomaterials* 26: 4985–4992.

Carpenedo, R.L., Bratt-Leal, A.M., Marklein, R.A., Seaman, S.R., Bowen, N.J., McDonald, J.F., and McDevitt, T.C. 2009. Homogeneous and organized differentiation with embryoid bodies induced my microsphere-mediated delivery of small molecules. *Biomaterials* 30(13): 2507–2515.

Cheng, K. and Kisaalita, W.S. 2010. Exploring cellular adhesion and differentiation in micro-/nano-hybrid polymer scaffolds. *Biotechnol. Prog.* (accepted).

Cheng, K., Lai, Y., and Kisaalita, W.S. 2008. Three-dimensional polymer scaffolds for high throughput cell-based assay systems. *Biomaterials* 29: 2808–2812.

Colombo, R. and Moll, J. 2008. Target validation to biomarker development. *Mol. Diag. Ther.* 12(2): 63–70.

Cukierman, E., Pankov, R., Stevens, D.R., and Yamada, K.M. 2001. Taking cell-matrix adhesions to the third dimension. *Science* 294: 1708–1712.

Damianova, R., Stefanova, N., Cukierman, E., Momchilova, A., and Pankov, R. Three-dimensional matrix induces sustained activation of ERK1/2 via Src/Ras/Raf signaling pathway. *Cell Biol. Int.* 32: 229–234.

Dayam, N., Aiello, F., Deng, J.X., Wu, Y., Garofalo, A., Chen, X.Y., and Neamati, N. 2006. Discovery of small molecule integrin alpha(v)beta(3) antagonists as novel anticancer agents. *J. Med. Chem.* 49(15): 4526–4543.

Ertel, S.I. and Kohn, J. 1994. Evaluation of tyrosine-derived polycarbonates as degradable biomaterials. *J. Biomed. Mater. Res.* 28: 919–930.

Escher, N., Kaatz, M., Melle, M., Hilper, C., Ziemer, M., Driesch, D., Wollina, U., and von Eggeling, F. 2007. Posttranslational modifications of transthyretin are serum markers in patients with mycosis fugiodes. *Neoplasia* 9: 254–259.

Görg, A., Postel, W., Gunter, S., and Weser, J. 1985. Improved horizontal two-dimensional electrophoresis with hybrid isoelectric-focusing in immobilized pH gradients in the 1st-dimension and laying-on transfer to the 2nd-dimension. *Electrophoresis* 6(12): 599–604.

Gosh, S., Elder, A., Guo, J.P., Mani, U., Patane, M., Sarson, K., Ye, Q., et al. 2006. Design, synthesis, and progress towards optimization of potent small molecule antagonists of CC chemokine receptor 8 (CCR8). *J. Med. Chem.* 49(9): 2669–2672.

Izzard, C.S. and Lochner, L.R. 1980. Formation of cell-to-substrate contacts during fibroblast motility: An interference-reflection study. *J. Cell Sci.* 42: 81–116.

Keough, T., Youngquist, R.S., and Lacey, M.P. 1999. A method for high-sensitivity peptide sequencing using postsource decay matrix-assisted laser desorption ionization mass spectrometry. *Proc. Natl. Acad. Sci. USA* 96: 7131–7136.

Klose, J. 1975. Protein mapping by combined isoelectric focusing and electrophoresis of mouse tissue. A novel approach to testing for induced point mutations in mammals. *Humangenetik* 26: 231–243.

Lo, S.-H., Janmey, P.A., Hartwig, J.H., and Chen, L.-B. 1994. Interaction of tensin with actin and identification of its three distinct actin-binding domains. *J. Cell Biol.* 125: 1067–1075.

Mackintosh, J.A., Choi, H.-Y., Bae, S.-H., Veal, D.A., Bell, P.J., Ferrari, B.C., Verrills, N.M., Paik, Y.K., and Karuso, P. 2003. A fluorescence natural product for ultra sensitive detection of proteins in one-dimensional and two-dimensional gel electrophoresis. *Proteomics* 3: 2273–2288.

Merril, C.R., Goldman, D., Sedman, S.A., and Ebert, M.H. 1981. Ultrasensitive stain for proteins in polyacrylamide gels shows regional variations in cerebrospinal-fluid proteins. *Science* 211: 1437–1438.

Neuhoff, V., Arold, N., Taube, D., and Ehrhardt, W. 1988. Improved staining of proteins in polyacrylamide gels including isoelectric-focusing gels with clear background at nanogram sensitivity using Coomassie brilliant blue G-250 and R-250. *Electrophoresis* 9(6): 255–262.

Pankov, R., Cukierman, E., Katz, B.Z., Matsumoto, K., Lin, D.C., Lin, S., Hahn, C., and Yamada, K.M. 2000. Integrin dynamics and matrix assembly: Tensin-dependent translocation of alpha (5)beta(1) integrins promotes early fibronectin fibrogenesis. *J.Cell Biol.* 148(5): 1075–1090.

Patel, S., Kurpinski, K., Quigley, R., Gao, H., Hsiao, B.S., Poo, M.-M., and Li, S. 2007. Bioactive nanofibers: Synergistic effects of nanotopography and chemical signaling on cell guidance. *Nano Lett.* 7(7): 212–2128.

Pienaar, H.S. and Daniels, W.M.U. 2008. Neuroproteomics as a promising toll in Parkinson's disease research. *J. Neural Transm.* 115: 1413–1430.

Rabilloud, T. and Chevallet, M. 2000. Solubilization of protein in 2D electrophoresis. In *Proteome Research: Two-Dimensional Gel Electrophoresis and Identification Methods*, T. Rabilloud (ed.), pp. 9–29, Berlin, Germany: Springer.

Sastry, S.K. and Burridge, K. 2000. Focal adhesions: A nexus for intracellular signaling and cytoskeletal dynamics. *Exp. Cell Res.* 261: 25–36.

Schindler, M., Ahmed, I., Kamal, J., Nur-E-Kamal, A., Grafe, TH., Chung, H.Y., and Meiners, S. 2005. A synthetic nanofibrillar matrix promotes *in vivo*-like organization and morphogenesis for cells in culture. *Biomaterials* 26: 5624–5631.

Simon, Jr., C.G., Stephens, J.S., Dorsey, S.M., and Becker, M.L. 2007. Fabrication of combinatorial polymer scaffold libraries. *Rev. Sci. Instrum.* 78: 072207-1–072207-7.

Simpson, R.J. 2003. *Purifying Proteins for Proteomics: A Laboratory Manual*. New York: Cold Spring Harbor Laboratory Press.

Smejkal, G.B., Robinson, M.H., and Lazarev, A. 2004. Comparison of fluorescence stains: Relative photostability and differential staining of proteins in two-dimensional gels. *Electrophoresis* 25: 2511–2519.

Tokarski, C., Cren-Olivé, C., Fillet, M., and Rolando, C. 2006. High-sensitivity staining of proteins for one- and two-dimensional gel electrophoresis using post migration covalent staining with a ruthenium fluorophore. *Electrophoresis* 27: 1407–1416.

Washburn, M.P., Ulaszek, R., Deciu, C., Schieltz, D.M., and Yates, J.R. 2002. Analysis of quantitative proteomic data generated via multidimensional protein identification technology. *Anal. Chem.* 74: 1650–1657.

Wolters, D.A., Washburn, M.P., and Yates, J.P. 2001. An automated multidimensional protein identification technology for shortgun proteomics. *Anal. Chem.* 73: 5683–5690.

Worth, D.C. and Parsons, M. 2008. Adhesion dynamics: Mechanisms and measurements. *Int. J. Biochem. Cell Biol.* 40: 2397–2409.

Wu, T.L. 2006. Two-dimensional difference gel electrophoresis. *Methods Mol. Biol.* 328: 71–95.

Yamada, K.M., Pankov, R., and Cukierman, E. 2003. Dimensions and dynamics in integrin function. *Braz. J. Med. Biol. Res.* 36: 959–966.

Yang, Y., Bolikal, D., Becker, M.L., Kohn, J., Zeiger, D.N., and Simon, C.G. 2008a. Combinatorial polymer scaffold libraries for screening cell-biomaterial interactions in 3D. *Adv. Mater.* 20: 2037–2043.

Yang, Y., Dorsey, S.M., Becker, M.L., Lin-Gibson, S., Schumacher, G.E., Flaim, G.M., Kohn, J., and Simon, C.G. 2008b. X-ray imaging optimization of 3D tissue engineering scaffolds via combinatorial fabrication methods. *Biomaterials* 29: 1901–1911.

Yim, E.K.F., Pang, S.W., and Leong, K.W. 2007. Synthetic nanostructures inducing differentiation of human mesenchymal stem cells into neuronal lineage. *Exp. Cell Res.* 313: 1820–1829.

Zhou, Y., Peng, H., Ji, G., Zhu, Z.P., and Yang, C.Z. 2006. Discovery of small molecule inhibitors of integrin alpha v beta 3 structure-based virtual screening. *Bioorg. Med. Chem. Lett.* 16(22): 5878–5882.

9 Readout Present and Near Future

9.1 READOUT PRESENT AND NEAR FUTURE

The need to screen more compounds in a given time is serving as a driving force for HTS laboratories to use high-density multiwell plates (e.g., 1536- or 3456-well plates) as well as liquid handling and plate reader formats that can support them. The 1536-well plate has become the industry workhorse and the feasibility of cell-based assays in these plates has been reported in the literature (Table 9.1). Two key observations can be made from the information presented in Table 9.1. First, the popularity of receptors and enzymes as targets is consistent with the two being the most common target classes for current drug therapies (Drews, 2000). Kinases have emerged as one of the most important target classes for drug discovery and development; this is because, as with GPCR targets, of their crucial role in cellular signaling (Cohen, 2002). Second, the most prevalent readout signal types are luminescence and fluorescence. Additional confirmation of this can be found from recent surveys of HTS Laboratory Directors (HighTech Business Decisions, 2007), where the top two readout technologies for both 2006 and 2008 surveys were time-resolved fluorescence resonance energy transfer (TR-FRET) and glow luminescence. In the same survey, the top planned purchase detection technology was label-free.

The possibility that in cell-based assays, labels may not only affect cell viability, but may also interfere with intracellular pathways (Stein, 2008) has created interest in implementing label-free technologies in HTS laboratories (Lee et al., 2008). The main advantage of label-free approaches is that the technology enables visualization of real-time cellular events in their true complexity, as opposed to endpoint measurements typical of assays with labels. Commercial platforms for measuring cell-based functional responses compatible with HTS formats are based on electrical impedance and refractive index (Galmann and Hansen, 2006; Fang et al., 2006; Minor, 2008) readout.

The purpose of this chapter is to provide an introductory understanding of the theory behind the most common readout technologies currently encountered or likely to be encountered in the near future in HTS laboratories. Any kit or instruments supplier's goal is to make it very easy for the client to implement the assay. This ease on the other hand may encompass a danger—inadequate understanding of the underlying principles and pitfalls, which may lead to all sorts of misinterpretations, errors, and frustrations. In the worst scenario, the technology may be perceived by the scientist to be "unreliable" or to be "producing unacceptable false negatives." It is in this spirit that this author found it fitting to provide an introductory theoretical background with the belief that the level of coverage will be more accessible and will

TABLE 9.1

Cell-Based Assay Types Implemented in High-Density Multiwell Plates Organized on the Basis of Detection Time or Temporal Resolution

Detection Timescale	Assay Type and/or Primary Signal	Secondary or Readout Signal Type	Biosensor/Assay Implementation	Typical Target Classes with References
Hours (<24 h)	Reporter gene; transcription regulation	Luminescence; fluorescence	Luciferase, β-lactamase as reporters	GPCRs (Hill et al., 2001; Chin et al., 2003; Cacace et al., 2003, Bovolenta et al., 2007) and other receptors (Zlokarnik et al., 1998; Golz, 2008)
Minutes to hours	Enzyme fragment complementation; protein–protein interaction; second messenger	Luminescence; fluorescence	Luciferase, β-galactosidase, or GFP	GPCR (Eglen, 2002; Weber et al., 2004) and tyrosine kinases (Issad et al., 2007)
Minutes	Second messenger accumulation; cAMP; IP_3 (IP_1)	Time-resolved fluorescence (TRF)	Homogeneous TRF with LANCE™	GPCR (Cacace et al., 2003; Reinscheid et al., 2003; Trinquet et al., 2006)
Seconds to minutes	Protein–protein interaction; protein conformation	Fluorescence resonance energy transfer (FRET)	GFP-derived fluorophores	GPCR (Milligan, 2004), tyrosine kinase (Issad et al., 2007), phosphodiesterase (PDE) (Herget et al., 2008), and guanylyl cyclase (Bader et al., 2001)

Seconds to minutes	Translocation	BRET	Renilla luciferase (RLUC) as donor and GFP as acceptor	GPCR (Milligan, 2004; Hamdan et al., 2005) and tyrosine kinase (Terrillon and Bouvier, 2004)
Seconds	pH	Fluorescence	pH-sensitive organic dyes (e.g., BCECF and SNARF)	Transporter (Simpson et al., 2000; Galietta et al., 2001)
Seconds	Membrane potential	Fluorescence	Voltage-sensitive organic dyes [e.g., $DiBAC_4(3)$] with background fluorescence masking	Transporters and ion channel (Gonzalez et al., 1999)
Seconds	Calcium release; calcium influx	Fluorescence; luminescence	Calcium-sensitive organic dyes (e.g., Fluo-4) with background fluorescence masking	GPCR (Chambers et al., 2003; Hodder et al., 2004) and calcium channel (Gonzalez et al., 1999; Cacace et al., 2003)
Seconds	cAMP, cGMP	Fluorescence; luminescence	Cyclic nucleotide-gated channels activation and monitoring of calcium via calcium indicator or photoprotein	GPCR (Cacace et al., 2003; Reinscheid et al., 2003) and PDE (Wunder et al., 2005)

Source: Adapted from Wunder, F. et al. 2008. *Comb. Chem. High Throughput Screen.*, 11: 497.

open the gate to seeking more advanced understanding elsewhere (considered beyond the scope of this book). For this reason, the coverage is restricted to fluorescence, luminescence, and label-free biosensing (impedance and resonant waveguide principles). Sample biosensor configurations of commercial interest are provided for illustrative purposes. There are textbooks or journal articles solely focusing on these readout principles. For the interested reader the following references may be useful: fluorescence and luminescence (Gadella, 2009; Albani, 2007; Inglese et al., 2007; Pfleger and Eidne, 2006; Pfleger et al., 2006; Lakowicz, 1999) and label-free biosensing (Cooper, 2009; Fang et al., 2006; Lo et al., 1995).

9.2 FLUORESCENCE-BASED READOUTS

9.2.1 JABLONSKI DIAGRAM AND FLUORESCENCE BASICS

Absorption of energy in the form of electromagnetic radiation by a population of molecules induces the passage of electrons from the singlet (an electronic state of a molecule that occurs when its spin state corresponds to a total spin of zero ground electronic level S_0 to an excited state S_n). Excited molecules return to the ground state S_0 via two successive steps (internal conversion and fluorescence/phosphorescence) depicted in the Jablonski diagram (Figure 9.1). The molecule at S_n may return to the lowest excited state S_1 via the internal conversion mechanism (dissipating a part of its energy in the surrounding environment). There are many competitive mechanisms

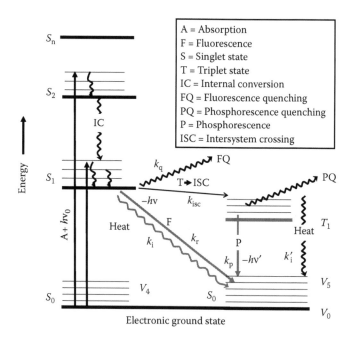

FIGURE 9.1 Jablonski (electronic transition), depicting the ground and excited states.

by which a molecule may reach the ground state S_0 from the excited state S_1, including (1) fluorescence (emission of a photon with a radiative rate constant k_r), (2) nonradiative heat dissipation (part of the absorbed energy may be dissipated in the medium as heat at a rate constant k_i), (3) energy release to nearby molecules at rate constants k_q (collisional quenching) and k_t (energy transfer at a distance to a donor—not shown in Figure 9.1), and (4) transient passage to the excited triplet [an electronic state of a molecule that occurs when its spin angular momentum quantum number S is equal to state one T_1 of energy lower than S_1 at a rate constant of k_{isc} (intersystem crossing)].

Excited molecules from the triplet state return to the ground state via competitive mechanisms similar to those of the excited singlet state: (1) phosphorescence emission of a photon with a radiative rate constant of k_p, (2) non-adiative heat dissipations at a rate constant of k_i', and (3) energy release to nearby molecules at rate constants k_q' (collision) and k_t' (energy transfer at a distance).

A chromophore that emits a photon is called a fluorophore. Many chromophores do not necessarily fluoresce. In such cases, the energy absorbed is dissipated within the environment as thermal or collisional energy and/or transfer to other molecules. Fluorescence intensity (I) is linearly related to the concentration of the fluorophore up to an absorption range of approximately 0.02–0.05 as follows:

$$I = QI_0(1 - 10^{\varepsilon cl}),\tag{9.1}$$

where Q is the fluorescence quantum yield, I_0 is the light source intensity, ε is the molar absorptivity, c is the fluorophore concentration, and l is the path length. The quantum yields (Q) and observed lifetime (τ) are dependent on summed rate constants for radiative (k_r) and nonradiative (k_i) processes as follows:

$$Q = \left(\frac{k_r}{k_i + k_r} \right),\tag{9.2}$$

that is, quanta emitted divided by quanta absorbed:

$$\tau = Q\tau_n,\tag{9.3}$$

where τ_n is the nonradiative process lifetime. The fluorescence lifetime observed is related to the average time spent in the excited state. Short lifetimes are associated with prompt fluorescence (FLINT) in the nanosecond range; however, some fluorophores such as Ln^{3+} ions exhibit long lifetimes in the millisecond ranges due to luminescence emission from the triplet state.

9.2.2 Fluorescence Readout Configurations

Although the use of nonfluorescence-based readouts in HTS is on the rise, currently the majority of commercial HTS assays are still using fluorescence-based readouts. This is mainly due to the sensitivity of fluorescence measurements, which now extend to the single molecule level, routinely. Additionally, the availability of different forms of fluorescence outputs, such as brightness, lifetime, anisotropy, and energy transfer, enables the construction of readouts designed to overcome undesirable interferences, that do not require separations step (true biosensors), and that have intrinsically higher information content (Pope et al., 1999; Heilker et al., 2009).

Fluorescence readout has been configured in commercial HTS systems in various formats (Pope et al., 1999; Gribbon and Sewing, 2003), including (1) prompt fluorescence intensity (FLINT), (2) fluorescence polarization/anisotropy (FP/FA), (3) time-resolved fluorescence (TRF), (4) fluorescence lifetime (FL), and (5) Förster (or fluorescence) resonance energy transfer (FRET). To illustrate these configurations, sample examples applicable in cell-based HTS formats are provided below.

For FLINT readouts, signals are changes in steady-state total light output. In a noncell-based example, Du et al. (2004) used FLINT in HTS format to identify novel small-molecule inhibitors for human transketolase (TK). TK is a metabolic enzyme that plays a crucial role in tumor cell nucleic acid synthesis, using glucose through the elevated nonoxidative pentose phosphate pathway (PPP). Effective anticancer therapeutic agents can hypothetically be developed by identifying inhibitors specifically targeting TK and preventing the nonoxidative PPP from generating RNA ribose precursor, ribose-5-phosphate. A purified human TK, cloned and expressed in *Escherichia coli*, was used in the assay reaction scheme shown in Figure 9.2, conducted in 384-well plates in 50 μL. FLINT was recorded as counts per second on an Analyst reader (Molecular Devices, Sunnyvale, CA). Fifty-four initial hits were identified from a library of 60,320 compounds.

FP and FA describe the same process with the exception that the units of measurement are different. In FP and FA, a sample is illuminated with light that is linearly polarized through an excitation polarizer as shown in Figure 9.3. The fluorescence intensity is measured at an angle relative to the excitation illumination through an

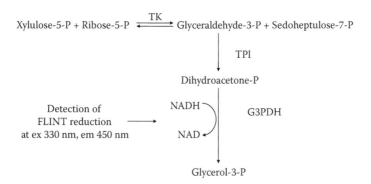

FIGURE 9.2 Reaction schematic of assay involving the transketolase (TK), coupled to reactions catalyzed by triose phosphate isomerase (TPI), and glyceraldehyde-3-phosphate (GDH3P). TK enzyme activity was proportional to FLINT (excitation at 330 nm and emission at 450 nm) reduction due to the resultant oxidation of NADH.

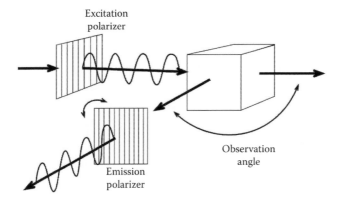

FIGURE 9.3 Schematic of optical path between the polarizers and sample for polarization and anisotropy readouts. In HTS, the observation angle is at or near 180° with the excitation beam directed from above.

emission polarizer that is configured to pass fluorescence that is polarized either parallel or perpendicular to the exciting illumination. The polarization signal (*P*) is defined as

$$P = \frac{(I_{par} - I_{per})}{(I_{par} + I_{per})},$$

(9.4)

where I_{par} and I_{per} are the fluorescence signals from the parallel and perpendicular channels of the readout instrument, respectively. The corresponding anisotropy values (*A*) are typically used in HTS applications and the relationship between *P* and *A* is as follows:

$$A = \frac{2P}{(3 - P)}.$$

(9.5)

A is a unitless value and is commonly represented as millianisotropy (mA). In multiple fluorophore mixtures, the anisoptropy is determined by the ratio of each fluorescent component in a linear manner as follows:

$$A = \varphi_1 A_1 + \varphi_2 A_2 + \cdots + \varphi_i A_i,$$

(9.6)

where φ represents the fractional fluorescence intensity of the corresponding fluorophore.

Owicki (2000) reviewed homogeneous anisotropy assay designs used in HTS and identified eight modes shown in Box 9.1. The Owicki review identified one report using intact cells for a cell-surface receptor application (Keating et al., 2000). A more recent literature search has hardly yielded cell-based biosensing configurations. Kashem et al. (2007) has recently screened a library of 10,208

BOX 9.1   FLUORESCENCE POLARIZATION/ ANISOTROPY ASSAY CONFIGURATIONS

1. Direct binding assay:

   F-ligand + Receptor $\leftrightarrow$ F-ligand:Receptor

2. Competitive binding assay:

   F-ligand + Receptor $\leftrightarrow$ F-ligand:Receptor

   I + Receptor $\leftrightarrow$ I:Receptor

3. Enzyme assay (direct immunodetection):

   F-Substr $\xrightarrow{\text{Enz}}$ F-Prod

   F-Prod + Ab $\leftrightarrow$ F-Prod:Ab or

   F-Substr +Ab $\leftrightarrow$ F-Substr:Ab

4. Enzyme assay (competitive immunodetection):

   Substr $\xrightarrow{\text{Enz}}$ Prod

   Prod + Ab $\leftrightarrow$ Prod:Ab

   F-Prod + Ab $\leftrightarrow$ F-Prod:Ab or

   Competition between Substr and F-Substr for Ab

5. Enzyme assay (transferase):

   F-SmallSubstr + LargeSubstr $\xrightarrow{\text{Enz}}$ F-SmallSubstr-LargeSubstr

6. Protease assay (size reduction):

   Protein-F$_n$ $\xrightarrow{\text{Enz}}$ nPeptide-F

7. Protease assay (detection by avidin binding):

   F-Oligopep1-Oligopep2-Biotin $\xrightarrow{\text{Enz}}$ F-Oligopep1 + Oligopep2-Biotin

   F-Oligopep1-Oligopep2-Biotin + Avidin $\rightarrow$ F-Oligopep1-Oligopep2-Biotin:Avidin

   F-Oligopep1 + Oligopep2-Biotin + Avidin $\rightarrow$ F-Oligopep1 + Oligopep2-Biotin:Avidin

8. Incorporation of labeled nucleotide triphosphate:

   F-NTP + Primer + Template $\xrightarrow{\text{Enz}}$ F-Oligonucleotide

Source: From Owicki, J.C. 2000. *J. Biomol. Screen.* 5(5): 297–306. With permission.

compounds to identify inhibitors of the nonreceptor protein kinase from the Tec family, interleukin-2-inducible T cell kinase (ITK). A competitive binding configuration was used with a fluorescently labeled small-molecule inhibitor probe on the Analyst in fluorescence polarization mode (excitation 485 nm, emission 530 nm). The FP of the probe is low and increases significantly on binding the ATP-binding pocket of the ITK. In the presence of a competitive inhibitor, the FP of the probe decreases. Together with two other assay types based on enzyme catalytic activity, the most comprehensive set of ITK inhibitors was identified (Kashem et al. 2007). Owing to the recent success of Imatinib Mesylate (Gleveec®) for chronic myeloid leukemia, Gefitinib (Iressa®) and Erlotinib (Tarceva®) for non-small-cell lung cancer, the search for effectors of protein kinases (as drug targets) is of great interest.

In TRF, the differences in emissions decay, as depicted by Equation 9.7, are taken advantage of. The decay of an excited state population of a fluorophore is a Poisson process. For a single emitting species, the probability of observing a photon at time t after the fluorophore is excited follows an exponential decay that fits the following relationship:

$$I = I_0 e^{-t/\tau}, \tag{9.7}$$

where I and I_0 are fluorescence intensities at time t and $t = 0$, respectively. The fluorescence lifetime is governed by the radiative and nonradiative decay rates of the excited state, k_r and k_i, respectively. For example, Ln^{3+} exhibits long-lived emission and can be monitored at a fixed time (e.g., 200 µs) after flash illumination, well after FLINT from other fluorescent molecules as well as other background signals (e.g., light scatter) have decayed. Implementation of TRF in HTS has been mainly accomplished in conjunction with FRET (Sharif et al., 2009; Engels et al., 2009; D'Souza et al., 2008; Schroeter et al., 2008).

To use "fluorescence" for the F in the FRET acronym has been rightly pointed out by Clegg (2009) to be a misnomer, although the usage is very common in the literature. The use of Förster makes it clear that the Förster mechanism is valid for the conditions in question. FRET involves the nonradiative energy transfer between an excited molecule (donor, D) and neighboring molecules in the ground state (acceptor, A). The transfer of energy usually takes place within D to A separations of 0.5–10 nm. It happens electrodynamically through space as a diplole–dipole interaction. If the spectral overlap between the emission and absorption spectra of D and A is sufficient and if the fluorescence quantum yield of D and the absorption coefficient of A are great enough, the excitation energy of D is transferred nonradiatively to A. In this process no photon emission or absorption takes place. The emission of A is detectable if A fluoresces; however, it is not a must that A fluoresces.

The rate constant of energy transfer (k_t) between D and A through a dipole–dipole interaction is (Selvin, 1995)

$$k_t = \left(\frac{1}{\tau_D} \right) \left(\frac{R_0}{R_{DA}} \right)^6, \tag{9.8}$$

where R_{DA} is the distance that separates D and A, R_0 is the distance between D and A when the energy transfer rate of the D molecule decays from the excited state as in the absence of the A molecule. This energy transfer rate equals $1/\tau_D$. For a detailed theoretical treatment of FRET, please see Selvin (1995), Wu and Brand (1994), and Cheung (1991).

FRET can be configured in several ways with the help of genetic tags (Giepmans et al., 2006) as shown in Figure 9.4. As an example, a cell-based biosensing system for the identification of small-molecule insulin mimetics was configured in a

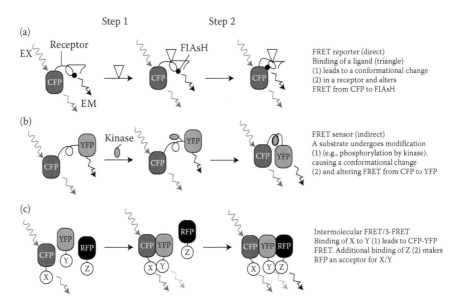

FIGURE 9.4 Advanced FRET configuration with genetic tags (fluorescent proteins, C, Y, and R represent cyan, yellow, and red fluorescent proteins): (a) Direct FRET where binding of a ligand (triangle) leads to a conformation change such that FRET occurs between the barrel (donor) and receptor label (loop); (b) indirect FRET where a substrate undergoes modification, such as phosphorylation (open dot, causing a conformation change that brings the barrels close enough for FRET); (c) intermolecular FRET, where binding of barrels X and Y leads to FRET and additional binding to barrel Z makes it the final acceptor for the XY donor. (Adapted from Giepmans, B.N.G. et al. 2006. *Science* 312(5771): 217–224.)

manner similar to the setup (b) of Figure 9.4 (see Figure 9.5 for schematic). Marine et al. (2006) used this system with 3456-well plates and screened 4195 hits confirmed from a biochemical assay in CHO clones overexpressing the insulin receptor. The screen removed false hits, mainly due to fluorescence artifacts and yielded 58 potential leads. The rationale for the campaign was based on previous work (Zhang et al., 1999; Qureshi et al., 2000) that identified a compound that induced insulin receptor tyrosine kinase activation, resulting in increased tyrosine phosphorylation of the receptor β subunit. The compound provided evidence in support of the possibility to obtain small molecules that function like insulin with efficacy in animal models.

9.3 BIOLUMINESCENCE-BASED READOUTS

Luciferins are a group of biological compounds with varied structure, which luminesce as a result of enzyme-catalyzed oxidation:

$$\text{Luciferin} \xrightarrow[\text{O}_2]{\text{Luciferase}} \text{Oxyluciferin} + h\nu(562\,\text{nm}) \tag{9.9}$$

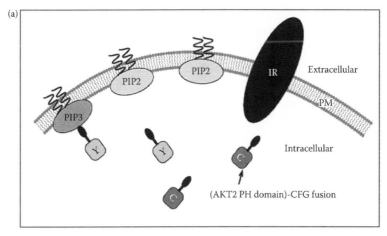

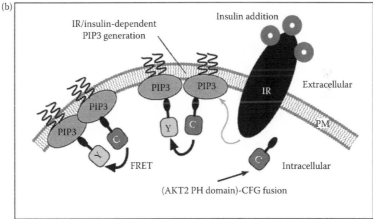

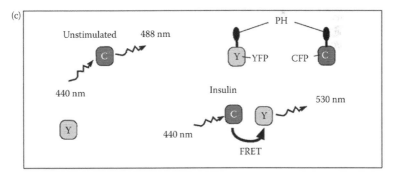

FIGURE 9.5 Schematic of indirect FRET based on insulin receptor signaling-dependent activation of phosphatidylinositol 3-kinase (a), which subsequently activates the serine/threonine AKT2 kinase through its PH domain (b). With the CFP- and YFP-labeled PH domain and excitation at 440 nm, YFP can only fluoresce if CFP and YFP are in close proximity to the cell membrane, which occurs when the insulin receptor is stimulated (c). The ratio of the 53 and 480 nm signals is an excellent readout for activation. (From Marine, S. et al. 2006. *Anal. Biochem.* 355: 267–277. With permission.)

As pointed out in Chapter 1, bioluminescence is a popular approach in developing bacterial biosensors for environmental monitoring. The typical reporter genes include *lux* (bacterial luciferase) and *luc* (firefly luciferase). The luciferase catalyzes the luciferin oxidation. Some of the luciferase reactions couple with cofactors such as ATP, FMN, FADH, and so on as shown below:

$$\text{ATP} + \text{luciferin} + O_2 \rightarrow \text{AMP} + \text{pyrophosphate} + \text{Oxyluciferin}$$
$$+ CO_2 + H_2O + h\nu. \qquad (9.10)$$

The reaction is highly sensitive to femtomole concentrations in some cases. It should be pointed out that bacterial luciferases do not involve luciferins but form excited complexes with reduced flavins such as FMNH as shown below:

$$\text{FMNH}_2 + O_2 + \text{RCHO} \rightarrow \text{FMN} + \text{RCOOH} + H_2O + h\nu(478-505 \text{ nm}) \qquad (9.11)$$

In cell-based biosensing, BRET configurations are popular. A typical configuration is shown in Figure 9.6. The difference between FRET and BRET is the illumination light source. The BRET signal strength is heavily influenced by the amount of overlap between the donor and acceptor emission spectra. The donor emission spectrum is substrate dependent. The donor and acceptor are typically proteins that are genetically fused to the protein of interest. In some cases nonprotein acceptors have been used (Yamakawa, et al., 2002). In most cases, BRET sensor configurations use a combination of a 35-kDa *Renilla luciferase* (Reluc) as the energy donor, a 27-kDa variant of GFP from *Aequrea* as the acceptor, and a coelenterazine as the substrate for the luciferase (in the presence of molecular oxygen). Coelenterazines are small, hydrophobic molecules that rapidly cross cell membranes, making real-time cell-based biosensing possible (Xu et al., 1999). Table 9.2 shows the rating of different GFP–substrate combinations. For example, use of DepBlueC (a coelenterazine derivative) with GFP10 is a good combination, while GFP10 and coelenterazine-h is not.

BRET is extensively used in HTS. Representative examples include Hamdan et al. (2005), who have described a BRET biosensor for screening of GPCR antagonists based on β-arrestin2 recruitment. GPCRs are a large family of cell surface proteins involved in signal transduction. GPCRs mediate diverse important physiological actions and their dysfunction has been associated with critical diseases such as cardiovascular diseases, asthma, and mental disorders, to name a few. They represent the most common target. Approximately 30% of the marketed drug act either directly or indirectly on GPCRs (Wise et al., 2002). See Box 9.2 for details on GPCRs as drug targets. GPCR signaling is terminated, in some instances, by receptor binding to intracellular adaptor proteins called arrestins. In response to exposure to an agonist, arrestin fused with *Reluc* binds to GPCR fused with GFP, resulting in a BRET signal. Any interference with arrestin–receptor binding results in reduced BRET signal. Four types of arrestins have been identified; arrestins 1 and 4 mediate the desensitization of photoreceptors (Gurevich and Gurevich, 2004).

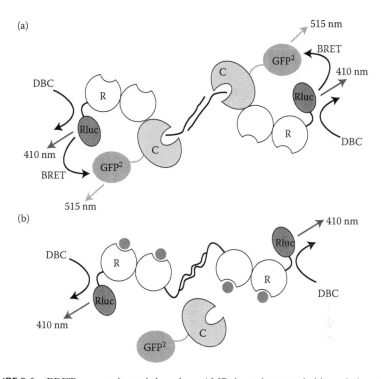

FIGURE 9.6 BRET sensor schematic based on cAMP-dependent protein kinase in intact cells. (a) Rluc is fused on the R subunit and GFP is fused on the C subunit. On addition of luciferase substrate, BRET occurs between Reluc (donor) and GFP (acceptor) depending on how close the two are (1–10 nm) via protein kinase A subunit interaction. (b) The binding of cAMP molecules (solid circles) to the regulatory, R, subunits results in a conformation change, releasing C subunits followed by decrease in BRET as the distance between Reluc and GFP increases. (From Prinz, A., Diskar, M., and Herberg, F.W. 2006. *ChemBioChem*. 7: 1007–1012. With permission.)

TABLE 9.2
GFPs in BRET

Mutation	GFP Example	Excitation Peak (nm)	Emission Peak (nm)	Suitability with Coelenterazine-h Substrate (BRET-1)	Suitability with DeepBlueC Substrate (BRET-2)
None	wtGFP or GFP	~395, ~475	~510	Poor	Good
Usually S65T	EGFP	~490	~510	Good	Suboptimal
Usually S202F and/or T203I	GFP10	~395	~510	Poor	Good
Usually S65G and T203Y	YFPs: EYFP, Topaz, Venus, Citrine	~515	~530	Good	Poor

Source: From Pfleger, K.D.G. and Eidne, K.A. 2006. *Nat. Methods* 3(3): 169. With permission.

BOX 9.2   TARGET DISTRIBUTION FOR COMMERCIALLY AVAILABLE DRUGS

Based on 483 therapeutical drugs on market then, Drews (2000) provided the following biochemical target classes:

45% Receptors
28% Enzymes
11% Hormones and factors
7% Unknown
5% Ion channels
2% DNA
2% Nuclear receptors

The majority of the receptors in the receptor category above are GPCRs. Approximately 30% of all marketed prescription drugs act on 20 GPCRs (Overington et al., 2006) making this family the pharmaceutically most successful target class (Heilker et al., 2009). GPCRs are heterotrimers ($\alpha\beta\gamma$) and are a central component of the primary mechanism used by virtually all eukaryotic cells to receive, interpret, and respond to a wide variety of structurally diverse extracellular stimuli (Cabrera-Vera et al., 2003). Activated GPCRs interact with G-protein, thereby catalyzing the exchange of guanosine triphosphate (GTP) for guanosine diphosphate (GDP) and the subsequent dissociation of the $G\alpha$-GTP complex from the $\beta\gamma$ complex (Cabrera-Vera et al., 2003), or the rearrangement of the G-protein, which enables both $G\alpha$-GTP and $\beta\gamma$ dimmers to interact with a variety of downstream effectors (Bünemann et al., 2003). In the human genome, there are 16 genes that encode α-subunits, five genes that encode β-subunits, and 14 genes that encode γ-subunits (Kostenis et al., 2005). The $G\alpha$-subunit nomenclature is commonly used to classify GPCRs; as such, they are referred to as G_s-, $G_{i/o}$-, or G_q-coupled, reflecting their primary signal transduction pathway. The primary G_s-subfamily stimulates adenylyl cyclases. The $G_{i/o}$-subfamily inhibits adenylyl cyclase; however, not all the isotypes share this property. The G_q-subfamily is linked to the stimulation of phospholipase Cβ (PLC-β) isoforms.

Arrestins 2 and 3, also known as β-arrestins 1 and 2, respectively, have been shown to interact with GPCRs (Shenoy and Lefkowitz, 2003). The biosensor was created by constitutively expressing optimal levels of β-arrestin 2-Reluc and GPCR fused to VENUS. For HTS, HEK293 cells stably coexpressing β-arrestin 2-Reluc and CCR5-VENUS were used with coelenterazine-h as the substrate (BRET1 assay). CCR5 is a CC chemokine receptor, the primary coreceptor for HIV-1. Agonist (RANTES)-promoted BRET was measured as the BRET ratio on the BMG FluoStar Optra plate reader. Only compounds that inhibited the agonist-promoted BRET signal by at least 50% and amounting to more than three standard deviations of the signal obtained in the presence of agonist alone were considered as potential hits. Of the 26,000 small molecules assayed against CCR5 antagonistic activity, 12 compounds were found to specifically inhibit the agonist-induced β-arrestin 2 recruitment to CCR5.

9.4   LABEL-FREE BIOSENSOR READOUTS

It has often been mentioned that the use of dyes or fluorescent labels in cell-based sensing or assays may interfere with the physiology of the cells. Labels may interfere with signaling pathways, through effects on molecular conformation, blocking of active binding epitopes, steric hindrance, and probably other ways that have not surfaced. Additionally, the majority of cell-based assays are endpoint types, wherein the cell is destroyed at the end of the assay. This precludes the opportunity to generate kinetic data that may offer deeper insights into the outcome of the interaction between the cell and the chemical compound being tested (Stein, 2008). In some situations it may not be possible to find an appropriate label or the labeling site may not be accessible. However, it is difficult to detect physical cellular changes such as mass, size, electrical impedance, or dielectric permittivity (Cunningham, 2009), which form the basis for label-free biosensing. The main advantage of label-free biosensing is offering the possibility to monitor cells in a more natural setting in time. This advantage has inspired engineers and scientists to develop platforms that can meet HTS requirements. Other important advantages that label-free biosensing offers are higher sensitivity and lower cost (Stein, 2008). Currently, the most common commercial cell-based instruments with HTS capability are CellKey (MDS Analytical Technologies, Sunnyvale, CA) based on impedance measurements and Epic (Corning, Inc.—Life Sciences, Lowell, MA), based on resonant waveguide measurements. Table 9.3 provides a sample of commercially available label-free platforms with an indication of HTS as well as cell-based application capabilities. The impedance and resonant waveguide principles are briefly reviewed below.

9.4.1   IMPEDANCE

Although measuring biological material impedance has long been the pursuit of biophysicists, validated physical principles that underlie biosensing instrumentation were only established a couple of decades ago (Giaever and Keese, 1991, 1993; Lo et al., 1995). To establish these principles, Giaever and Keese simplified adherent cells as discs of radius r_c and height h as illustrated in Figure 9.7. Additional assumptions were: (1) current flows radially in the space formed between the ventral surface of the cell and the substratum, (2) the current density under the cells does not change in the vertical direction, (3) the electrode potential V_c as constant that is independent of position, (4) the potential in solution of the dorsal side of the cell, V_s, is likewise treated as a constant and for convenience set to zero, (5) the electrical potential inside the cell, V_i, is constant, (6) the presence of the cell does not affect the electrode

TABLE 9.3
Other Commercial Label-Free Platforms

Instrument	Principle	Company and Website	HTS Capability[a,b]	Cell-Based Capability[a] and/or References
Biacore (products include A100, T100, and chip)	Optical (SPR)	GE Healthcare http://www.gehealthcare.com	No	No
Epic™	Optical (resonant waveguide)	Corning http://www.corning.com	Yes	Yes
RAP + id4	Acoustic	Akubio http://www.akubio.com	No	No
RAPid4	Resonant quartz crystal	TTP LabTech http://www.ttplabtech.com	No	No
ProteOn XPR36	Optical (SPR optical biosensor)	Bio-Rad Laboratories http://www.bio-rad.com/evportal/evolutionPortal.portal	No	No
AutoLab Espirit	Optical (SPR)	EcoChemie http://www.ecochemie.nl	Yes	No
RAISE platform	Optical (SPR) for fragment library screening	Graffinity http://www.graffinity.com	Yes	No
SPRImager®II	Optical (SPR)	GWC technologies http://www.gwctechnologies.com	No	Yes
SensiQ	Optical (SPR)	Icx Nomadics http://www.nomadics.com	No	No
Proteomic processor	Optical (SPR)	Lumera http://www.lumera.com	No	No
Octet System	Interferometry	ForteBio http://www.fortebio.com	Yes	No

Product	Description	Company / URL		
No product available, only at prototype stage	Microparticles and grating-coupled SPR	Toshiba Cambridge Research Lab http://www.toshiba-europe.com/research/crl	No	Yes
EVOM epithelial voltohmmeter	Impedance (confluent monolayer of cells between parallel plate electrodes)	World Precision Instruments http://www.wpiinc.com	No	Yes
ECIS Z, ECIS Zθ (**Theta**)	Impedance (cells in contact with—even growing on—asymmetric electrodes)	Applied Biophysics http://www.biophysics.com	Yes	Yes
CellKey	Impedance (cells in contact—even growing on—symmetric electrodes)	MDS Analytical Technologies http://www.mdssciex.com/home/	Yes	Yes
RT-CES®	Impedance (monitor cellular events by measuring the electronic impedance of sensor electrodes integrated on the bottom of microtiter E-plates. Based on measured impedance, a dimensionless parameter, "Cell Index," is derived and reported to provide quantitative information about the biological status of the cells, including cell number, viability, morphology, and cytoskeletal dynamics)	ACEA Biosciences http://www.aceabio.com	Yes	Yes
VeriScan 3000	Microcantilever	Protiveris http://www.protiveris.com	Yes	Yes
CantiLab3D and CantiLabPro	Mass (a microcantilever imprinted with antibody deflects on antigen binding; the cantilever deflection is optically detected)	Cantion http://www.cantion.com	No	No
AMBRI ICS™	Electrochemical (sodium channels are embedded into a lipid bilayer near a gold electrode, and when sodium ions pass through the channel a signal is registered at the electrode—with an antibody bound to the outside of the channel, a lower signal is registered when the antibody is bound by an antigen, blocking the channel opening)	Ambri http://www.ambri.com	No	Yes

continued

TABLE 9.3 (continued)
Other Commercial Label-Free Platforms

Instrument	Principle	Company and Website	HTS Capability[a,b]	Cell-Based Capability[a] and/or References
High-performance capillary electrophoresis with label-free intrinsic imaging	Electrophoretic separation coupled to algorithms to identify biomolecules	DeltaDOT http://www.deltadot.com	No	No
VP-ITC system	Mass (differential scanning calorimetry)	MicroCal http://www.microcal.com	No	No
AutoiTC200	Isothermal titration calorimetry (the instrument passively measures the heat produced during a chemical reaction)	MicroCal http://www.microcalorimetry.com	Yes	No
Not known	ITC (as above)	Thermometric http://www.thermometric.com	Not known	Yes

a Based on company literature and/or website.
b Can accommodate at least 96-well format or equivalent.

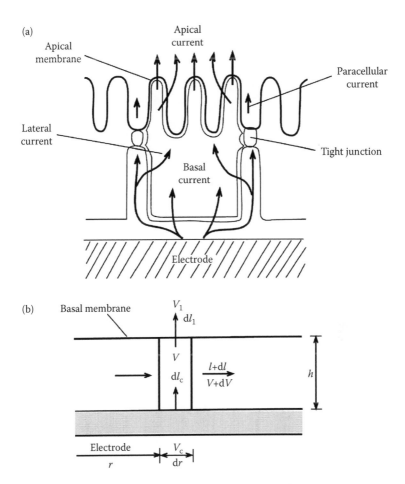

FIGURE 9.7 Schematic of a cell in a monolayer on top of an electrode substratum with current flow paths (a). When regarded as a simple disc shape from the top view, a side schematic diagram can be constructed (b) to enable the development of differential equations. (From Lo, C.-M., Keese, C.R., and Giaever, I. 1995. *Biophys. J.* 69: 2802. With permission.)

polarization, and (7) the monolayer is confluent. The approach to modeling was to calculate the current flowing from the area of a single cell as a function of V_c. The impedance of the cell-covered electrode Z_c is

$$Z_c = \frac{(\text{electrode area})(V_c - V_s)}{I_c} = \frac{\pi r_c^2 V_c}{I_{ct}}, \qquad (9.12)$$

where I_c is the total current flowing from the electrode and I_{ct} is the current flowing from the area of a single cell. Lo et al. (1995) using the schematic in Figure 7.7 used Ohm's law together with a differential equation and boundary condition to obtain a matrix equation with numeric constants A, C, D, and V_i that were found by solving the equation. From Ohm's law, I_{ct} was found to be

$$I_{ct} = \frac{-2\pi r_c^2 A}{Z_n \gamma r_c} I_1(\gamma r_c) + \frac{\pi r_c^2}{Z_n + Z_b}(V_c - V_i) \tag{9.13}$$

where I_{ct} is total current from the area of a single cell in amperes, r_c is cell radius in micrometer, A is a numerical constant, Z_n is impedance per unit area of the cell-free electrode in Ω cm^2, $\gamma = [(\rho/h)/(1/Z_n + 1/Z_b)]^{-0.5}$, Z_b is the specific (per unit area) impedance through the basal cell membrane in Ω cm^2, h is the height between the basal cell surface and the substratum in nanometer, ρ is the resistivity of the cell culture medium in Ω cm, I_l is the paracellular current through the intercellular lateral path in Amperes, and V_c is the applied voltage across the cell in Volts, V_i is the potential inside the cell in Volts. The impedance was obtained by dividing V_c by I_{ct}. The model was used to calculate the resistance and capacitance of the impedance and the results were compared to experiments data with and without cells (MDCK). The calculated values were based on impedance of the naked electrode as the solution resistance (constriction resistance) is a significant part of the measured impedance; it was first subtracted from the measured impedance before the calculation was done and then added back for comparison with experimental results. The Lo et al. (1995) results are presented in Figure 9.8, showing the frequency dependence and excellent agreement between theory and measured values.

There are numerous practical arrangements of impedance biosensors. McGuinness and Verdonk (2009) have described four electrode geometries including (1) parallel electrodes for dilute cell suspension, (2) parallel electrodes for suspended monolayers,

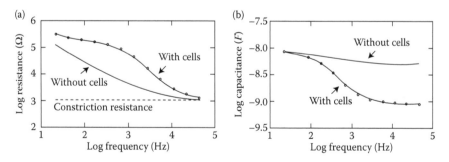

FIGURE 9.8 Frequency dependence of resistance (a) and capacitance (b) of an electrode with and without a monolayer of MDCHK cells. The black circles are calculated values based on measured impedance of the cell-free electrode and the model. The resistance is a sum of the electrode resistance and the frequency-independent solution resistance (constriction resistance). At low frequency, the electrode resistance dominates, and at high frequency, the constriction resistance dominates. The measured values of the cell-covered electrode can be normalized by dividing with the corresponding values of the cell-free electrode. (From Lo, C.-M., Keese, C.R., and Giaever, I. 1995. *Biophys. J.* 69: 2803. With permission.)

(3) monolayers on asymmetric electrodes, and (4) cells growing on symmetric electrodes. Examples of commercial instruments with these configurations are provided in Table 9.3. The configuration from the ACEA RT-CES™ system is highlighted in Figure 9.9. A parameter termed cell index (CI) as defined below is used to quantify the cell status with respect to the measured cell-electrode impedance:

$$CI = \max(i = 1,...,N)\left(\frac{R_{cell}(f_i)}{R_0(f_i)} - 1 \right), \tag{9.14}$$

where $R_{cell}(f_i)$ and $R_0(f_i)$ are frequency-dependent electrode resistances with or without cells in the well, respectively, at three different frequencies (e.g., 10, 25, and 50 kHz; $N = 3$). N is the number of frequency points at which the impedance is measured. Under RT-CES software, control wells are automatically selected and each frequency is applied for about 100 ms to each well followed by replicate measurements at specific times (usually 2–10 min). As suggested by the formula, CI is a

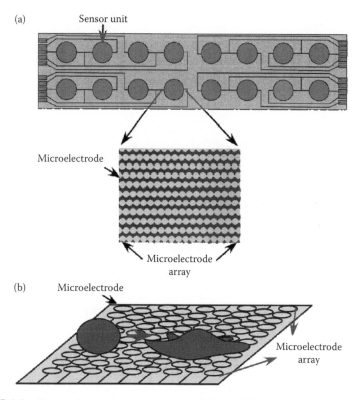

FIGURE 9.9 Schematic electrode arrays of ACEA RT-CES™ high-density well plate for impedance measurement (a). Upon cell seeding, the round cells spread, making contact with many electrodes, resulting in higher impedance readout (b). The effect of a compound that would affect cell spreading is easily detectable in this system. (From Atienza, J.M. et al. 2009. *J. Biomol. Screen.* 10(8): 797. With permission.)

measure of total cell contact area on the electrodes as well as the strength of cell adhesion to the electrode (Ge et al., 2009; Glamann and Hansen, 2006; Xing et al., 2005; Wegener et al., 2000).

Atienza et al. (2005) have used RT-CES™ and demonstrated that the integrity of the actin cytoskeleton is required for meaningful cell adhesion and spreading on the ECM-coated sensors of the system. The majority of the applications have so far been in cytotoxicity (Huang et al., 2009; Xia et al., 2008; Zhu et al., 2006). However, a differently configured impedance system (CellKey, MDS Sciex, South San Francisco, CA) has been used on GPCRs. Peters et al. (2007) used agonists and antagonists for the G_i-coupled GPCRs with CHO cell lines stably transfected to express human recombinant receptors and compared the CellKey performance with the classical methods of [^{35}S]GTPγS binding (scintillation proximity assay, SPA) and cAMP production (CatchPoint™ cAMP Fluorescent Kit form Molecular Devices Corporation). The CellKey results correlated with data obtained with the classical methods, suggesting the platform's suitability for structure–activity relationship (SAR) studies. McGuinness and Verdonk (2009) reported receptor panning studies with primary and immortalized prostate, Jukart, and U-2 OS cells, treated with panels of ligands (1–10 μM doses). Active endogenous receptors were identified from impedance responses that were greater than the buffer response. Also, cellular impedance (CellKey) confirmed cell-specific G_s and G_q coupling for the endogenous melanocortin-receptor and dual G_i and G_s signaling with endogenous cannabinoid (CB-1) receptor (Peters and Scott, 2009), which is valuable signaling information that is not available in end-point type of biosensing. Taken together, these results demonstrated the importance of dynamic impedance measurements and high sensitivity generally associated with label-free technologies that in most cases should render the standard need for overexpressed cellular systems unnecessary (McGuinness and Verdonk, 2009).

9.4.2 SURFACE PLASMON RESONANCE

SPR is the most common readout principle utilized among current cell-free platforms as shown in Table 9.3. SPR is one of three variations of reflectance methods (internal reflectance spectroscopy). The other two variations include attenuated total reflectance (ATR) and total internal reflection fluorescence (TIRF) (Eggins, 1996). The basic principle in internal reflectance spectroscopy is illustrated in Figure 9.10. With media of two refractive indices (n_1 and n_2), light striking the interface of the media undergoes total internal reflection when θ obeys the following equation:

$$\sin \theta_c = \frac{n_2}{n_1} \quad \text{and} \quad n_1 > n_2. \tag{9.15}$$

If $\theta > \theta_c$, an evanescent wave refracted through the interface in the z-direction penetrates the n_2 medium (Figure 9.10a) a distance d_p and the electrical field vector

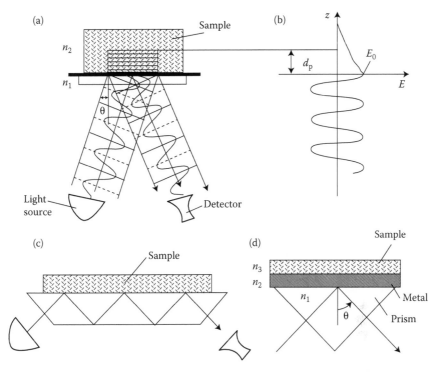

FIGURE 9.10   Attenuation of total reflection, generation of the evanescent wave at the interface between two optical media (a and b), and multiple reflections showing enhanced sensitivity (c). Kretschmann prism arrangement where n_1, n_2, and n_3 are refractive indices of the glass prism, metal layer, and sample, respectively (d).

of this wave (E) is largest at the interface (E_0). The electric field decays with the distance z (Figure 9.10b) according to

$$E = E_0 \exp\frac{z}{d_p}. \tag{9.16}$$

In addition to the indices of the media, d_p is dependent on the light wavelength (λ) and θ and the larger the d_p value, the larger is the attenuation of the reflection. Other factors that attenuate the reflection include polarization-dependent electric field intensity at the reflecting surface, the sampling area, and the matching of the two refractive indices. The sensitivity of the signal can be increased by multiple reflections as shown in Figure 9.10c, by the use of a waveguide. The longer and thinner the waveguide is, the longer the number of reflections and consequently the higher the sensitivity. In ATR practical configurations, predominantly used in immunosensors, the absorbing material (e.g., antigen–antibody binding) is placed in contact with the reflecting surface of the waveguide, causing the attenuation of the internally reflected light.

For a waveguide of length L and thickness T and angle of incidence θ, the number of reflections is given by

$$N = \frac{L}{T}\cot\theta. \tag{9.17}$$

As previously mentioned, many factors influence the attenuation of the reflection. If d_e is defined as the effective depth penetration, accounting for all the factors, then reflectivity R is given by

$$R = 1 - \alpha d_\theta, \tag{9.18}$$

where α is the absorption coefficient. For N reflection,

$$R^N = 1 - N\alpha d_\theta. \tag{9.19}$$

Surface plasmon is described as an oscillation of electrons on the surface of a metal or a semiconductor (Earp and Dessy, 1998). Figure 9.10d illustrates the Krestschmann prism arrangement, which is the classical SPR coupling method. Light that is perpendicularly (p) polarized with respect to the metal surface is launched into the prism and coupled into the surface plasmon mode on the metal film. Only perpendicularly polarized light can be coupled because this perpendicular polarization has the electric field vector oscillating normal to the plane that contains the metal film. When the p-polarized incident field has an angle θ such that the photon momentum along the surface matches the plasmon frequency, SPR occurs (Earp and Dessy, 1998). The intensity of totally reflected light exhibits a sharp decrease with an increase in θ depending on the depth, the width, the characteristic absorbance, and the thickness of the metal.

The Corning resonant waveguide (RWG) biosensor in their Epic system is an excellent example of practical arrangement sensing based on internal reflection principles. A schematic of the biosensor is shown in Figure 9.11. In this sensor the waveguide is grated; diffraction grating is of considerable importance in spectroscopy, due to its ability to separate (disperse) polychromatic light into its constituent monochromatic components (Loewen and Popov, 1997). Broad band light, incident on the grated (nanoscale) waveguide, passes through the glass at the sensor bottom and is selectively coupled through the 2.36 refractive index material. A resulting evanescence electromagnetic field of the guided wave penetrates into the medium above the sensor and changes in the refractive index above are detected. The effective thickness (d_e) above the sensor is approximately 150 nm. Changes in the refractive index on the sensor surface manifest into spectral selection. The ligand-induced change in effective refractive index is related to various system variables as follows:

$$\Delta N = S(C)\alpha d_e \sum_i \Delta c_i(t)\left[e^{\frac{-z_i}{\Delta z_c}} - e^{\frac{-z_{i+1}}{\Delta z_c}}\right], \tag{9.20}$$

where ΔN is the change in effective refractive index, $S(C)$ is the sensitivity to the cell layer, α is the specific refractive increment, d_e is the arbitrary thickness of a slice

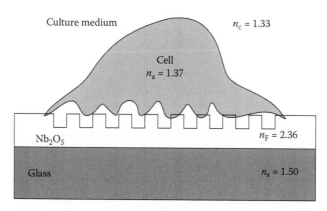

FIGURE 9.11 Schematic of a biosensor used in the Corning Epic system. The sensor was made up of a thin waveguide film of Nb_2O_5 (refractive index, $n_F = 2.36$) on a glass substrate (refractive index, $n_s = 1.50$), surrounded by culture media (refractive index, $n_c \sim 1.33$). The overall cell monolayer refractive index $n_a = 1.37$. The sensors are incorporated in each well of the 384-well plate. (From Fang, Y. et al. 2005. *Anal. Chem.* 77: 5721. With permission.)

within the cell layer, $\Delta C_i(t)$ is the change in local concentration of biomolecules in space and time, ΔZ_C is the penetration depth into the cell layer, and z_i is the distance where the mass redistribution occurs. This mathematical model of the sensor, published by Fang et al. (2006), has been verified by experimental studies (e.g., Lee et al., 2008; Fang and Ferrie, 2007); these studies have revealed that the resulting shift in incident angles was primarily due to stimulation-triggered dynamic mass redistribution perpendicular to the sensor surface. A summary of real-time cell-based studies conducted with the Epic by Corning Scientists has been presented by Fang et al. (2009). These include panning of endogenous receptors in cells, receptor signaling, ligand pharmacology evaluation, and screening of compounds. One of the most interesting results was the ability of the biosensor to discriminate between GPCR signaling in the human cancer cell line A431 of G_q (protease-activated receptor subtype 1, 40 U/mL thrombin), G_s signaling (β_2-adrenergic receptor, 25 nM epinephrine), and G_i signaling (lysophosphatic acid receptor [LPA] subtype 1, 200 nM LPA) that is not possible with calcium flux signaling measurements (Fang et al., 2007) (Figure 9.12). The Epic has been recently used by Dodgson et al. (2009) as a rapid orthogonal screening Fluorometric Imaging Plate Reader (FLIPR), in CHO K1 cells to indentify antagonists of the muscarinic M3 receptor and the Epic was able to identify FLIPR false positive.

9.5 CONCLUDING REMARKS

Cell-free sensing offers the opportunity to use cells in assays in their native form, without any genetic modification. Additionally, platforms like the Epic offer pharmacological evaluation of cellular responses to GPCR activation differentiating GPCR signaling, whereas traditional biosensing such as in the FLIRR calcium assay does not. These two features highlight the niche label-free technology is offering. On the other hand,

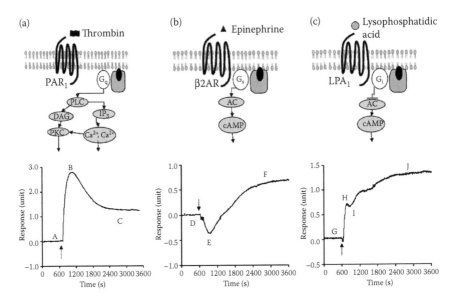

FIGURE 9.12 Mass redistribution due to GPCR signaling. (a) G_q signaling stimulated by thrombin (40 units/mL) in CHO cells. (b) G_s signaling stimulated by epinephrine (25 nm) in A431 cells. (c) G_i signaling stimulated by lysophosphatic acid (200 nM) in A431 cells. Arrows indicate the beginning of stimulation. The signals exhibited different characteristics, for example, while G_q displays an initial rapid increase (a-AB), G_s displays a decrease (b-DE). Although both G_q and G_i displayed rapid signal increases, their profiles after the increase went in different directions (a-BC and c-IJ). (From Fang, Y., Li, G., and Ferrie, A.M. 2007. *J. Pharmacol. Toxicol. Methods* 55: 316. With permission.)

there is a general need to address low specificity issues with many of the label-free biosensors platforms, a sentiment echoed by many current users. The challenges multiply when 3D cultures are added in the mix. For example, the detection zone limitation of 150 nm from the biosensor surface limits the application of the Epic System and probably other label-free technologies with 3D cell cultures. Most of these limitations can be addressed by going back to the drawing board and reconfiguring the sensor for 3D cultures. A possible design change could involve using microwell scaffolds such as those presented in Chapter 6. Scaffolding typically presents two main challenges of light scattering and absorption. However, these limitations are not insurmountable; for example with judicious selection of scaffolding material, Cheng et al. (2008) successfully monitored voltage-gated calcium channel gating in response to high potassium depolarization with primary and neural progenitor cells in a 3D scaffold. This author is confident that as 3D cultures begin to take their place in HTS laboratories, readout instrumentation suppliers will rise up to the occasion.

REFERENCES

Albani, J.R. 2007. *Principles and Applications of Fluorescence Spectroscopy*. Ames, IA: Blackwell Publishing.

Atienza, J.M., Zhu, J., Wang, X., Xu, X., and Abassi, Y. 2005. Dynamic monitoring of cell adhesion and spreading on microelectronic sensor arrays. *J. Biomol. Screening* 10(8): 795–805.

Bader, B., Butt, E., Palmetshofer, A., Walter, U., Jarchau, T., and Druecker, P. 2001. A cGMP-dependent protein kinase assay for high throughput screening based on time-resolved fluorescence resonance energy transfer. *J. Biomol. Screen.* 6(4): 255–264.

Bovolenta, S., Foti, M., Lohmer, S., and Corazza, S. 2007. Development of a Ca^{2+}-activated photoprotein, photina®, and its application to high-throughput screening. *J. Biomol. Screen.* 12: 694–704.

Bünemann, M., Frank, M., and Lohse, M.J. 2003. Gi protein activation in intact cells involves subunit rearrangement rather than dissociation. *PNAS* 100(26): 16077–16082.

Cabrera-Vera, T.M., Vauhauwe, J., Thomas, T.O., Medkova, M., Preininger, A., and Mazzoni, M.R. 2003. Insights into G protein structure, function, and regulation. *Endocrine Rev.* 24(6): 765–781.

Cacace, A., Banks, M., Spicer, T., Civoli, F., and Watson, J. 2003. An ultra-HTS process for the identification of small molecule modulators of orphan G-protein-coupled receptors. *Drug Discov. Today* 8(17): 785–792.

Chambers, C., Smith, F., Williams, C., Marcos, S., Liu, Z.H., Hayter, P., Ciaramella, G., Keighley, W., Gribbon, P., and Sewing, A. 2003. Measuring intracellular calcium fluxes in high throughput mode. *Comb. Chem. High Throughput Screen.* 6: 355–362.

Cheng, K., Lai, Y., and Kisaalita, W.S. 2008. Three-dimensional polymer scaffolds for high throughput cell-based assay systems. *Biomaterials* 29: 2802–2812.

Cheung, H. 1991. Resonance energy transfer. *Topics Fluoresc. Spectr.* 3: 127–176.

Chin, J., Adams, A.D., Bouffard, A., Green, A., Lacson, R.G., Smith, T., Fischer, P.A., Menke, J.G., Sparrow, C.P., and Mitnaul, L.J. 2003. Miniaturization of cell-based β-lactamase-dependent FRET assays to ultra-high throughput formats to identify agonists of human liver X receptors. *ASSAY Drug Dev. Technol.* 1(6): 777–787.

Clegg, R.M. 2009. Förster resonance energy transfer—FRET: What is it, why do it, and how it is done. In *Laboratory Techniques in Biochemistry and Molecular Biology, Vol 33: FRET and FLIM Techniques,* T.W.J. Gadella (ed.), pp. 1–57. New York: Elsevier.

Cohen, P. 2002. Protein kinases—the major drug targets of the twenty-first century? *Nat. Rev. Drug Discov.* 1: 309–315.

Cooper, M.A. (ed.) 2009. *Label-Free Biosensors: Techniques and Applications.* Cambridge: Cambridge University Press.

Cunningham, B.T. 2009. Label-free optical biosensors: An introduction. In *Label Free-Biosensors: Techniques and Application*, M.A. Cooper (ed.), pp. 1–28. Cambridge: Cambridge University Press.

Dodgson, K., Gedge, L., Murray, D.C., and Coldwell, M. 2009. A 100K well screen for muscarinic receptor using the Epic® label-free system—a reflection on the benefits of the label-free approach to screening seven-transmembrane receptors. *J. Recep. Signal. Transduc.* 29(3–4): 163–172.

Drews, J. 2000. Drug discovery: A historical perspective. *Science* 287: 1960–1964.

Du, M.X., Sim, J., Fang, L., Yin, Z., Koh, S., Stratton, J., Pons, J., Wang, J.-X., and Carte, B. 2004. Identification of novel small-molecule inhibitors for human transketolase by high-throughput screening with fluorescence intensity (FLINT) assay. *J. Biomol. Screen.* 9(5): 427–433.

D'Souza, P.C.G., Ally, N., Agarwal, A., Rani, P., Das, M., and Sarma, D. 2008. Demonstration of LanthanScreen® TR-FRET-based nuclear receptor coactivator recruitment assay using PHERAstar, a multi-detection HTS microplate reader. *Ind. J. Pharmacol.* 40(2): 89–90.

Earp, D.L. and Dessy, R.E. 1998. Surface plasmon resonance. In *Commercial Biosensors*, G. Ramsey (ed.), pp. 99–164. New York: Wiley.

Eggins, B. 1996. *Biosensors: An Introduction.* New York: Wiley.

Eglen, R.M. 2002. Enzyme fragment complementation: A flexible high throughput screening assay technology. *ASSAY Drug Dev. Technol.* 1: 97–104.

Engels, I.H., Daguia, C., Huynh, T., Urbina, H., Buddenkotte, J., Schumacher, A., Caldwell, J.S., and Brinker, A. 2009. A time-resolved fluorescence energy transfer-based assay for EDN1 peptidase activity. *Anal. Chem.* 390(1): 85–87.

Fang, Y., Ferrie, A.M., Fontaine, N.H., Mauro, J., and Balakrishnan, J. 2006. Resonant waveguide grating biosensor for living cell sensing. *Biophys. J.* 91: 1925–1940.

Fang, Y., Li, G., and Ferrie, A.M. 2007. Non-invasive optical biosensor for assaying endogenous G protein-coupled receptors in adherent cells. *J. Pharmacol. Toxicol. Methods* 55: 314–322.

Fang, Y. and Ferrie, A.M. 2007. Optical biosensor differentiates signaling of endogenous PAR1 and PAR2 in A431 cells. *BMC Cell Biol.* 8(24).

Fang, Y., Fung, J., Tran, E., Xie, X., Hallstrom, M., and Frutos, A. 2009. High-throughput analysis of biomolecular interactions and cellular responses with resonant waveguide grating biosensor. In *Label Free-Biosensors: Techniques and Application*, M.A. Cooper (ed.), pp. 206–222. Cambridge: Cambridge University Press.

Fung, Y., Ferrie, A.M., Fontaine, N.H., Mauro, J., and Balakrishnan, J. 2006. Resonant waveguide grating biosensor for living cell sensing. *Biophys. J.* 91: 1925–1940.

Gadella, T.W.J. (ed.) 2009. *Laboratory Techniques in Biochemistry and Molecular Biology, Vol 33: FRET and FLIM Techniques.* New York: Elsevier.

Galietta, L.V.J., Jayaraman, S., and Verkman, A.S. 2001. Cell-based assay for high-throughput quantitative screening of CFTR chloride transport agonists. *Am. J. Physiol. Cell Physiol.* 281: C1734–C1742.

Galmann, J. and Hansen, A.J. 2006. Dynamic detection of natural cell-mediated cytotoxicity and cell adhesion by electrical impedance measurements. *Assay Drug Dev. Technol.* 4(5): 555–563.

Ge, Y.K., Deng, T.L., and Zheng, X.X. 2009. Dynamic monitoring of changes in endothelial cell–substrate adhesiveness during leukocyte adhesion by microelectrical impedance assay. *Acta Biochim. Biophys. Sinica* 41(3): 256–262.

Giaever, I. and Keese, C.H. 1993. A morphological biosensor for mammalian cells. *Nature* 366: 591–592.

Giaever, I. and Keese, C.R. 1991. Micromotion of mammalian cells measured electrically. *Proc. Natl. Acad. Sci. USA* 88: 7896–7900.

Giepmans, B.N.G., Adams, S.R., Ellisman, M.H., and Tsien, R.Y. 2006. The fluorescent toolbox for assessing protein location and function. *Science* 312(5771): 217–224.

Glamann, J. and Hansen, A.J. 2006. Dynamic detection of natural killer cell-mediated cytotoxicity and cell adhesion by electrical impedance measurement. *Assay Drug Dev. Technol.* 4(5): 555–564.

Gonzalez, J.E., Oades, K., Leychkis, Y., Harootunian, A., and Negulescu, P. 1999. Cell-based assays and instrumentation for screening ion-channel targets. *Drug Discov. Today* 4: 431–439.

Gribbon, P. and Sewing, P. 2003. Fluorescence readouts in HTS: No gain without pain? *Drug Discov. Today* 8(22): 1035–1043.

Gurevich, V.V. and Gurevich, E.N. 2004. The molecular acrobats of arrestin activation. *Trends Pharmacol. Sci.* 25(2): 105–111.

Hamdan, F.E., Audet, M., Garneau, P., Pelletier, J., and Bouvier, M. 2005. High-throughput screening of G protein-coupled receptor antagonists using a bioluminescence resonance energy transfer 1-based {szligbeta}-arrestin2 recruitment assay. *J. Biomol. Screen.* 10: 463–475.

Heilker, R., Wolff, M., Tautermann, S., and Bieler, M. 2009. G-protein-coupled receptor-focused drug discovery using a target class platform approach. *Drug Discov. Today* 14(5/6): 231–240.

Herget, S., Lohsen, M.J., and Nikolaev, V.O. 2008. Real-time monitoring of phosphodiesterase inhibition in intact cells. *Cell. Signal.* 20(8): 1423–1431.

HighTech Business Decisions. 2007. *High Throughput Screening 2007: New Strategies, Success Rates, and Use of Enabling Technologies*, vols. I and II. A market study report by HighTech Business Decisions (www.hightechdecisions.com), Can Jose, CA.

Hill, S.J., Baker, J.G., and Rees, S. 2001. Reporter-gene systems for the study of G-protein-coupled receptors. *Curr. Opin. Pharmacol.* 1: 526–532.

Hodder, P., Mull, R., Cassaday, J., Berry, K., and Struluvici, B. 2004. Miniaturization of intracellular calcium functional assays to 1536-well plate format using a fluorometric imaging plate reader. *J. Biomol. Screen.* 9: 417–426.

Huang, D.Y., Mock, M., Hagenbuch, B., Chan, S., Dmitrovic, J., Gabos, S., and Kinnburgh, D. 2009. Dynamic cytotoxic response to microcysteines using microelectronic sensor array. *Environ. Sci. Technol.* 43(20): 7803–7809.

Inglese, J., Johnson, R.L., Simeonov, A., Xia, M., Zheng, W., Austin, C.P., and Auld, D.S. 2007. High-throughput screening assays for the identification of chemical probes. *Nat. Chem. Biol.* 3(8): 466–479.

Isaad, T., Blanquart, C., and Gonzalez-Yanes, C. 2007. The use of bioluminescence resonance energy transfer for the study of therapeutic targets: Application to tyrosine kinase receptors. *Expert Opin. Ther. Targets* 11: 541–556.

Jayaraman, S. Golz, S. 2008. Reporter genes in cell based ultra high throughput screening. In *Bioengineering in Cell and Tissue Research*, G.M. Artmann and S. Chien (eds), pp. 3–22. Berlin Heidelberg: Springer.

Kashem, M.A., Nelson, R.M., Yingling, J.D., Pullen, S.S., Prokopowcz, III, A.S., Jones, J.W., Wolak, J.P., et al. Three mechanistically distinct kinase assays compared: Measurements of intrinsic ATPase activity identified the most comprehensive set of inhibitors. *J. Biomol. Screen.* 12(1): 70–83.

Keating, S., Marsters, J., Beresini, M., Lander, C., Zioncheck, K., Clark, K., Arellano, F., and Bodary, S. 2000. Putting the pieces together: Contribution of fluorescence polarization assay to small molecule lead optimization. *Prog. Biomed. Optics Imaging (SPIE)* 3913: 128–137.

Kostenis, E., Waelbroeck, M., and Milligan, G. 2005. Techniques: Promiscuous Gα proteins in basic research and drug discovery. *TRENDS Pharmacol. Sci.* 26(11): 595–602.

Lakowicz, J.R. 1999. *Principles of Fluorescence Spectroscopy.* New York: Kluwer Academic/Plenum.

Lee, P.H., Gao, A., van Staden, C., Ly, J., Salon, J., Fang, Y., and Verkleeren, R. 2008. Evaluation of dynamic mass redistribution technology for pharmacological studies of recombinant and endogenously expressed G protein-couples receptors. *Assay Drug Dev. Technol.* 6(1): 83–94.

Lo, C.-M., Keese, C.R., and Giaever, I. 1995. Impedance analysis on MDCK cells measured by electrical cell–substrate impedance sensing. *Biophys. J.* 69: 2800–2807.

Loewen, E.G. and Popov, E. 1997. *Diffraction Gratings and Applications.* New York: Marcel Dekker.

Marine, S., Zamiara, E., Smith, S.T., Stec, E.M., McGarvey, J., Kornienko, O., Jiang, G. et al. 2006. A miniaturized cell-based fluorescence resonance energy transfer assay for insulin-receptor activation. *Anal. Biochem.* 355: 267–277.

McGuinness, R.P. and Verdonk, E. 2009. Electrical technology applied to cell-based assays. In *Label Free-Biosensors: Techniques and Application*, M.A. Cooper (ed.), pp. 251–277. Cambridge: Cambridge University Press.

Milligan, G. 2004. Applications of bioluminescence- and fluorescence resonance energy transfer to drug discovery at G protein-coupled receptors. *Eur. J. Pharm. Sci.* 21: 397–405.

Minor, L.K. 2008. Label-free cell-based functional assays. *Comb. Chem. High Throughput* 11(7): 573–580.

Overington, G., Fredriksson, R., and Schioth, H. 2006. How many drug targets are there? *Nat. Rev. Drug Discov.* 5: 993–996.

Owicki, J.C. 2000. Fluorescence polarization and anisotropy in high throughput screening: Perspectives and primer. *J. Biomol. Screen.* 5(5): 297–306.

Peters, M.F. and Scott, C.W. 2009. Evaluating cellular impedance assays for detection of GPCR pleiotropic signaling and functional sensitivity. *J. Biomol. Screen.* 14(3): 246–255.

Peters, M.F., Knappenberger, K.S., Wilkins, D., Sygowski, L.A., Lazor, L.A., Liu, J., and Scott, C.W. 2007. Evaluation of cellular dielectric spectroscopy, a whole-cell label-free technology for drug discovery on G_i-couples GPCRs. *J. Biomol. Screen.* 12(3): 312–319.

Pfleger, K.D.G. and Eidne, H.A. 2006. Illuminating insights into protein–protein interactions using bioluminescence resonance energy transfer (BRET). *Nat. Methods* 3(3): 165–174.

Pfleger, K.D.G., Seeber, R.M., and Eidne, K.A. 2006. Bioluminescence resonance energy transfer (BRET) for the real-time detection of protein-protein interactions. *Nature Protocols* 1(1): 337–345.

Pope, A.J., Haupts, U.M., and More, K.J. 1999. Homogeneous fluorescence readouts for miniaturized high-throughput screening: Theory and practice. *Drug Discov. Today* 4(8): 350–362.

Prinz, A., Diskar, M., and Herberg, F.W. 2006. Application of bioluminescence resonance energy transfer (BRET) for bimolecular interaction studies. *ChemBioChem* 7: 1007–1012.

Qureshi, S.A., Ding, V., Li, Z., Scalkowski, D., Biazzo-Ashnault, D.E., Xie, D., Saperstein, R., Brady, E., Huskey, S., Shen, X., Liu, K., Xu, L., Salituro, G.M., Heck, J.V., Moller, D.E., Jones, A.B., and Zhang, B.B. 2000. Activation of insulin signal transduction pathways and anti-diabetic activity of small molecule insulin receptor activators. *J. Biol. Chem.* 275: 36590–36595.

Reinscheid, R.K., Kim, J., Zeng, J., and Civelli, O. 2003. High-throughput real-time monitoring of G_s-coupled receptor activation in intact cells using cyclic nucleotide-gated channels *Eur. J. Pharmacol.* 478: 27–34.

Schroeter, T., Minond, D., Weiser, A., Dao, C., Habel, J., Spicer, T., Chaser, P., et al. 2008. Comparison of miniaturized time-resolved fluorescence resonance energy transfer and enzyme-coupled luciferase high-throughput screening assays to discover inhibitors of Rho-kinase II (ROCK-II). *J. Biomol. Screen.* 13(1): 17–28.

Selvin, P. 1995. *Fluorescence Resonant Energy Transfer.* San Diego: Academic Press.

Sharif, O., Hu, H., Klock, H., Hampton, E.N., Nigoghossian, E., Knuth, M.W., Matzen, J. et al. 2009. Time-resolved fluorescence resonance energy transfer and surface plasmon resonance-based assays for retinoid and transthyretin binding to retinol-binding protein 4. *Anal. Biochem.* 392(2): 162–168.

Shenoy, S.K. and Lefkowitz, R.J. 2003. Trafficking patterns of beta-arrestin and G protein-coupled receptors determined by the kinetics of beta-arrestin deubiquitination. *J. Biol. Chem.* 278(16): 14498–14506.

Simpson, P.B., Woollacott, A.J., Pillai, G.V., Maubach, K.A., Hadingham, K.L., Martin, K., Choudhury, H.I., and Guy, R. 2000. Pharmacology of recombinant human GABAA receptor subtypes measured using a novel pH-based high-throughput functional efficacy assay. *Seabrook J. Neurosci. Method* 99: 91–100.

Stein, R.A. 2008. Advantages of label-free cell-based assays. *Genetic Eng. Biotechnol. News Suppl.* to 28(18): 4–5.

Terrillon, S. and Bouvier, M. 2004. Receptor activity-independent recruitment of beta arrestin2 reveals specific signaling modes. *EMBO J.* 23(20): 3950–3961.

Trinquet, E., Fink, M., Bazin, H., Grillet, F., Maurin, F., Bourrier, E., Ansanay, H., et al. 2006. D-myo-Inositol 1-phosphate as a surrogate of D-myo-inositol 1,4,5-tris phosphate to monitor G protein-coupled receptor activation. *Anal. Biochem.* 358: 126–135.

Weber, M., Ferrer, M., Zheng, W., Inglese, J., Strulovici, B., and Kunapuli, P. 2004. A 1536-well cAMP assay for G_s- and G_i-coupled receptors using enzyme fragmentation complementation. *Assay Drug Dev. Technol.* 2(1): 39–49.

Wegener, J., Keese, C.R., and Giaever, I. 2000. Electric cell–substrate impedance sensing (ECIS) as a noninvasive means to monitor the kinetics of cell spreading to artificial surfaces. *Exp. Cell Res.* 259: 158–166.

Wise, A., Gearing, K., and Rees, S. 2002. Target validation of G protein-coupled receptors. *Drug Discov. Today* 7(4): 235–246.

Wu, P. and Brand, L. 1994. Resonant energy transfer: Methods and applications. *Anal. Biochem.* 218: 1–13.

Wunder, F., Tersteegen, A., Rebmann, A., Erb, C., Fahrig, T., and Hendrix, M. 2005. Characterization of the first potent and selective PDE9 inhibitor using a cGMP reporter cell line. *Mol. Pharmacol.* 68: 1775–1781.

Wunder, K., Kalthof, B., Muller, T., and Huser, J. 2008. Functional cell-based assays in microliter volumes for ultra-high throughput screening. *Comb. Chem. High Throughput Screen.* 11: 495–504.

Xia, M., Huang, R., Witt, K.L., Southal, N., Foster, J., Cho, M.H., Jadhav, A. et al. 2008. Compound cytotoxic profiling using quantitative high-throughput screening. *Environ. Health Perspect.* 116(3): 284–291.

Xing, J.Z., Zhu, L.J., Jackson, J.A., Gabos, S., Sun, X.J., Wang, X.B., and Xu, X. 2005. Dynamic monitoring of cytotoxicity on microelectronic sensors. *Chem. Res. Toxicol.* 18(2): 154–161.

Xu, Y., Piston, D.W., and Johnson, C.H. 1999. A bioluminescence resonant energy transfer (BRET) system: Applications to interacting circadian clock proteins. *Proc. Natl. Acad. Sci. USA* 96: 151–156.

Yamakawa, Y., Ueda, H., Kitayama, A., and Nagamune, T. 2002. Rapid homogeneous immunoassay of peptides based on bioluminescence romance energy transfer firefly luciferase. *J. Biosci. Bioeng.* 93: 537–542.

Zhang, B.B., Salituro, G.M., Szalkowski, D., Li, Z., Zhang, Y., Royo, I., Vilella, D., et al. 1999. Discovery of small molecule insulin mimetic with antidiabetic activity in mice. *Science* 284: 974–977.

Zhu, J., Wang, X.B., Xu, X., and Abassi, Y.A. 2006. Dynamic and label-free monitoring of natural killer cell cytotoxic activity using electronic cell sensor arrays. *J. Immunol. Methods* 309(1–2): 25–33.

Zlokarnik, G., Negulescu, P.A., Knapp, T.E., Mere, L., Burres, N., Feng, L., Whitney, M., Roemer, K., and Tsien, R.Y. 1998. Quantitation of transcription and clonal selection of single living cells wit h b-lactamase as reporter. *Science* 279: 84–88.

10 Ready-to-Use Commercial 3D Plates

10.1 INTRODUCTION

There are numerous commercially available products or kits with which an investigator can create their own single or multiwell 3D plates. For example, Millipore, Invitrogen, and 3DM are marketing a 3D collagen culture kit, laminin/collagen hydrogel, and a synthetic peptide hydrogel, respectively, for this purpose. These products offer a lot of flexibility for the investigator to tailor their 3D construct to their needs. However, in most cases the preparation process is time consuming and well-to-well and plate-to-plate variations are likely to be high. They are therefore most suitable for research in academic laboratories or for tissue engineering and/or regenerative medicine applications.

Founded in 2005 with the technology, personnel and intellectual property spun out of the disbanded Tissue Sciences Inc., RegeneMed Inc. started shipping engineered 3D cocultures of primary human liver stromal and parenchymal cells in 96-well plates in 2008. RegeneMed's focus is on delivering physiologically relevant cells to enable absorption, distribution metabolism, excretion, and toxicity (ADMET) studies to be performed earlier in drug discovery. It is not yet possible for customers to obtain a separate plate to create 3D cultures with their own cells. There are other companies with a similar 3D business model, some offering high and others offering low-throughput products. For example, CellnTec is marketing Cell Estrous—a vaginal cell system, and is developing 3D bladder and epidermal models. Another example is EpiDerm™, a 3D skin model based on keratinocytes that is marketed by MatTe Corporation. Another 3D product available in 96-well plates marketed by MatTe is EpiAirway™, a model for respiratory track tissue.

The scope of this chapter has been restricted to those products that are delivered from the vendor in a ready-to-use format for HTS, where the customer can incorporate in-house cells. Also, there are many products in which a coating such as Matrigel or a low-aspect-ratio surface modification is applied, for example, BD Biosciences' BioCoat™ cell ware products and SCIVAX NanoCulture®Plate marketed by InfiniteBio, Inc., of San Jose, California. These types of products are also not considered in this chapter because they do not provide the z-direction spatial dimension that is considered a prerequisite for true three-dimensionality, as discussed in Chapter 5. However, the NanoCulture®Plate is somewhat unique in that it is designed to promote spheroid formation. Apart from company publicity literature, no information is available at the time of writing from independent users to enable full coverage here. Additionally, to be consistent with the HTS theme of the book, only product examples with well densities of 96 or higher are featured, excluding for example, BD Biosciences' calcium phosphate, collagen composite, and open-cell-poly-lactic acid scaffolds that are offered in 24- and 48-well plates. Commercially available ready-to-use 3D plates for HTS cell-based assay can be grouped into "wet" and "dry" categories. The only wet product on the market is Extracel™ hydrogel from Glycosan. The dry products include AlgiMatrix™ from Invitrogen; Extracel™ sponge from Glycosan; and Ultra-Web from Corning. These products are available in the 96- or higher density well format. The Ultra-Web plates marketed by Corning fall into the "low-aspect ratio surface modification" category. However, an exception was made and Ultra-Web was included in this chapter because it is possible to fabricate nanofibrous electrospun multilayered constructs that provide the z-dimension (Srouji et al., 2008) or nanofibrous "mats" with large enough pores that cells can infiltrate deeper into the scaffold (Ekaputra et al., 2008).

The purpose of this chapter was to give the readers a deeper insight into the fabrication process of the products, a closer look at the resultant ultrastructure, and user experiences based on published works—where available, as well as the market entry opportunities and challenges—where applicable. The information on user experience is rather limited based on the fact that, at the time of writing, some of the products had been on the market for less than a year. More user experience will be available in a future book edition.

10.2 ALGIMATRIX™

On June 28, 2007, Invitrogen Corporation (http://www.invitrogen.com/3D-cellculture) launched the GIBCO(R) AlgiMatrix™ 3D sponge in 96-well plates for cell culture. The plate scaffold (sponge) is fabricated from pharmaceutical-grade alginate from brown seaweed. The alginate extracted from brown algae is made up of two uronic acids: D-mannuronic acid and L-guluronic acid. Polyvalent cations are responsible for interchain and intrachain reticulations because they are tied to the polymer when two guluronic acid residues are close. The reticulation process consists of the simple substitution of sodium ions with calcium ions (Gombotz and Wee, 1998). The relatively mild gelation process has enabled not only proteins, but also cells (Murtas et al., 2005) and DNA (Douglas and Tabrizian, 2005) to be incorporated into alginate matrices with retention of full biological activity.

10.2.1 FABRICATION

AlgiMatrix™ fabrication is a four-step process involving the preparation of sodium alginate stock solutions of 1–3% (w/v), crosslinking of alginate by adding, dropwise, the bivalent crosslinker (e.g., calcium gluconate), freezing the crosslinked alginate in a homogeneous, cool (−20°C) environment, and lyophilization to produce a sponge-like scaffold. The scaffold can be sterilized with ethylene oxide gas. Plates can be stored, aseptically, at room temperature until use. Figure 10.1 shows alginate chemical and the structure (a) and the ultrastructure (b and c). The structure shown was characterized by 90% porosity and pore size in the range of 50–200 μm in diameter (Shapiro and Cohen, 1997; Zmora et al., 2002). The wide range of pore sizes is

(a)

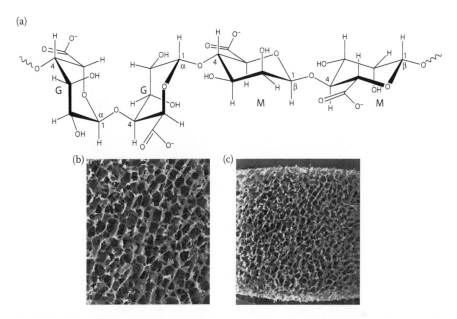

(b)

(c)

FIGURE 10.1 (a) Shows alginate copolymer chemical structure made up of two uronic acids: D-mannuronic acid (M) and L-guluronic acid (G), (b) and (c) show AlgiMatrix™ porous ultrastructure and pore-interconnectivity, respectively.

claimed to facilitate AlgiMatrix™'s superior cell loading, nutrient delivery, and potential for cell-to-cell interaction.

10.2.2 COMPLEX PHYSIOLOGICAL RELEVANCE

At the time of writing, this product had been on the market for approximately one year and no peer-reviewed literature was found about the use of this product. However, results are available from the work of Itskovits-Eldor's group at the Israel Institute of Technology that inspired the development of AlgiMatrix™. Gerecht-Nir et al. (2004) used larger scaffolds (5 mm diameter × 2 mm height) in 12-well plates and cultured human embryonic stem cells (hESCs) with the goal of forming human embryoid bodies (hEBs). The confining environment of the alginate scaffold pores enabled the efficient formation of hEBs with a relatively high degree of cell proliferation, and differentiation, encouraged round, small-sized hEBs and most importantly induced vasculogenesis in the forming hEBs to a greater extent in comparison to 2D or rotating wall vessel culture (Figure 10.2). Results provided in Invitrogen promotion literature, C3A human hepatocarcinoma, rat cardiomyocytes, and bovine aorta epithelial cells have been successfully cultured in the plate. C3A human hepatocarcinoma cells cultured in the plate expressed a significantly higher cytochrome P450 (CYP) 1A2 isoform relative to 2D control, 2% alginate gel, PuraMatrix, and Matrigel. CYP1A2 was induced by 24 h exposure to methylcholanthrene.

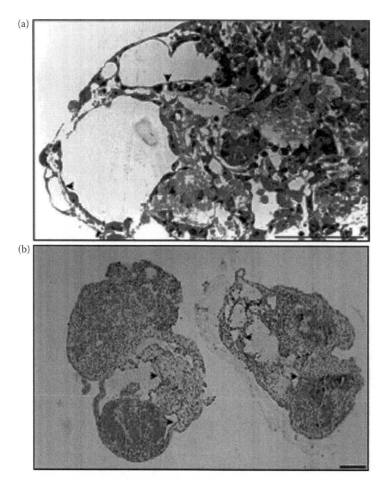

FIGURE 10.2 Vasculogenesis characterized by different voids surrounded by cells with elongated endothelial morphology indicated by arrowheads (a) and CD34+ cells surrounding the voids indicated by arrowheads (b). (Adapted from Gerecht-Nir, S., et al., 2004. *Biotechnol. Bioeng.* 88(4): 313–320.) CD34+ is a marker/antigen for endothelial cells and is also found on immature hematopoietic cells.

10.2.3 UNIQUE FEATURES

A key unique feature of AlgiMatrix™ is that it is fabricated from a natural material (alginate) that is animal-free. Some animal-sourced materials have been found to be contaminated with lactate dehydrogenase-elevating virus (LDEV), which is becoming an issue for Matrigel (Carlson et al., 2008), a solubilized basement membrane preparation, extracted from Engelbreth–Holm–Swarn mouse sarcoma tumor cells. Also, alginate scaffolds are biodegradable and this feature is highly desirable for tissue engineering and regenerative medicine application, but may not be a factor in HTS application depending on how long the culture needs to be maintained.

10.3 EXTRACEL™

Extracel™ or synthetic extracellular matrix (sECM) was originally developed by Glenn D. Prestwich's research group at the Center for Therapeutic Biomaterials (CTB) at the University of Utah (Shu et al., 2003; Liu et al., 2004). The hydrogel is a simple and effective biocompatible material that mimics the natural EMC with respect to stiffness and short-range chemistry (see Section 10.3.1). Glycosan BioSystems (Salt Lake City, UT; http://www.glycosan.com), a privately held life sciences technology company, was founded by Glenn Prestwich, William Tew, and Anna Scott in January 2006. The company received startup support from the Centers of Excellence Program for innovative research in Utah and from the University of Utah, and then closed its first round of investment in May 2007 to commercialize the technology for applications including pseudo 3D cell culture or regenerative medicine and controlled release of growth factors and antiproliferative drugs.

10.3.1 SYNTHESIS

Crosslinking the semisynthetic thiol-modified hyaluronan-like glycosaminoglycans (GAGs) with thiol-modified gelatin produces the hydrogel, now available as Extracel™. The hydrogel may be formed *in situ* in the presence of cells or tissue to provide an injectable cell-seeded hydrogel for regenerative medicine applications. For HTS applications, the hydrogel is aliquoted in 96-well plates and lyophilized to create a stiffer macroporous scaffold (Extracel™ sponge). Figure 10.3 shows the two-component model (a), the sponge microstructure (b), and the chemical reaction (c).

10.3.2 COMPLEX PHYSIOLOGICAL RELEVANCE

In the context of HTS applications, the physiological relevance of the scaffold has been demonstrated with primary hepatocytes (Prestwich et al., 2007). Freshly harvested hepatocytes from perfused rat liver were cultured under standard 2D, on collagen-coated plates, and on Extracel™ sponge. As a marker for physiological relevance, cytochrome P450 isoform CYP1A1 activity was monitored. As shown in Figure 10.4, hepatocytes cultured in 2D lost the enzyme activity, while those in the sponge demonstrated a cyclical activity, which is normal for healthy hepatocytes.

10.3.3 UNIQUE FEATURES

Extracel™ uniqueness includes mimicking the natural EMC with respect to stiffness and short-range chemistry. As presented in Chapter 7, substrate stiffness can modulate tissue phenotype. To be able to create an environment that very closely mimics the actual *in vivo* conditions requires the ability to independently manipulate the physical properties while keeping the chemical environment the same. This is a challenge for most scaffold constructs. Following a synthesis similar to the Extracel™ process, but using fibronectin in place of gelatin, Ghosh et al. (2007) have been able to precisely control the hydrogel's sheer modulus by altering the

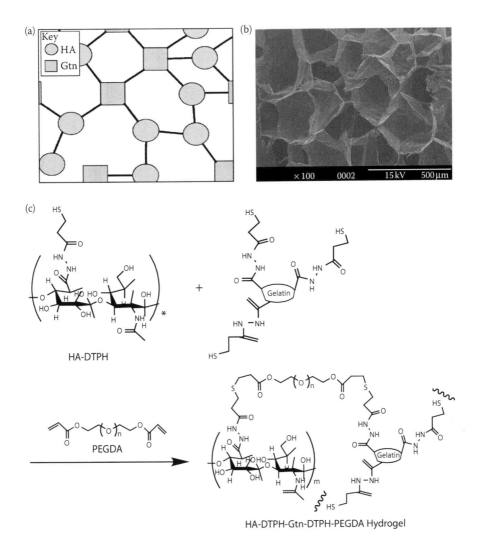

FIGURE 10.3 Schematic of the two-component matrix (a) and the 3D matrix ultrastructure (b) resulting from crosslinking of thio-modified hyaluronan (HA-DTPH), gelatin-DTPH, and poly(ethylene glycol) diacrylate (PEGDA) (c). (From Prestwich, G.D., et al. *Enzyme Regul.* 47: 196–207. With permission.)

crosslinker bulk density. Thus the hydrogel construct provides a single system in which both substrate mechanics (a function of crosslinker bulk density) and biochemical signaling (a function of ligand type and bulk density) can be independently altered and their effects on genotype and phenotype monitored. Using the system, Ghosh et al. (2007) found that human dermal fibroblast migrated faster on softer substrates while proliferating preferentially on stiffer ones. This probably offers the explanation why a single Extracel™ composition is not optimal for every cell type application. For example, stem cells prefer lower gelatin percentage in the

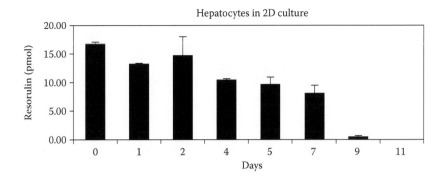

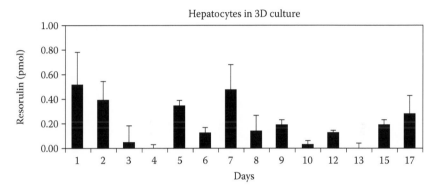

FIGURE 10.4 Comparison of cytochrome P1A1 activity measured as 7-ethoxyresorufin-*O*-deethylase activity during culture for 2D (collagen-coated polystyrene dish, top) and 3D (Extracel™ sponge, bottom). (From Prestwich, G.D., et al. 2007. *Adv. Enzyme Regul.* 47: 196–207. With permission.)

hyaluronan–gelatin mixture (Duflo et al., 2006). CTB is in the process of developing tools to assist in selecting the optimal composition and compliance for a given application.

Extracel™'s second unique feature lies in its synthesis that is modeled after natural ECM. The ECM is a heterogeneous collection of covalent and noncovalent molecular interactions consisting primarily of proteins and GAGs. In the ECM, covalent interactions connect chondroitin sulfate, heparin sulfate, and other sulfated GAGs to core proteins forming proteoglycans. Noncovalent interactions include binding of link modules of proteoglycans to hyaluronan, electrostatic association with ions, hydration of the polysaccharide chains, and triple helix formation to generate collagen fibrils. The synthetic ECM consisting of hyaluronan and gelatin provides resemblance to natural ECM.

10.4 ULTRA-WEB™

A commercial Ultra-Web™ fabrication process was invented by Chung and coworkers, and then used at the Donaldson Co. of Minneapolis, Minnesota, which specializes in filtration. However, the use of electrospun nanofibrous scaffolds in cell culture was first published by Yang et al. (2004), who successfully cultured and differentiated neural stem cells culture on the structures. The use of Ultra-Web™ structures as cell culture scaffolds was simultaneously published by Schindler et al. (2005) and Nur-E-Kamal et al. (2005). Products based on this technology are currently being marketed by SurModics Inc. (Eden Prairie, MN) and Corning (Lowell, MA).

10.4.1 FABRICATION

The Ultra-Web™ nanofibers (Figure 10.5b and c) are electrospun using the industrial electrospinning process developed by Chung et al. (2004). A schematic of the process is shown in Figure 10.5a. A field strength of 30 kV was originally used and the fibers were electrospun on glass coverslips (18 mm, No. 1, in a controlled thickness and fiber density). Polyamide polymers, $(C_{28}O_4N_4H_{47})_n$ and $(C_{28}O_4 \cdot 4N_4H_{47})_n$, crosslinked in the presence of an acid catalyst, were used. As shown in Figure 10.5b and c, the resulting nanofibers are organized into fibrillar networks reminiscent of the architecture of the basement membrane, a form of extracellular matrix.

10.4.2 COMPLEX PHYSIOLOGICAL RELEVANCE

Schindler et al. (2005) cultured NIH 3T3 fibroblasts, NRK, and breast epithelial cells (3T3) on Ultra-Web. NRK cells displayed morphology and characteristics similar to

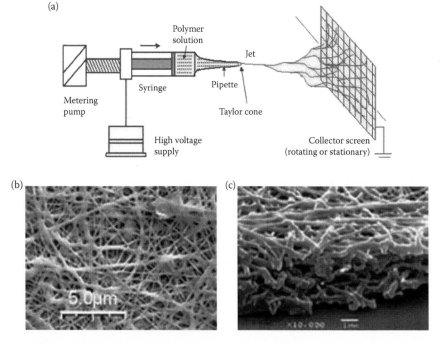

FIGURE 10.5 Schematic of a typical electrospinning setup (a) and scanning electron microscope images showing the random orientation of the polyamide nanofibers of the Ultra-Web surface (b) and cross-section (c). (SEM images provided by Corning.)

their counterparts *in vivo*. Breast epithelial cells formed multicellular spheroids containing lumens (Figure 10.6), a morphogenesis not observed with standard 2D substrates. In a more recent publication, Ahmed et al. (2007) cultured mouse embryonic fibroblast and observed a dynamic redistribution of myosin II-B previously observed with nanofibers composed of collagen I but not for cells on standard 2D substrates coated with monomeric collagen.

10.4.3 Unique Features

One of the unique features is that the scaffold is very easy to fabricate and to incorporate into standard high-density plastic plates. The scaffold does not seem to require

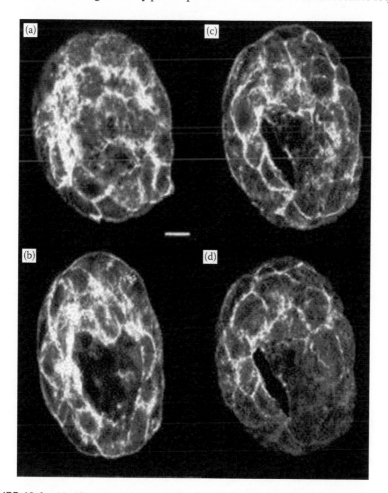

FIGURE 10.6   (a)–(d) are a series of confocal sections of T47D breast epithelial multicellular spheroids cultured on Ultra-Web nanofibers and stained with phalloidin–Alexa Fluor and taken at 0 (a), 20 (b), 34 (c), and 48 (d) μm steps from the top. The lumen extending through the spheroid is observable that was absent in cells cultured on 2D glass (not shown). (Adapted from Schindler, M. et al. 2005. *Biomaterials* 26: 5624–5631.)

the *z*-dimension to yield cells that emulate *in vivo* situation. Like the NanoCulture® Plate, introduced at the beginning of the chapter, Ultra-Web plates may in some cases yield three dimensionality through the promotion of spheroid formation as described above. The question that confronts investigators with structures like Ultra-Web, which only emulate *in vivo* nanostructure surface topology, is whether differences in cell phenotype produced more accurately emulate *in vivo* or they are just different. In the absence of validated three-dimensionality markers, it is a challenge to answer this question. This kind of question provides justification for the search for biomarkers. This topic has already been discussed in detail in Chapter 8.

10.5 MARKET OPPORTUNITIES

10.5.1 THE OPPORTUNITY

The target-driven drug discovery approach created around HTS laboratories was first implemented in the pharmaceutical/biotechnological industries in the early 1990s. On average, HTS laboratories have been in place at these industries for eight years. Three drugs resulting from this approach (Gleevec®, Paretin®, and Tagretin®) are now on the market. Although it takes 5–10 years to advance drug candidates from screening through launch, the results of the target-driven approach have been disappointing, with few commercial therapies that can be traced back to HTS leads. As pointed out in Chapter 2, the disappointing results have been attributed to low-quality leads, which has led to a high attrition rate of investigational compounds in the preclinical, clinical, and postapproval phases. Use of more physiological relevant screening platforms, based on 3D cell-based assays, early in drug development will likely yield quality leads, reducing the overall drug attrition rate. According to a market study conducted by HighTech Business Decisions (http://www.hightechdecisions.com) in 2007, approximately 50% of HTS assays conducted are 2D cell based and more HTS laboratories are expected to increase screening with cells. The adoption of 3D cultures represents an interesting opportunity; however, there are challenges—discussed in detail in Chapter 11—which are not insurmountable.

10.5.2 POTENTIAL CUSTOMERS

The potential customers are pharmaceutical and biotechnological laboratories that can be subdivided into three groups. The first group (high-throughput group) has average weekly throughput that involves the reading of 100,000 wells or more per week. High-end reads are in several million wells/week. A typical customer belonging to this group is one of the "big pharma" companies like Pfizer. The second group (medium throughput) has average weekly throughputs that involve reading fewer than 100,000 wells per week. A typical customer belonging to this group would have about five people in the laboratory in a biotechnological company like Tularik Inc. (San Francisco, CA). The third group is noncommercial laboratories (academic centers and government institutes). These laboratories read a wide range of wells per week that may vary from year to year depending on funding. Ten such centers have been funded through the NIH Roadmap—Molecular Libraries and Imaging at

National Human Genomic Research Institute, University of Pennsylvania, Emory University, University of Pittsburg, Southern Research Institute, The Burhman Institute, The Scripps Institute, Columbia University Medical Center, University of New Mexico—Albuquerque, and Vanderbilt University.

10.5.3 MARKET SIZE

In 2006, the worldwide HTS market size by product/service category totaled US$1.4B (12% consumables, 22% reagents and cell lines, 31% equipment, and 35% other). The "other" category includes compound libraries, data management software, bioinformatics, custom cloning, target generation, assay development and optimization, protein production, tissue automation, structure determination, outsourced screening, compound library service contracts, instrument training services, lead optimization, secondary screening, and *in vitro* absorption–distribution–metabolism–excretion/toxicology (ADME/Tox) services. Market drivers for HTS in 2008, identified by a HighTech Business Decisions (2007) market study that included 87 respondents from pharmaceutical, biotechnological, and institute HTS laboratories, and HTS supplier companies, are outlined in Box 10.1.

In their market study, HighTech Business Decisions asked HTS laboratory directors to report on the number of 96- or higher-density well plates they used in 2006. Twenty-four directors provided this number—12 from high-throughput group, 8 from medium-throughput group, and 4 from noncommercial group. In total, the 24 respondents use approximately 0.64 M plates in 2006. It should be noted that only the 24 respondents mentioned above constituted approximately half of the total that returned the survey. Some of the other half indicated that they were unsure of the exact numbers. These figures provide a basis for estimating the total number of plates used in 2006 and, together with the 7.8% predicted growth in Research and Development spending by pharmaceutical and biotechnology companies (HighTech Business Decisions, 2007), a reasonable market size of the proposed plates can be estimated as outlined in the paragraph below.

10.5.4 MARKET SIZE ESTIMATION

(1) Given that approximately half of the respondents in the HighTech Business Decisions study reported plate usage, the number of plates for the study participants is doubled to 1.28 M. (2) Assuming a 10% response—typical of mail surveys—from the number of surveys sent out, the above figure is multiplied by 10 to arrive at an estimate of the number of plates used in 2006 in the United States of 12.8 M plates. (3) Using the 7.8% growth rate, five-year plate use projection comes to the following: 2008 (1.50 M), 2009 (1.62 M), 2010 (1.75 M), 2011 (1.88 M), and 2012 (2.03 M). (4) Assuming a conservative market capture of 30% by the 3D plates and pricing of $100 per plate yields a 3D market value of $0.3 \times 2.03 \text{ M} \times \$100 = \$60.9 \text{ M}$ in 2012. In a separate market study by Frost and Sullivan (cited in *Genetic Engineering & Biotechnology News*, April 15, 2008, issue), US cell-based assay revenue in 2007 was estimated at $112.5 M and is projected to reach $232.5 by 2014. In another Frost and Sullivan study (http://www.drugdiscovery.frost.com), European cell-based assay

---

BOX 10.1   MARKET DRIVERS FOR HTS IN 2008

HTS laboratories are pushing to increase screening capacity and through-puts through the use of more cell-based assays and high content screening (and/or 3D) technologies for more physiologically relevant results and to find higher quality leads

HTS laboratories are shifting to novel targets and target classes and are working with more diverse compounds

HTS laboratories are seeking more robust assays, including sensitivity to inhibitors, biologically relevant assays, and an efficient ability to assess targets

HTS laboratories are looking for ways to make *in vitro* assay amenable to HTS

HTS laboratories drive for efficiency through the use of improved liquid handling tools and automated cell tools

HTS laboratories have expanding roles in target validation, assay development, secondary screening, ADME/Tox, and lead optimization

HTS laboratories area seeking label-free technology to avoid interferences and improve results

HTS laboratories are seeking more data analysis software to obtain meaningful information, especially for HCS (and/or 3D), and *in silico* tools

HTS laboratories are continuing to miniaturize to reduce screening time and cost

HTS laboratories are pushing to profile compound libraries, cherry-pick compounds, and conduct focused screens, when possible

HTS laboratories are increasing the use of outsourcing to access expertise and capacity, save time, and lower costs

HTS laboratories rely on collaborations and partnering to increase success

---

market kits earned revenues of $66.2 M in 2007 and were estimated to reach $220.1 in 2014. The proposed 3D plate market value of $60.9 M is considered reasonable in light of these projections.

10.6 CONCLUDING REMARKS

The three products featured are compared side by side in Table 10.1. A quick glance at the table immediately identifies a key challenge of 3D cell-based systems' entry into the HTS market. Given that the most popular HTS instrumentation readouts are optical and in particular fluorescent in nature (Chapter 9), the paucity of optically transparent 3D plates presents a challenge and a product development opportunity, which a number of laboratories including the author's Cellular Bioengineering laboratory are currently addressing. The last three chapters (Chapters 11 through 13)

TABLE 10.1
Comparison of 3D Cell Culture Products

Attribute	AlgiMatrix™	Extracel™	Ultra-Web™
Company	Invitrogen	Glycosan	Corning
Material	Natural—alginate from brown seaweed	Natural—hyaluronic acid and denatured collagen	Synthetic— polyamide (nylon)
Pore size range	30–300 μm	N/A	300–500 nm
Fiber size	N/A	N/A	~280 nm
Mechanical strength	Medium	Low	Medium
Optical property	Opaque	Transparent	Opaque
Ready-to-use	Yes	Yes	Yes
Cell recovery	Trypsinization	Recovery solution provided	Trypsinization
Well-density availability	96-well plate	96-well plate	96-well plates

explore the major challenges and opportunities for developing and deploying 3D cell-based biosensing platforms in HTS. The discussion of challenges is expanded in Chapter 11. The question of evidence in support of the value 3D cultures provide is presented in Chapter 12 in a study case format and, lastly, an ideal case study design for those that might be interested in generating more supportive evidence concludes the book (Chapter 13).

REFERENCES

Ahmed, I., Ponery, A.S., Nur-E-Kamal, A., Kamal, J., Meshel, A.S., Sheetz, M.P., Schindler, M., and Meiners, S. 2007. Morphology, cytoskeleton organization, and myosin dynamics of mouse embryonic fibroblasts cultured on nanofibrillar surfaces. *Mol. Cell Biochem.* 301: 241–249.

Carlson, J., Garg, R., Compton, S.R., Zeiss, C., and Uchio, E. 2008. Poliomyelitis in SCID mice following injection of basement membrane matrix contaminated with lactate dehydrogenase-elevating virus. *J. Am. Assoc. Lab. Anim. Sci.* 47(5): 80–81.

Chung, H.Y., Hall, J.R.B., Cogins, M.A., Crofoot, D.G., and Wiek, T.M. 2004. Polymer, polymer microfiber, polymer nanofiber, and applications including filter structures. United States Patent No. 6,743,273 B2; June 1.

Douglas, K.L. and Tabrizian, M. 2005. New approach to the preparation of alginate-chitosan nanoparticles as gene carriers. *J. Biomater. Sci. Polym. Edn.* 16(1): 43–56.

Duflo, S., Thibeault, S.L., Li, W., Shu, X.Z., and Prestwich, G.D. 2006. Vocal fold tissue repair *in vivo* using a synthetic extracellular matrix. *Tissue Eng.* 12: 1–10.

Ekaputra, A.K., Prestwich, G.D., Cool, S.M., and Hutmacher, D.W. 2008. Combining electrospun scaffolds with electrosprayed hydrogel leads to three-dimensional cellularization of hybrid constructs. *Biomacromolecules* 9: 2097–2103.

Gerecht-Nir, S., Cohen, S., Ziskind, A., and Itskovitz-Eldor, J. 2004. Three-dimensional porous alginate scaffolds provide a conducive environment for generation of well vascularized embryoid bodies from human embryonic stem cells. *Biotechnol. Bioeng.* 88(4): 313–320.

Gombotz, W.R. and Wee, S.F. 1998. Protein release from alginate matrices. *Adv. Drug Deliv. Rev.* 31: 267–285.

Ghosh, K., Pan, Z., Guan, E., Ge, S., Liu, Y., Nakamura, T., Ren, X.-D., Rafailovich, M., and Clark, R.A.F. 2007. Cell adaptation to a physiologically relevant ECM mimic with different viscoelastic properties. *Biomaterials* 28: 6671–6679.

HighTech Business Decisions. 2007. *High Throughput Screening 2007: New Strategies, Success Rates, and Use of Enabling Technologies*, vols. I and II. A market study report by HighTech Business Decisions (www.hightechdecisions.com), Can Jose, CA.

Liu, Y., Shu, X.Z., Gray, S.D., and Prestwich, G.D. 2004. Disulfide-crosslinked hyaluronan-gelatin sponge: growth of fibrous tissue *in vitro*. *J. Biomed. Mater. Res. A* 68: 142–149.

Murtas, S., Campuani, G., Dentini, M., Manetti, C., Masci, G., Massimi, M., Miccheli, A., and Crescenzi, V. 2005. Alginate beads as immobilization matrix for hepatocytes perfused in a bioreactor. *J. Biomater. Sci. Polym. Edn.* 16(7): 829–846.

Nur-E-Kamal, A., Ahmed, I., Kamal, J., Schindler, M., and Meiners, S. 2005. Three dimensional nanofibrillar surface induces activation of Rac. *Biochem. Biophys. Res. Commun.* 331(2): 428–434.

Prestwich, G.D., Liu, Y., Yu, B., Shu, X.Z., and Scott, A. 2007. 3-D culture in synthetic extracellular matrices: New tissue models for drug toxicology and cancer drug discovery. *Adv. Enzyme Regul.* 47: 196–207.

Schindler, M., Ahmed, I., Kamal, J., Nur-E-Kamal, A., Grafe, T.H., Chung, H.Y., and Meiners, S. 2005. A synthetic nanofibrillar matrix promotes *in vivo*-like organization and morphogenesis for cell culture. *Biomaterials* 26: 5624–5631.

Shapiro, L. and Cohen, S. 1997. Novel alginate sponges from cell culture and transplantation. *Biomaterials* 18: 583–590.

Shu, X.Z., Liu, Y., Palumbo, F.S., and Prestwich, G.D. 2003. Disulfide-crosslinked hyaluronan-gelatin hydrogel films: A covalent mimic of extracellular matrix for *in vitro* cell growth. *Biomaterials* 24: 3825–3834.

Srouji, S., Kizhner, T., Suss-Tobi, E., Livne, E., and Zussman, E. 2008. 3-D nanofibrous electrospun multilayered construct in an alternative ECM mimicking scaffold. *J. Mater. Sci: Mater. Med.* 19: 1249–1255.

Yang, F., Murugan, R., Ramakrishna, S., Wang, X., Ma, Y.-X., and Wang, S. 2004. Fabrication of nano-structured porous PLLA scaffold intended for nerve tissue engineering. *Biomaterials* 25: 1891–1900.

Zmora, S., Glicklis, R., and Cohen, S. 2002. Tailoring the pore architecture in 3-D alginate scaffolds by controlling the freezing regime during fabrication. *Biomaterials* 23: 4087–4094.

Part IV

---

*Technology Deployment
Challenges and Opportunities*

11 Challenges to Adopting 3D Cultures in HTS Programs

11.1 TYPICAL HTS LABORATORY AND ASSAY CONFIGURATIONS

In a typical drug discovery program, HTS laboratory activities are usually one of many components that make up the overall process from target identification to hit confirmation. For illustrative purposes, we present the Emory University Molecular Library Screening Center (MLSC) process flow (Figure 11.1), in which HTS activities are placed in relation to the overall center approach to drug discovery. Not all the elements shown in Figure 11.1 will be present in other academic laboratories, let alone pharmaceutical/biopharmaceutical companies. But the central themes of assay development, namely compound libraries, HTS, structure optimization, and secondary bioassay, are most likely common. Because such centers are state or federally funded, the flowchart begins with securing funding and ends with uploading data to PubChem (http://pubchem.ncbi.nih.gov/), a free database of chemical structures of small organic molecules and information on their biological activity. The HTS laboratory is highlighted at the center of the overall process flow. Inspired by the NIH Roadmap Molecular Libraries and Imaging Program, academic institutions worldwide have established centers to facilitate the discovery of small molecules that might eventually be developed into therapeutics. Emory University is one of many such institutions that have established HTS core facilities. A list compiled by the Society for Biomolecular Sciences can be found at http://web.memberclicks.com/mc/directory/viewallmembers.do. When last accessed (January 2009), the directory comprised 57 screening centers. Most centers list types of libraries (e.g., small molecules) and size (number of compounds in the library), therapeutic or target focus, as well as readout (transduction) technologies available at the facility.

A functional HTS laboratory is typically equipped with stand-alone products (workstations) and/or robotic systems. For illustrative purposes, Figure 11.2 presents a schematic of a customized HTS screening laboratory. In general, a workstation is a device that performs multiple different functions on multiwell plates. As such, plate washers, dispensers, and readers are considered stand-alone devices, while liquid handlers, which combine multiple tools or which can perform multiple actions on a multiwell plate, are considered workstations. Robotic systems are a collection of stand-alone devices, workstation, and one or two robotic arms integrated into a functional environment controlled by system management software. As discussed in Chapter 10, commercial multiwell plates have been developed in which scaffolds have been successfully integrated in standard multiwell plates. Given that the

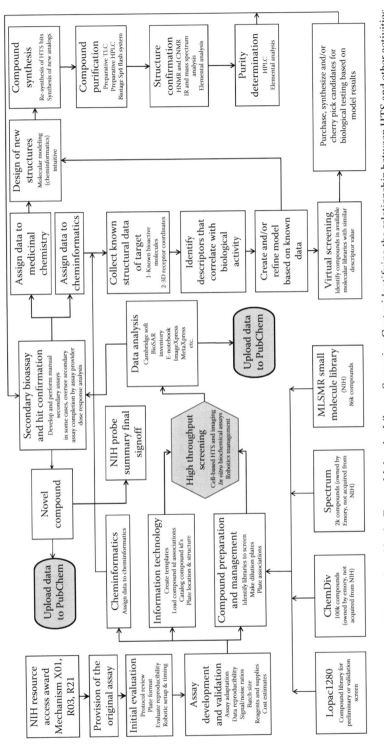

FIGURE 11.1 Process flow employed in the Emory Molecular Library Screening Center identifying the relationship between HTS and other activities. (Adapted from http://www.emory.edu/chemical-biology/workflow.html. Last accessed in January 2009. With permission.)

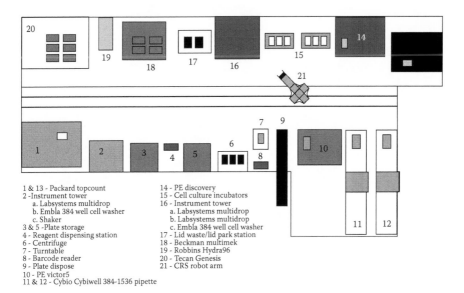

1 & 13 - Packard topcount
2 -Instrument tower
   a. Labsystems multidrop
   b. Embla 384 well cell washer
   c. Shaker
3 & 5 -Plate storage
4 - Reagent dispensing station
6 - Centrifuge
7 - Turntable
8 - Barcode reader
9 - Plate dispose
10 - PE victor5
11 & 12 - Cybio Cybiwell 384-1536 pipette

14 - PE discovery
15 - Cell culture incubators
16 - Instrument tower
   a. Labsystems multidrop
   b. Labsystems multidrop
   c. Embla 384 well cell washer
17 - Lid waste/lid park station
18 - Beckman multimek
19 - Robbins Hydra96
20 - Tecan Genesis
21 - CRS robot arm

FIGURE 11.2 Schematic illustration of a customized HTS Laboratory equipped with a 6-m linear track, **Packard Topcount** (1 and 13), instrument tower (2: **Labsystems Multidrop; Embla 384 well cell washer; and Shaker**), **plate storage** (3 and 5), reagent dispensing station (4), centrifuge (6), **turntable** (7), **barcode reader** (8), **plate dispose** (9), Perkin Elmer Victor5 (10), **Cybio Cybiwell 384-1536 pipette** (11 and 12), PE Discovery (14), **cell culture incubators** (15), instruments tower (16: **Labsystems Multidrop; Labsystems Multidrop; Embla 384 well cell washer**), **lid washer/lid park station** (17), Beckman Multimek (18), Robbins Hydra96 (19), Tecan Genesis (20), and **CRS robot arm** (21). Instruments in bold were used in a scintillation proximity assay (SPA) for the discovery of inositol phosphatase inhibitors. (From Zheng, W. et al. 2004. *J. Biomol. Screen.* 9(2): 132. With permission.)

handling of multiwell plates is very well developed in laboratory automation, incorporation of 3D plates into HTS operations does not present unique challenges from a physical handling standpoint.

Unlike automation, signals from cells (biosensor) in response to test compounds to readout instruments in HTS assays require a closer examination. In the 57 academic screening core facilities from the Society for Biomolecular Sciences Directory, most HTS readout technologies are predominantly based on fluorescence (including genetically encoded fluorescence sensors) or radioactive labels. Although no reliable information is available from HTS laboratories in bio/pharmaceutical industry, the readout technology distribution in the academic laboratories is probably an accurate reflection of the industry situation. Figures 11.3 and 11.4 provide schematics of representative fluorescence and radiolabel HTS assays, respectively. There are numerous variations in biosensor/bioassay configurations from the formats presented above, but the common themes in all these assays are (1) engineered cells to overexpress targets and/or biosensor (reporter) moieties, and (2) use of fluorescent dyes to tag or label targets of interest.

Engineering of cells is a time-consuming process. Moreover, overexpression of receptor and/or reporter genes as well as use of fluorescent dyes can potentially

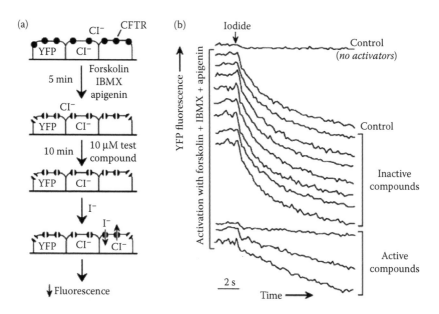

FIGURE 11.3   Schematic illustration of a fluorescence HTS assay for identifying potentiators of a defective (ΔF508 mutation) cystic fibrosis transmembrane conductance regulator (CFTR) channel. (a) Fisher rat thyroid cells coexpressing ΔF508-CFTR and a halide (chloride or iodide) yellow fluorescence protein biosensor (YFP-H148Q/I148L) were grown for 24 h at 27°C to allow plasma membrane ΔF508-CFTR expression. Cells were washed and forskolin (agonist stimulator; 20 µM) and test compounds (collection of 100,000 diverse drug-like small molecules; 2.5 µM) were added. After adding I⁻ to the external solution, I⁻ influx was monitored from the time course of the biosensor fluorescence. (b) Representative time courses of biosensor fluorescence in control wells (saline, negative control; 50 µM genistein, positive control) with examples of inactive and active compounds. Over 30 compounds that potentiate ΔF508-CFTR Cl⁻ channel activity were found. (Adapted from Verkman, A.S., Lukacs, G.L., and Galietta, L.J.V. 2006. *Curr. Pharm. Des.* 12: 2235–2247.)

disrupt the intricate movements and interactions of intracellular proteins that determine signaling pathways (Atienza et al., 2006). Also, the inherent assumption in monitoring a single molecule as is the case in most fluorescence and radiolabel assays is that one signaling pathway corresponds to one ligand–receptor complex. Although the assumption has served the drug discovery process in the bio/pharmaceutical industries well, recent results suggest that it does not capture well the true complex and dynamic intracellular signaling (Sun et al., 2007). The development of label-free biosensors is being fueled by the need to address this traditional biosensor limitation, as outlined before in Chapter 9. Biacore, now part of GE Healthcare, pioneered the use of SPR as a label-free means of quantifying molecular interactions. Biacore, like many label-free biosensors that are either in development (Morrow, 2008) or commercially available (Table 9.3), is not configured for cell-based functional assays, but is serving well in a complementary manner (Abdiche et al., 2008). The Epic System, covered in detail in Chapter 9, was launched in September 2006 by Corning Life Sciences. Epic is a label-free screening platform that identifies hits and leads based on SPR

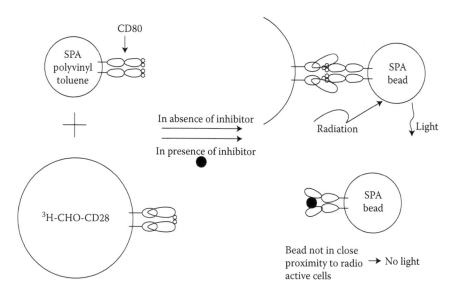

FIGURE 11.4 Schematic illustration of a radiolabel (scintillation proximity) HTS assay for identifying selective small-molecule CD80 inhibitors. The costimulatory pathway involving CD80–CD28 interaction plays a critical role in regulation of the immune response, making it an attractive target for therapeutic manipulation of autoimmune diseases. CD80 was presented in a multivalent manner on SPA beads by capture through the IgG domain. QCD28 was expressed in CHO cells that were metabolically labeled ([4,5-^{3}H]leucine). CHO cells were incubated for 18–20 h at 37°C in a CO_2 incubator and were dislodged with nonenzymatic cell dissociation solution and resuspended in assay buffer prior to use in SPA assay. As shown, in the presence of an inhibitor, binding to CD80, the bead is not in close proximity to cells and, on the contrary, in the absence of inhibitors, radio emission from cells interacts with the bead and detectable light is emitted. The assay yielded a hit rate of 0.03% from a total of approximately 4000 test compounds. (Adapted from Uvebrant, K. et al. 2007. *J. Biomol. Screen.* 12(4): 464–472.)

principles or the refractive index of light, also covered in Chapter 9. Lee et al. (2008) have demonstrated the utility of the system with pharmacological evaluation of cellular responses to GPCR activation and the results suggested that as many as 85% of GPCR compounds thought to be antagonists really behave as inverse agonists.

The challenge of interfacing label-free readouts with 3D cultures has been discussed in Chapter 9 to include limited specificity and detection zone. The requirement to suspend cells in radiolabel assays is not compatible with the benefits 3D cell cultures are purported to offer as it is not possible for beads to penetrate deep into 3D culture tissue. However, it may be possible for mass produced cells in 3D cultures (Sanchez-Bustamante et al., 2005) to maintain their desirable 3D attributes soon after dissociation from 3D tissue and suspension in assay buffer. Studies to confirm or rule out this possibility are under way in the author's laboratory. In the absence of such results, fluorescence-based readout remains the main potential target for 3D cultures. However, as discussed in the concluding remarks of Chapter 9, re-engineering the biosensor with 3D in mind has potential to improve readout performances. Despite the benefits that 3D cultures offer, adoption in HTS programs is slow. The possible reasons, including the emerging just-in-time preparation of the reagents model, perceived limited value-addition from 3D culture physiological relevance, and general paucity of data in support of 3D culture superiority are further explored below.

11.2   JUST-IN-TIME REAGENTS PROVISION MODEL

The increasing use of cell-based assays in HTS laboratories has resulted in exponential demand for cells. For example, a typical cell-based HTS at Pfizer, Sandwich, UK, uses 1×10^{10} plated cells (Cawkill and Eaglestone, 2007) and this number does not include cell stock maintenance and assay development requirements. The traditional manual production of cells has failed to meet such a demand and in addition the need to increase quantity has resulted in lower quality (Terstegge et al., 2007), making cell provision a limiting factor in HTS. The challenge to meet this demand is being met through several recent innovations reflected in the Cawkill and Eaglestone (2007) proposal for a desired optimal distribution of reagents provision with respect to robust and timely screening in the bio/pharmaceutical industry (Figure 11.5). These innovations include (1) decoupling cell production from screening through use of cryopreserved (frozen) cells (Zaman et al., 2007), (2) increasing cell quantity while maintaining quality through automated production (Terstegge et al., 2007), and (3) outsourcing to increase overall capacity.

Cryopreserved cells have been successfully used with many cell types and assays widely used in HTS (Table 11.1), yielding results that are comparable to assay with cells that are continuously maintained in culture. The three main advantages of being frozen in HTS and other downstream drug discovery processes are (1) increased flexibility—when compounds are available, testing can start without having to wait for cells to be produced, (2) improved consistence—the same cell batch can be used for all testing, and (3) reduced cost due to not having to maintain cells in culture and the maintenance of the cell culture facility. Thus, by separating the cell-culturing from drug screening activities through the use of frozen cells, the logistics of cell-based assays are approaching the logistics of biochemical assays with respect to the

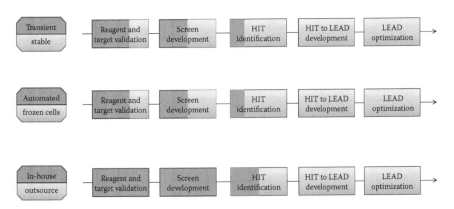

FIGURE 11.5   Reagent provision distributions to facilitate robust and timely cell production for primary screening in the biopharmaceutical industry. (From Cawkill, D. and Eaglestone, S.S. 2007. *Drug Discov. Today* 12(19/20): 824. With permission.)

TABLE 11.1

Examples of Successful Use of Cryopreserved Cells

Target	Target Class	Host Cell Line	Cell Supply	Readout	Reference
Very late antigen-4 (VLA-4)	Adhesion molecule	Ramos	Frozen	Fluorescence	Weetall et al. (2001)
Hepatitis C virus replication	Replication proteins	Huh-7	Frozen	β-lactamase reporter	Zuck et al. (2004)
Gαq-coupled receptor[a]	GPCR	CHO	Frozen and division-arrested[b]	Ca²⁺ (FLIPR)	Kunapuli et al. (2005)
Gαs-coupled receptor[a]	GPCR	CHO	Frozen and division-arrested	Enzyme fragment complementation	Kunapuli et al. (2005)
Potassium channel	Ion channel	HEK293	Frozen and division-arrested	Ca²⁺ (FLIPR)	Kunapuli et al. (2005)
hERG	Ion channel	CHO	Frozen	Rb⁺ efflux	Ding et al. (2006)
Pregnane X receptor	Nuclear receptor	HepG2	Frozen transients[c]	Luciferase reporter	Zhu et al. (2007)
Vanilloid-1 receptor	Ion channel	NK[d]	Frozen transients	Ca²⁺	Bianchi et al. (2007)
Vanilloid-4 receptor	Ion channel	NK[d]	Frozen transients	Ca²⁺	Bianchi et al. (2007)
Melastatin-8 receptor	Ion channel	NK[d]	Frozen transients	Ca²⁺	Bianchi et al. (2007)

Source: Modified from Zaman, G.J.R. et al. 2007. *Drug Discov. Today* 12: 523.

[a] Target name was not disclosed.

[b] For large screening jobs, there can be a considerable time lag between processing of the first and the last assay plates and cell doubling may cause significant plate-to-plate viability. Cells can be division-arrested by suspension in serum-free medium or treated with mitomycin C (a natural antitumor antibiotic) to arrest mitosis (Digan et al., 2005; Fursov et al., 2005).

[c] Transiently transfected cells.

[d] Not known.

just-in-time provision of reagents and the validation of large reagent batches. It is not yet clear how willing industry might be to outsource the proprietary cell lines for frozen cell production and delivery when needed.

Stable cell line development comprises three stages: (1) transfection and selection of a heterogeneous pool of cells, (2) isolation of cell clones and screening for suitability of use, and (3) clone characterization in assay formats of choice. Down- and upstream drug discovery activities rely on stable cell lines, the development of which is time consuming (Herman, 2004). Numerous studies have recently appeared with the goal of reducing the time for stable cell line development (Goetz et al., 2000; Carroll and Al-Rubeai, 2004; Burke and Mann, 2006). The transient nature of transient transfection has rendered the approach unsuitable for generating cells for use over a number of years. The use of frozen cells has changed the status quo, enabling the rapid construction and provision of cells as a reagent, reducing the lead time from months to weeks (Zhu et al., 2007; Chen et al., 2007).

Fitting 3D cultured cells in the just-in-time reagent provision model has not yet happened, but is not out of the question. It has not happened because, for cells to maintain their 3D phenotype, they require maintaining in the microenvironment in which they were grown. Studies have shown gene expression changes not long after cells are transferred from 3D to 2D microenvironments (Duggal et al., 2009; Parreno et al., 2008; also see Chapter 3). Fitting 3D cultures in the just-in-time reagent provision is not out of the question because the next intuitive step in the model is to combine the benefits of automation and frozen cells by cryopreserving 2D cultures directly in assay plates. The automated cell culture system will directly produce plates for freezing. Successfully deploying this strategy will require addressing the storage of large numbers of plates (Cawkill and Eaglestone, 2007).

11.3 LIMITED VALUE-ADDITION FROM 3D CULTURE PHYSIOLOGICAL RELEVANCE: TRANSEPITHELIUM DRUG TRANSPORT AND INDUCTION OF DRUG METABOLIZING ENZYME CASES

11.3.1 TRANSEPITHELIUM DRUG TRANSPORT: CACO-2 ASSAY

Oral delivery is without question the most desirable route of drug administration (van de Waterbeemd et al., 2001). The fraction of an orally administered dose of a drug that reaches the systemic circulation unchanged, referred to as "oral bioavailability," is an important property in determining drug suitability for oral administration. Oral bioavailability is mainly determined by solubility of the drug in the gastrointestinal (GI) fluids as well as its permeability through the intestinal barrier. Traditionally, drug permeability studies have been conducted upstream of the discovery process. But one of the common areas of failure in clinical trials is drug absorption, leading many laboratories to currently implement screening for ADME early in the discovery process to eliminating drug candidates with, for example, poor oral bioavailability (Kellard and Englestein, 2007).

Cell lines are now routinely cultivated as monolayers on permeable filters for screening for transepithelial transport of drugs. Caco-2 cells have been extensively used for this purpose (for reviews, see Artursson et al., 2001; Press and Di Grandi, 2008). Caco-2 cells were originally derived from human colon adenocarcinoma and model enterocytes of the small intestine with respect to expression of various efflux and microvillar transporters, Phase II conjugation enzymes, and membrane peptidases and disaccharidases (Artursson, 1991). When cultured over a 21-day period, Caco-2 cells are able to fully polarize into differentiated monolayers displaying brush microvilli regions and cell–cell tight junctions. Press and Di Grandi (2008) have summarized the Caco-2 assay (reproduced here with permission, Box 11.1). More detailed

BOX 11.1 LOW-THROUGHPUT CACO-2
ASSAY PROCEDURE SUMMARY

Caco-2 cells are cultured in 75 cm^2 flasks at a density of approximately 30,000 cells/cm^2 in Dulbecco's modified Eagle's medium (DMEM), containing 20% fetal bovine serum. Cells between passages 30 and 40 (from ATCC original passage number 18) should be utilized for up to months in culture. For transport studies, following an initial washing step with 10 mL of PBS, cells are harvested from the flasks with trypsin–EDTA solution for 10 min at 37°C and seeded at a density of approximately 100,000 cells/cm^2 on a high-density polyethylene terephthalate (PET) or polycarbonate membrane insert (Figure 11.6). Culture media should be changed at least twice weekly, and cell monolayers are suitable for use between 21 and 28 days postseeding. To certify cell monolayers before each experiment, monolayer integrity is determined by measuring the transepithelial electrical resistance (TEER). TEER values should be measured in the permeability buffer (PUB), comprising Hank's Balanced Salt Solution containing 10 mM glucose and 20 mM HEPES, pH 7.4, using an EVOMX voltmeter with an STX100F electrode (World Precision Instruments, Inc.) or equivalent. For each assay plate, the TEER value of each well intended to be used in permeability studies is measured prior to inclusion in the study. The quality control of assay plates used in permeability studies is typically conducted in two steps. The first step consists of measuring the suitability of each assay well of the plate and the second step consists of testing both high- and low-permeability reference compounds. Common reference compounds used to assess the suitability are atenolo or mannitol (for low permeability), and metoprol or naproxen (for high permeability). For each assay plate tested, the permeability coefficient for atenolo/mannitol should be within a specified range to be certified as "acceptable." Additionally, apparent permeability coefficient (P_{app}) values for low- and high-permeability control compounds must meet their respective classifications to be certified as "acceptable." P_{app} values for atenolol/mannitol and metoprol/naproxen must be less and greater than 1×10^{-6} cm/s, respectively.

For each drug studied, both "apical" to "basal" and "basal" to "epical" transports are routinely investigated in order to determine if a carrier-mediated transport or a carrier-mediated efflux mechanism is involved. For transport studies, compounds are diluted to 0.1% in PAB in order to achieve a final concentration of 10 μM and a final solvent concentration of 0.1%. Lower drug concentrations should be used when aqueous solubility is a problem (or, to measure efflux substrate potential), and changes must be noted in the calculations. When using [^{14}C]- or [^{3}H]-radiolabeled compounds, final specific activity should be approximately 1.0 μCi/mL. The "apical" to "basal" assay is initiated by 0.25 mL of drug solution to the apical side of the monolayer. The "basal" to "epical" assay is performed in a similar manner; except that 0.25 mL of transport buffer is added to the epical side and 1.0 mL of drug solution is added

continued

to the basal side. Permeability experiments should be conducted in triplicate ($n = 3$), and aliquots of the dosing solution from the donor well should be taken at the beginning and end of the assay, to allow for recovery measurements. P_{aap} values are calculated according to the following equation:

$$P_{app} = \frac{V}{AC_i} \cdot \frac{C_f}{T}$$

where V is the solution of the receptor chamber (1.0 mL), A is the area of the membrane insert (0.31 cm^2), C_i is the initial concentration of the drug (10 μM), C_f is the final concentration of the drug (μM), and T is the assay time in seconds.

Source: From Press, B. and Di Grandi, D. 2008. *Curr. Drug Metab.* 9: 894. With permission.

assay procedures can be found in the following references: Irvine et al. (1999), Artursson et al. (2001), Youdim et al. (2003), and Volpe (2008). Interestingly, Justice et al. (2009) have pointed out that well inserts, such as those used in the Caco-2 assay (Figure 11.6), were the first technologies that begun to approach 3D-like structures where all cell membrane sides are exposed to the same microenvironment.

The Caco-2 *in vitro* model effectiveness in predicting human absorption has been found to be excellent (see Figure 11.7). This effectiveness has been attributed to the fact that the major mechanism of drug uptake across the intestinal mucosa is passive diffusion, driven by concentration differences. Because of its effectiveness, the assay has been recognized by FDA for establishing permeability assessment. In 2000, FDA issued "Guidance for Industry, Waiver of *In Vivo* Bioavailability and Bioequivalence Studies for Immediate-release Solid Oral Dosage Forms Based on the Biopharmaceutics Classification System" in which new drugs' apparent permeabilities (P_{app}) are tested together with 20 model drugs in a system like Caco-2. Based on the results from this *in vitro* assay, biowaivers may be requested for high-solubility high-permeability compounds (Class I) formulated in immediate release solid oral-dosage form that exhibit rapid *in vitro* dissolution.

Traditionally, Caco-2 assays have been conducted in 12- or 24-well formats. However, in efforts to increase throughput, 96-well formats have successfully been developed and deployed (Alsenz and Haenel, 2003; Marino et al., 2005). The morphology and function of Caco-2 cells mirror that of primary enterocytes. However, monolayers of Caco-2 cells differ from human intestinal epithelium in several ways: (1) Caco-2 cells form less permeable tight junctions in comparison to the human intestinal epithelium, resulting in low levels of paracellular permeability of the Caco-2 model; (2) permeability (P)-glycoprotein, a cellular efflux pump encoded by the MDR1 gene, is overexpressed in Caco-2 cells; and (3) levels of intestinal CYP 3A4, 3A5, 3A7, 2C9, 2C19, and 2B6 enzymes are low or nonexistent in Caco-2 cells. The underestimation or overestimation of permeation of some compounds (e.g., efflux substrates, esters, amides, and acids) is attributable to a mix of the above limitations.

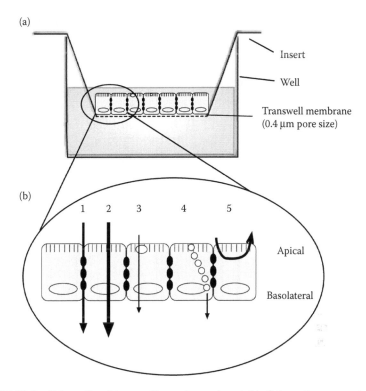

FIGURE 11.6 Schematic of transwell membrane insert (a). Schematic representation of routes and mechanisms for transport of molecules across the intestinal epithelium (b) including: (1) passive diffusion via paracellular transport between cells, (2) transcellular transport across the cell membrane, (3) active carrier-mediated uptake, (4) transcytosis mechanisms, and (5) protective barrier—recycling xenobiotics back out to the intestinal lumen, an "antitransport" process in which substances can enter the cytoplasm of enterocytes, but are then recycled back to the intestinal lumen, across the apical plasma membrane. The efflux process in intestinal cells is carried out predominantly by transporters (efflux pumps) that act as barriers to absorption. Screening for efflux liability is another application that utilizes Caco-2 assay. (From Press, B. and Di Grandi, D. 2008. *Curr. Drug Metab.* 9: 893. With permission.)

To improve intestinal absorption prediction of different classes of drug compounds, the accepted view is to use additional cell-based screens, for example, MDCK (Madin-Darby Canine Kidney) cell-based assay (Irvine et al., 1999), artificial membrane assays (described below), combined with *in silico* predictions (Press and Di Grandi, 2008). It is interesting to note that the cellular monolayer, including Caco-2, tends to be "flat," whereas the human jejunum is highly folded. See Figure 11.8 for a comparison schematic. Despite the availability of nano- and microfabrication technologies (Michel et al., 2001; Kang et al., 2008), with which high-aspect-ratio scaffolds that would provide scaffolding to support cells mimicking the jejunum folding can be developed, no study has been found taking this route to improve intestinal adsorption by developing a more physiologically relevant culture through replication *in vitro* of the *in vivo* folding. Interestingly, Wang et al. (2009)

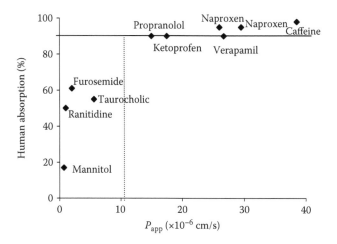

FIGURE 11.7 Comparison of Caco-2 permeability to human absorption for commercially available reference drugs. P_{app} values of new drugs are used to predict the drug absorption with reference to the curve established by reference drugs of known human absorption exhibiting a wide range of permeability coefficients. The curve profile is typical of *in vitro* permeability assays—asymptotic for high P_{app} values (>10) and steep for drugs with lower P_{app} values. (From Press, B. and Di Grandi, D. 2008. *Curr. Drug Metab.* 9: 899, 2008. With permission.)

has shown that nerve cells cultured in high-aspect-ratio (microwell) structures are physiologically different and are not merely a folded 2D monolayer.

The view that Caco-2 monolayers function as a mere membranes for the majority of drugs tested inspired the development of an alternative permeability assay based on a membrane composed of egg lecithin dissolved in *n*-dodecane, referred to as parallel artificial membrane permeability assay (PAMPA) (Kansy et al., 1998). For a

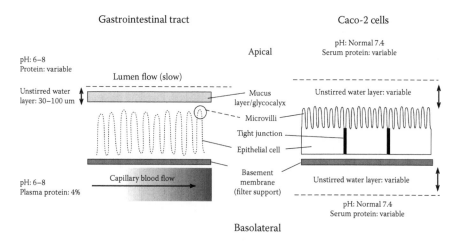

FIGURE 11.8 Comparison of the *in vivo* and *in vitro* barriers that drugs must pass to reach the other side. (From Youdim, K.A., Avdeef, A., and Abbott, N.J. 2003. *Drug Discov. Today* 8(21): 998. With permission.)

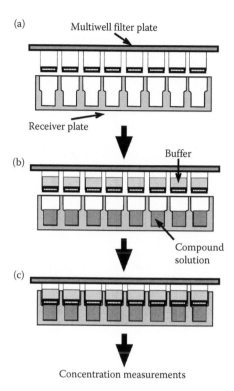

FIGURE 11.9   Schematic representation of a PAMPA assay: (a) illustrates the filter (pre-coated with an artificial membrane) and receiver plates in 96-well format; (b) compound solutions are added in the plate (permeation donor) and buffer is added in the well of the filter plate (permeation acceptor); (c) the plates are coupled together and incubated for a certain duration followed by separation and measurement of the compound concentration in each plate. (From Chen, X. et al. 2008. *Pharm. Res.* 25(7): 1512. With permission.)

schematic representation of a typical PAMPA assay, see Figure 11.9. A key advantage of the PAMPA technology is that a number of membrane compositions developed after the pioneering Kansy study can easily be prepared and used to mimic different biological barriers as shown in Table 11.2. In a comparative permeability study (apical to basolateral) with 800 compounds, 60% of the compounds were identically classified, less than 5% were classified high in one assay and low in the other or vice versa (Faller, 2008). Compounds with less than 30% recoveries in the Caco-2 assay were excluded from the analysis. Recent improvements in artificial membrane design (Chen et al., 2008) and automation (Kellard et al., 2007; Flaten et al., 2009) have been reported to provide improvements in the assay. It should be pointed out that PAMPA assays unlike cell-based assays are limited to absorption studies only and cannot be applied to studying effects like efflux liability due to absence of pumps in the artificial membranes.

Although the Caco-2 assay is strictly not a biosensor, since the monolayer does not biologically (no biological reaction) interact with the test compounds to produce a quantifiable signal, it serves to underscore a key challenge to incorporating 3D

TABLE 11.2
Characteristic of PAMPA Assays

Assay Name	Target Barrier	Filter	Solvent	Membrane Composition	pH	Sink	Incubation Time (h)	Reference
Egg—PAMPA	GI tract	Hydrophobic PVDF (125 μm)	n-Dodecane	10% egg lecithin	Iso 6.5 and 7.4	No—bile salt in donor	15	Kansy et al. (1998)
HDM—PAMPA	GI tract	10 μm PC	n-hexadecane	n-hexadecane	Iso 4.0, 6.8–8.0	No	4	Wohnsland et al. (2001)
BM—PAMPA	GI tract	Hydrophobic PVDF (125 μm)	1.7-Octadiene	3% of a phospholipid mixture[a]	Iso 6.5	No	15	Sugano et al. (2001)
BM—PAMPA	GI tract	Hydrophobic PVDF (125 μm)	n-Dodecane	1% egg lecithin	Iso 5.5 and 7.4	No	2	Zhu et al. (2002)
DOPC—PAMPA	GI tract	Hydrophobic PVDF (125 μm)	n-Dodecane	2% DOPC		No	15	Avdeef et al. (2001)
DS—PAMPA	GI tract	Hydrophobic PVDF (125 μm)	n-Dodecane	20% (phospholipid mixture)	Gradient 6–7.4	Res—anionic surfactant in acceptor	15	Bermejo et al. (2004)
BBB—PAMPA	BBB	Hydrophobic PVDF (125 μm)	n-Dodecane	2% PBL	Iso 7.4	No	18	Di et al. (2003)
PAMPA—Skin	Skin	Hydrophobic PVDF (125 μm)	70% silicone oil, 30% IPM	70% silicone oil, 30% IPM	Variable (target neutral species)	No	7	Ottavian et al. (2006)

Source: From Faller, B. 2008. *Curr. Drug Metab.* 9: 888. With permission.

[a] Lipid mixture made of 33% cholesterol, 2.7% phosphatidylcholine, 2.7% phosphatidylserine, and 7% phosphatidylinositol.

cultures in HTS laboratories. First, it models the *in vivo* situations very well as far as absorption goes, so well that the results generated by the assay are acceptable by a regulatory agency–FDA. Why then would anybody "adventurously tamper" with the assay? Second, the success of Caco-2 possibly creates a mindset in some that with the exception of a few cases all 2D cell-based assays can perform as well and therefore the extra effort needed to implement 3D cultures is difficult to justify. Simply put, "why fix it when it is not broken." The overall take-home message from the Caco-2 assay case is that the physiological relevance exhibited by some 2D cultures is sufficient and as such the development of 3D cultures needs to target those applications where the value added is evident or easily justifiable.

11.3.2 INDUCTION OF DRUG METABOLIZING ENZYMES: HEPATOCYTE ASSAYS

A second case with which one can make the perceived "limited value-addition" is found in the FDA guidelines on carrying out and interpreting *in vitro* induction studies using hepatocytes. The induction of drug metabolism was described almost 50 years ago. The induction of drug metabolizing enzymes can have serious implications on the pharmacokinetics and toxicity of drugs. It is paramount to establish the induction potential of a new chemical entity before it moves ahead through the development pipeline. The FDA publishes guideline documents to help researchers design enzyme induction studies.

Hewitt et al. (2007) published the results of their survey in which they wanted to establish how well companies complied with the FDA recommendations. Thirty investigators from 27 companies completed a questionnaire of 19 questions between November 2005 and February 2006 on (1) *in vitro* methods and study design, (2) data analysis and interpretation, and (3) action as a consequence of the results. The methodology results are presented in Table 11.3. What is striking is that because the FDA document is a guideline and not a mandate, 73% of the respondents used monolayer cultures as opposed to the sandwich. The sandwich hepatocyte culture is a 2D culture that maintains the polarized state of the hepatocytes, bile acids are secreted across the apical surfaces, and albumin is secreted across the basolateral surface, recapitulating the *in vivo* situation of exhibiting the complex physiological relevance introduced in Chapter 4 (Dunn et al., 1989). It is interesting to note that the 2006 FDA guideline does not even mention sandwich cultures, yet they place a much higher emphasis on drug–transporter interactions. It would thus be wise for the companies to use the sandwich 3D cultures that are ideal for studying transporters (Li et al., 2009; Bi et al., 2006). In a way, the availability of credible evidence in support of superiority of 3D is not enough, as long as the regulatory agency does not mandate its use.

11.4 PAUCITY OF CONCLUSIVE SUPPORT OF 3D CULTURE SUPERIORITY

At the time of writing this section (March 2010), a search of the Web of Knowledge or Medline databases yielded over sixfold more number of citations with "high-throughput" than "physiological relevance" search phrases, underscoring the relentless pursuit of improvements in throughput. From a business standpoint, it is easy to

TABLE 11.3

Compliance with the 2004 FDA Methodology Guidelines Recommendation for Enzyme Induction Studies[a]

Condition/Parameter	2004 FDA Guidelines	Survey
Fresh or cryopreserved hepatocytes?	Fresh or cryopreserved	23% used only fresh hepatocytes
		37% used only cryopreserved hepatocytes
		40% used both fresh and cryopreserved
Number of donors	Greater or equal to 3	1 for screening
		3 or more
Culture type	Sandwich	73% conventional monolayer
		23% sandwich culture
Use of sub-μM dexamethasone in medium	No preference	70% used dexamethasone
		30% did not supplement
Preincubation time	48–72	3% incubated for 3 h
		52% incubated for 24 h
		73% incubated for 48 h (some specified 24–48 h)
Induction period	3–4 days	7% induced for 24 h (mRNA)
	24–72 h for mRNA	59% induced for 48 h
		34% induced for 3 days
		14% induced for up to 8 days
Number of concentrations used	Greater than three concentrations	10% used two concentrations
		60% used three concentrations
		6% used 2–6 concentrations
		23% used more than four concentrations
Endpoints	Enzyme activities (mRNA and protein for confirmation only)	100% measure enzyme activity
		40% measure mRNA
		23% measure protein
		40% determine concomitant cytotoxicity
		67% measure a combination of endpoints

Source: From Hewitt, N.J., de Kanter, R., and LeCluyse, E. 2007. *Chem. Biol. Interact.* 168: 55. With permission.

[a] Induction of CYP3A4 was used in this study as the example because it is the most cited of all the P450 cytochrome P enzymes and in most cases rifampin was used as the control.

appreciate the return from throughput investment in terms of more compounds or wells screened in a given time, while the benefits of investing in physiologically more relevant screening systems, despite hypothetically making sense, have not been clearly demonstrated. To look for "circumstantial evidence," we first turn to 3D culture work in bioprocessing where the superiority has been conclusively established, outlined in the following paragraph.

Biopharmaceutical manufacturing has been defined as the science and/or art of producing large amounts of high-quality protein pharmaceuticals (Wurm, 2004). Until recently, it has been accepted that unrestricted proliferation of monodispersed cell lines is the key to manufacturing biopharmaceuticals. In a recent study, Sanchez-Bustamante et al. (2005) used a gravity-enforced hanging drop method to culture 3D tissues for transient gene expression and production of recombinant proteins. Numerous cell lines, including those commonly used in drug discovery (e.g., CHO, BHK, and HEK) cultivated as 3D, showed a 20-fold increase in specific productivity when compared to their monolayer counterparts. The poor productivity in 2D cultures was reasoned to be a consequence of the missing tissue-specific inter-cellular crosstalk and the productivity of extracellular matrix (ECM) components (Sanchez-Bustamante et al., 2005; Wong et al., 2007). It is also well known that cells in suspension or in monolayers actively proliferate and thus devote most of their metabolic energy on reproduction as opposed to recombinant protein production. Based on this understanding, controlled proliferation technologies have been used to increase protein productivity. Examples include nutrient withdraw or addition of DNA synthesis inhibitors (Suzuki and Ollis, 1990), temperature-induced growth arrest (Jenkins and Hovey, 1993), and overexpressing tumor suppressor genes (Watanabe et al., 2002). While these strategies have potential to increase specific productivity by several 10-fold, they have in the past led to significant cell viability reduction (Mercille and Massie, 1994) and increased apoptosis (Ko and Prives, 1996). The physiological relevance nature of 3D cultures seems to enhance specific productivity without the above undesirable effects. The argument that 3D culture can do for HTS what they are doing for biopharmaceutical manufacturing is attractive.

In the bio/pharmaceutical drug discovery component of the industry and in HTS laboratories in particular, convincing evidence of similar quality to that described above from the biopharmaceutical manufacturing industry has to show failed compounds (leads) that passed the 2D, but now fail the 3D cell-based assay. Current literature has provided only three cases that come close to this requirement and these cases are the subject of the following chapter (Chapter 12). These cases are somewhat similar to CPR examples described in Chapter 4, with the main difference that the focus in Chapter 12 is on drug therapeutical outcomes as opposed to biological functional outcomes. The three cases presented involve monoclonal antibody therapy in breast cancer, tumor cell metastasis, and cancer drug resistance.

REFERENCES

Abdiche, Y.N., Malashock, D.D., and Pons, J. 2008. Probing the binding mechanism and affinity of tanezumab, a recombinant humanixed anti-NGF monoclonal antibody, using a repertoire of biosensors. *Protein Sci.* 17(8): 1326–1335.

Alsenz, J. and Haenel, E. 2003. Development of a 7-day, 96-well Caco-2 permeability assay with high-throughput direct UV compound analysis. *Pharm. Res.* 20(12): 1961–1969.

Artursson, P. 1991. Cell cultures as models for drug absorption across the intestinal mucosa. *Crit. Rev. Ther. Drug Carrier Syst.* 8: 305–330.

Artursson, P., Palm, K., and Luthman, K. 2001. Caco-2 monolayers in experimental and theoretical predictions of drug transport. *Adv. Drug Deliv. Rev.* 46: 27–43.

Atienza, J.M., Yu, N.C, Kirstein, S.L., Xi, B., Wang, X.B., Xu, X., and Abassi, A. 2006. Dynamic and label-free cell-based assays using the real-time cell electronic sensing system. *Assay Drug Dev. Technol.* 4(5): 545–553.

Avdeef, A., Strafford, M., Block, E., Balogh, M.P., Chambliss, W., and Khan, W. 2001. Drug absorption in vitro model: Filter-immobilized artificial membrane 2 > Studies of the permeability properties of lactones in Piper methysticum Forst. *Eur. J. Pharm. Sci.* 14(4): 271–280.

Bermejo, M., Avdeef, A., Ruiz, A., Nalda, R., Ruell, J.A., Tsinman, O., Gonzalez, L., et al. 2004. PAMPA—a drug absorption in vitro model 7. Comparing rat in situ, Caco-2, and PAMPA permeability of fluoroquinolones. *Eur. J. Pharm. Sci.* 21(4): 429–441.

Bi, Y.A., Kazolias, D., and Duignan, D.B. 2006. Use of cryopreserved human hepatocytes in a sandwich culture to measure hepatobiliary transport. *Drug Metab. Dispos.* 34(9): 1658–1665.

Bianchi, B.R., Moreland, R.B., Faltynek, C.R., and Chen, J. 2007. Application of large-scale transiently transfected cells to functional assays on ion channels: Different targets and assay formats. *Assay Drug Dev. Technol.* 5(3): 417–424.

Burke, J.F. and Mann, C.J. 2006. Rapid isolation of monoclonal antibody producing cell lines: Selection of stable, high-secreting clones. *BioProcess Int.* 4: 48–51.

Carroll, S. and Al-Rubeai, M. 2004. The selection of high-producing cell lines using flow cytometry and cell sorting. *Expert Opin. Biol. Ther.* 4: 1821–1829.

Cawkill, D. and Eaglestone, S.S. 2007. Evolution of cell-based reagent provision. *Drug Discov. Today* 12(19/20): 820–825.

Chen, J., Lake, M.R., Sabet, R.S., Niforatos, W., Pratt, S.D., Cassar, S.C., Xu, J. et al. 2007. Utility of large-scale transiently transfected cells for high throughput screens to identify transient receptor potential channel A1 (TRPA1) antagonists. *J. Biomol. Screen.* 12 (1): 61–69.

Chen, X., Murawski, A., Patel, K., Crespi, C.L., and Balimane, P.V. 2008. A novel design of artificial membrane for improving the PAMPA model. *Pharm. Res.* 25(7): 1511–1520.

Di, L., Kerns, E.H., Fan, K., McConnell, O.J., and Carter, G.T. 2003. High throughput artificial membrane permeability assay for blood-brain barrier. *Eur. J. Med. Chem.* 38(3): 223–232.

Digan, M.E., Pou, C., Niu, H.L., and Zhang, J.L. 2005. Evaluation of division-arrested cells for cell-based high-throughput screening and profiling. *J. Biomol. Screen.* 10(6): 615–623.

Ding, M., Stjernborg, L., and Albertson, N. 2006. Application of cryopreserved cells to hERC screening using a non-radioactive Rb$^+$ efflux assay. *Assay Drug Dev. Technol.* 4(1): 83–88.

Duggal, S., Fronsdal, K.B., Szoke, K., Shahdadfar, A., Melvik, J.E., and Brinchmann, J.E. 2009. Phenotype and gene expression of human mesenchymal stem cells in alginate scaffolds. *Tissue Eng.* 15(7): 1763–1773.

Dunn, J.C., Yarmush, M.L., Koebe, H.G., and Tompkins, R.G. 1989. Hepatocyte function and extracellular matrix geometry: Long-term culture in sandwich configurations. *FASEB J.* 176(3): 174–177.

Faller, B. 2008. Artificial membrane assay to assess permeability. *Curr. Drug Metab.* 9: 886–892.

Flaten, G.E., Awoyemi, O., Luthman, K., Brandl, M., and Massing, U. 2009. The phospholipid vesicle-based drug permeability assay: 5. Development toward an automated procedure for high-throughput permeability screening. *JALA* 14: 12–21.

Fursov, N., Cong, M., Federich, M., Platchek, M., Haytko, P., Tacke, R., Livelli, T., and Zhong, Z. 2005. Improving consistency of cell-based assays by using division-arrested cells. *Assay and Drug Develop. Technol.* 3(1): 7–15.

Goetz, A.S., Andrews, JL., Littleton, T.R., and Ignar, D.M. 2000. Development of a facile method for high throughput screening with reporter gene assays. *J. Biomol. Screen.* 5(5): 377–384.

Herman, E. 2004. Generation of model cell lines expressing recombinant G-protein coupled receptors. *Methods Mol. Biol.* 259: 137–154.

Hewitt, N.J., de Kanter, R., and LeCluyse, E. 2007. Induction of drug metabolizing enzymes: A survey of *in vitro* methodologies and interpretations used in the pharmaceutical industry—do they comply with FDA recommendation? *Chem. Biol. Interact.* 168: 51–65.

Irvine, J.D., Takahashi, L., Lockhart, K., Cheong, J., Tolan, J.W., Selick, H.E., and Grove, J.R. 1999. MDCK (Madin-Derby Canine Kidney) cells: A toll for membrane permeability screening. *J. Pharm. Sci.* 88(1): 28–33.

Jenkins, N. and Hovey, A. 1993. Temperature control of growth and productivity in mutant Chinese Hamster Ovary Cells synthesizing a recombinant protein. *Biotechnol. Bioeng.* 42: 1029–1036.

Justice, B.A., Badr, N.A., and Felder, R.A. 2009. 3D cell culture opens new dimensions in cell-based assays. *Drug Discov. Today* 14(1/2): 102–107.

Kang, L., Chung, B.-G., Langer, R., and Khademhosseini, A. 2008. Microfluidics for drug discovery and development: From target selection to product lifecycle management. *Drug Discov. Today* 13(1/2): 1–13.

Kansy, M., Senner, F., and Gubernator, K. 1998. Physicochemical high throughput screening: Pararell artificial membrane permeation assay in the description of passive absorption processes. *J. Med. Chem.* 41(7): 1007–1010.

Kellard, L. and Engelsteine, M. 2007. Automation of cell-based and noncell-based permeability assays. *JALA* 12: 104–109.

Ko, L.J. and Prives, C. 1996. p53: Puzzle and paradigm. *Genes Dev.* 10(9): 1054–1072.

Kunapuli, P., Zheng, W., Weber, M., Solly, K., Mull, R., Platcherk, M., Cong, M., Zhong, Z., and Strulovici, B. 2005. Application of division-arrest technology to cell-based HTS: Comparison with frozen and fresh cells. *Assay Drug Dev. Technol.*, 3(1): 17–26.

Lee, P.H., Bao, A., van Staden, C., Ly, J., Xu, A., Fang, Y., and Verkleeren, R. 2008. *Assay Drug Dev. Technol.* 6(1): 83–94.

Li, N., Bi, Y.A., Duignan, D.B., and Lai, Y.R. 2009. Quantitative expression profile of hepatobiliary transporters in sandwich cultured rat and human hepatocytes. *Mol. Pharm.* 6(4): 1180–1189.

Marino, A.M., Yarde, M., Patel, H., Chong, S.H., and Balimante, P.V. 2005. Validation of the 96 well Caco-2 cell culture model for high throughput permeability assessment of discovery compounds. *Int. J. Pharm.* 297(1–2): 235–241.

Mercille, S. and Massie, B. 1994. Induction of apoptosis in nutrient-deprived cultures of hybridoma and myeloma cells. *Biotechnol. Bioeng.* 44: 1140–1154.

Michel, B., Bernard, A., Bietsch, A., Delamarche, E., Geissler, M., Juncker, D., Kind, H. et al. 2001. Printing meets lithography: Soft approaches to high-resolution patterning. *IBM J. Res. Dev.*, 45(5): 697–719.

Morrow, K.J. 2008. Innovations mark biosensor development. *Genetic Eng. Biotechnol. News* 28(5): 36, 38 and 40.

Ottaviani, G., Martel, S., and Carrupt, P.A. 2006. Parallel artificial membrane permeability assay: A new membrane for the fast prediction of passive human skin permeability. *J. Med. Chem.* 49(13): 3948–3954.

Parreno, J., Buckley-Herd, G., de-Hemptinne, I., and Hart, D. 2008. Osteoblastic MG-63 cell differentiation, contraction, and mRNA expression in stress-relaxed 3D collagen I gels. *Mol. Cell Biochem.* 317: 21–32.

Press, B. and Di Grandi, D. 2008. Permeability for intestinal absorption: Caco-2 assay and related issues. *Curr. Drug Metab.* 9: 893–900.

Sanchez-Bustamante, C.D., Kelm, J.M., Mitta, B., and Fussenegger, M. 2005. Heterologous protein production capacity of mammalian cells cultivated as monolayers and microtissues. *Bitechnol. Bioeng.* 93(1): 169–180.

Sugano, K., Hamada, H., Machida, M. Ushio, H., Saitoh, K., and Terada, K. 2001. Optimized conditions of bio-mimetic artificial membrane permeation assay. *Int. J. Pharm.* 228(1–2): 181–188.

Sun, Y., Huang, J., Xiang, Y., Bastepe, M., Juppner, H., Kobilka, B.K., Zhang, J.J., and Huang, X.-Y. 2007. Dose-dependent switch from G protein-coupled to G protein-independent signaling by a GPCR. *EMBO J.* 26(1): 53–63.

Suzuki, E. and Ollis, D.F. 1990. Enhanced antibody production at slowed growth rates: Experimental demonstration and a simple structured model. *Biotechnol. Prog.* 6(3): 231–236.

Terstegge, S., Laufenberg, I., Pochert, J., Schenk, S., Itskovits-Eldor, J., Endl, E., and Brustle, O. 2007. Automated maintenance of embryonic stem cell cultures. *Biotechnol. Bioeng.* 96(10): 195–201.

Uvebrant, K., Thrige, D.D.G., Rosen, A., Akesson, M., Berg, H., Walse, B., and Bjork, P. 2007. Discovery of selective small-molecule CD80 inhibitors. *J. Biomol. Screen.* 12(4): 464–472.

van de Waterbeemd, H., Smith, D.A., Beaumont, K., and Walker, D.K. 2001. Property-based design: Optimization of drug absorption and pharmacokinetics. *J. Med. Chem.* 44(9): 1313–1333.

Verkman, A.S., Lukacs, G.L., and Galietta, L.J.V. 2006. CFTR chloride channel drug discovery—inhibitors as antidiarrheals and activators for therapy of cystic fibrosis. *Curr. Pharm. Des.* 12: 2235–2247.

Volpe, D.A. 2008. Variability in Cao-2 and MDCK cell-based intestinal permeability assays. *J. Pharma. Sci.* 97(2): 712–725.

Wang, L., Wu, Z.-Z., Xu, B., Zhao, Y., and Kisaalita, W.S. 2009. SU-8 microstructure for quasi-three-dimensional cell-based biosensing. *Sens. Actuators B Chem.* 140: 349–355.

Watanabe, S., Shttleworth, J., and Al-Rubeai, M. 2002. Regulation of cell cycle and productivity in NS0 cells by the over-expression of p21. *Biotechnol. Bioeng.* 77: 1–7.

Weetall, M., Hugo, R., Friedman, C., Maida, S., West S., Walttanasin, S., Bouhel, R., Weitz-Schmidt, G., and Lake, P. 2001. A homogeneous fluorimetric assay for measuring cell adhesion to immobilized ligand using V-well microtiter plates. *Anal. Biochem.* 293(2): 277–287.

Wohnsland, F. and Faller, B. 2001. High-throughput permeability pH profile and high-throughput alkane/water lop P with artificial membranes. *J. Med. Chem.* 44(6): 923–930.

Wong, H., Wang, M., Cheung, P., Yao, K., and Chan, B.P. 2007. A 3D collagen microsphere culture system for GDNF-secreting HEK293 cells with enhanced protein productivity. *Biomaterials* 28(35): 5369–5380.

Wurm, F.M. 2004. Production of recombinant protein therapeutics in cultivated mammalian cells. *Nat. Biotechnol.* 22: 1393–1398.

Youdim, K.A., Avdeef, A., and Abbott, N.J. 2003. *In vitro* trans-monolayer permeability calculations: Often forgotten. *Drug Discov. Today* 8(21): 997–1003.

Zaman, G.J.R., de Roos, J.A.D.M., Blomenrohr, M., van Koppen, C.J., and Oosterom, J. 2007. Cryopreserved cells facilitate cell-based drug discovery. *Drug Discov. Today* 12(13/24): 521–526.

Zhu, C.Y., Jiang, L., Chen, T.M., and Hwang, K.K. 2002. A comparative study of artificial membrane permeability assay for high throughput profiling of drug absorption potential. *Eur. J. Med. Chem.* 37(5): 399–407.

Zhu, Z.R., Puglisi, J., Connors, D., Stewart, J., Herbs, J., Marino, A., Sinz, M., O-Connell, J., Dicknson, K., and Cacace, A. 2007. Use of cryopreserved transiently transfected cells in high-throughput pregnane X receptor transactivation assay. *J. Biomol. Screen.* 12(2): 248–254.

Zuck, P., Murray, E.M., Steck, E., Grobler, J.A., Simon, A.J., Strulovici, B., Inglese, J., Flores, O.A., and Ferrer, M. 2004. A cell-based β-lacatmase reporter gene assay for the identification of inhibitors of hepatitis C virus replication. *Anal. Biochem.* 334: 244–355.

12 Cases for 3D Cultures in Drug Discovery

12.1 THREE CASES

In looking for previous studies that best make the case for 3D cultures in drug discovery programs, the author had two criteria. First, the drug and target involved are either commercial or belong to a class of drugs from which commercial products exist. Second, the outcome of the drug interaction with the target in 3D cultures is directly linked to the success or failure *in vivo*, but such an outcome is absent in 2D cultures. Three groups of studies that meet the above criteria were found. The first comes from Mina J. Bissell's group at the Lawrence Berkeley National Laboratory, Berkeley, California, involving a β1-integrin monoclonal antibody that only successfully treated breast cancer cells in 3D cultures and tumors *in vivo*. The second comes from Peter Friedel's group at the University of Würzburg, Würzburg, Germany, involving identification of alternative tumor cell migration after blocking the pericellular proteolysis; a potential partial explanation of the failure of metalloproteinase inhibitors in cancer therapy. The third comes from a study published almost a decade ago by Robert Kerbel's group, at the Sunnybrook Health Science Center of the University of Toronto, in which the expression of drug resistance observed *in vivo* was fully recapitulated *in vitro* with multicellular spheroids. More detailed background of the target(s), drug(s), cell culture system(s), and results for each case are presented below.

12.2 THE β1-INTEGRIN MONOCLONAL ANTIBODY CASE

12.2.1 INTEGRINS

The integrin superfamily has already been discussed in Chapter 5 with respect to the role these glycoproteins play in cell adhesion. The discussion is briefly extended here from a drug target perspective. Given the diversity of interactions in the ECM, deviations from precise regulation present many druggable opportunities and a number of diseases have been implicated (Huxley-Jones et al., 2008). Integrin-specific examples include but are not limited to cancer (Ria et al., 2002; Dayam et al., 2006), inflammatory bowel disease (Mosnier et al., 2006), vascular diseases (Hu et al., 2007; Lal et al., 2009), thrombosis (Furlan et al., 1997), and multiple sclerosis (Landry and Gies, 2008). Numerous small molecules or biologicals with integrin inhibitory activities are in production or development pipelines (Landry and Gies, 2008; Huxley-Jones et al., 2008).

12.2.2 Monoclonal Antibodies

Hybridoma technology, proposed by Kohler and Milstein (1975), involved the fusion of an antibody-producing white cell with a myeloma (skin cancer) cell of a mouse (or other animals) resulting in the formation of a hybridoma cells. Each hybridoma cell synthesizes a single molecular species of antibody. Cells cultured from such a hybridoma produce a single monoclonal antibody (mAb). Antibodies, which are primary agents in immunological defenses of vertebrates, are glycoproteins, members of a class of proteins called immunoglobulins that have the common structural features indicated in Figure 12.1 (Loo et al., 2008). The most common class of immunoglobulins is the IgG. The primary differentiation, within a class of antibodies, occurs at the amino-terminal ends of the chains, named the variable region (Figure 12.1). The mAbs exhibit high affinity for antigens, remarkable specificity, long half-life, and potentially low toxicity, which makes them promising for clinical applications such as delivering cytotoxic agents to tumors. But early attempts to use mAbs in clinical trials used murine molecules, which often elicited human anti-mouse antibody responses that limited the number of treatments (Kunq et al., 1979; Tjandra et al., 1990).

A field of "antibody engineering" emerged in the mid 1980s to address the clinical use limitations outlined above. For excellent antibody engineering reviews, see Stockwin and Holmes (2003), Loo et al. (2008), Liu et al. (2008), and Stiehm et al. (2008). As an example, these efforts have yielded chimeric mAb (designed to contain the variable region of mouse antibody and the constant regions of human antibody), humanized mAb (only the complementary determining regions are of murine origin), and small antibody fragments as shown in Figure 12.1. At the time of writing, mAbs represented over 30% of biological proteins undergoing clinical trials and are the second largest class of biological drugs after vaccines (da Silva et al., 2008). More than 20 mAb therapeutics are licensed for treatment of various diseases (Jefferis, 2009) and nine of these are for cancer therapy (Table 12.1).

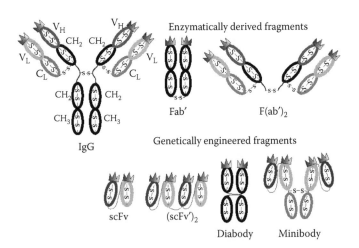

FIGURE 12.1 Schematic diagram of intact antibody (IgG) and various enzymatically derived and genetically engineered antibody fragments.

continued

TABLE 12.1
Monoclonal Antibodies (mAbs) Approved by FDA for Cancer Therapy

Generic (Trademark)	Target Antigen	Isotype	Species	Payload	Mechanism of Action	Antitumor Therapeutic Activity	FDA Year of Approval	References
Rituximab (Rituxan™)	CD20	IgG1 k	Chimeric	—	Introduction of apoptosis, ADCC, CDC, chemosensitization	Low-grade B-cell NHL	1997	Reff et al. (1994), Pescovitz (2006), Glennie et al. (2007), Jazirehi et al. (2004), Jazirehi et al. (2005)
Trastuzumab (Herceptin™)	Her-2/neu	IgG1 k	Humanized	—	ADCC, chemosensitization, CCA, inhibition of angiogenesis	Her-2 overexpressed metastatic breast cancer	1998	Carter et al. (1992), Hudis (2007), Izumi et al. (2002), Barok et al. (2007), Baselga et al. (2005)
Alemtuzumab (Campath-IH™)	CD52	IgG1 k	Humanized	—	ADCC, CDC	B-cell CLL	2001	Hale et al. (1983, 1988), Ravandi and O'Brien et al. (2005), Alinari et al. (2007)
Cetuximab (Erbitux™)	EGFR (Her-1)	IgG1 k	Chimeric	—	Chemosensitization, and radiosensitization, CCA, ADCC	Metastatic colorectal cancer, head and neck cancers	2004	Mendelsohn (1997), Galizia et al. (2007), Kawagichi et al. (2007)
Bevacizumab (Avastin™)	VGEF	IgG1 k	Humanized	—	Inhibition of angiogenesis	Colorectal cancer	2004	Kim et al. (1993), Presta et al. (1997), Kramer and Lipp (2007), Gerber et al. (2007)

TABLE 12.1 (continued)
Monoclonal Antibodies (mAbs) Approved by FDA for Cancer Therapy

Generic (Trademark)	Target Antigen	Isotype	Species	Payload	Mechanism of Action	Antitumor Therapeutic Activity	FDA Year of Approval	References
Panitumumab (Vectibix™)	EGFR	IgG2 k	Human	—	Inhibition of cell growth, induction of apoptosis, decreased proinflammatory cytokines and VEGF production	Metastatic colorectal cancer	2006	Cohenuram and Saif (2007), Jakobavits et al. (2007)
Gemtuzumab ozogamicin (Mylotarg™)	CD33	IgG4 k	Humanized	Calicheamicin	Double-stranded DNA breaks and cellular death induced by payload after intracellular hydrolysis	CD33+ relapsed AML	2000	Hamann et al. (2002), Pagano et al. (2007)
Ibritumomab tiuxetan (Zevalin™)	CD20	IgG1 k	Murine		Cellular death induced by β-emitter, induction of apoptosis, ADCC, CDC	Rade or follicular, relapsed or refractory, CD20+ B-cell NHLs; Rituximab-refractory follicular NHL	2002	Chinm et al. (1999), Davies (2007)
Tositumomab (Bexxar™)	CD20	IgG 2a λ	Murine	^{131}I-iodine	Cellular death induced by γ-emitter, induction of apoptosis, ADCC, CDC	Relapsed CD20+ B-cell NHL; Rituximab-refractory NHL	2003	Davies (2007), Nadler et al. (1981), Zalutsky et al. (1990), Kaminski et al. (2001)

Abbreviations: FDA, Food and Drug Administration; IgG, immunoglobulin G; ADCC, antibody-dependent cell cytotoxicity; CDC, complement dependent cytotoxicity; NHL, non-Hodgkin's lymphoma; CLL, chronic lymphocytic leukemia; Her-2, human-2 epidermal growth factor receptor-2; CCA, cell cycle arrest; EGFR, epidermal growth factor receptor, Her-1, human-1 epidermal growth factor receptor-1; VEGF, vascular endothelial growth factor; and AML, acute myeloid leukemia.

Source: From Liu, X.-Y., Pop, L.M., and Vietta, E.S. 2008. *Immunol. Rev.* 222: 11. With permission.

12.2.3 EXPERIMENTAL SYSTEM: BREAST CANCER CELLS IN MATRIGEL

In their pioneering experiment, Weaver et al. (1997) used a nonmalignant ("normal") S-1 and a malignant T4-2 (became spontaneously tumorigenic after a certain passage) cell lines. Both were sublines of the HMT-3522 breast cancer series established from a breast biopsy of a woman with nonmalignant breast lesion (Briand et al., 1987; Nielsen and Briand, 1989). The two cell lines, one originating from the other by spontaneous genetic events, provided a unique tool for the goals of the study, which was to better understand the specific mechanisms involved in malignant conversion in the breast.

For 3D preparations, cells were embedded in EHS extract currently known as laminin-rich extracellular matrix (lrECM) or Matrigel—a soluble and sterile extract of the basement membrane protein derived from EHS tumor (Kleinman and Martin, 2005). Matrigel is commercially available from several sources including BD Biosciences. The EHS mouse tumor, named so to acknowledge J. Englbreath-Holms of Denmark, who discovered it, and Richard Swarm, who maintained and characterized it, was originally identified as a poorly differentiated chondrosarcoma, but further analysis suggested that the tumor matrix was distinct from cartilage and instead resembled basement membrane (Kleinman and Martin, 2005). For 2D preparations, cells were grown as monolayers on plastic dishes coated with collagen I.

12.2.4 TREATMENT WITH β1-INTEGRIN INHIBITORY ANTIBODY REDUCED MALIGNANCY IN *IN VITRO*-3D AND *IN VIVO*, BUT NOT IN *IN VITRO*-2D SYSTEMS

Because T4-2 cells exhibited a higher level of β1- to β2-integrins, Weaver et al. (1997) reasoned that it was possible that malignancy was a reflection of changes in these integrins. Accordingly, they treated the 3D preparations with a rat β1-integrin inhibitory mAb. While the treatment caused massive apoptosis in "normal" S-1 cells, T4-2 cells were refractory. Most importantly, T4-2-treated cells reverted to a "non-malignant" phenotype and were not distinguishable from S-1. Culture in 3D was required for complete expression of reverted phenotype. As shown in Figure 12.2, cryosections of "normal" and treated cells revealed acinar formation, uniform and polarized nuclei, well-organized filamentous actin (12.2a and a″). On the other hand, untreated or mock-treated cells revealed polymorphic nuclei and a grossly disorganized actin cytoskeleton, visualized as random, hatched bundles (12.2a′). The structure resulting from acinus formation in 3D cultures resembled an *in vivo* mammary acinus (Bissell et al., 2003). These results were replicated *in vivo*, by injecting "normal" and tumor cells treated in suspension (β1-integrin inhibitory mAb, mock mAb, and no treatment). Actively growing tumors were observed in greater than 75–90% of the mock mAb or no treatment mice.

Follow-up experiments in Mina Bissell's lab conclusively showed that there is a bidirectional cross-modulation of β1-integrin and epidermal growth factor receptor (EGFR) signaling through the mitogen-activated protein kinase (MAPK) pathway in 3D, which does not occur in monolayer (2D) cultures (Wang et al., 1998). The breast epithelial tumor cell reversion described above has been used to understand the multiple ways a given malignant population can be reverted to near-normal phenotype

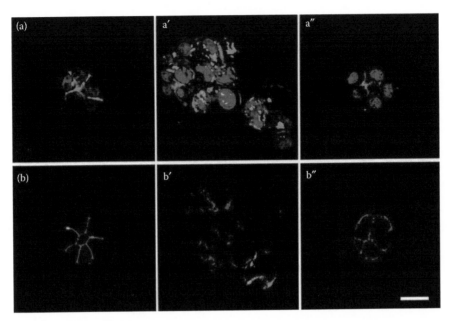

FIGURE 12.2 Confocal microscopy images of stained F-actin and nuclei for 3D S-1 ("normal") cells (a), 3D T4-2 (malignant) mock treated cells (a′), and 3D T4-2 cells treated with β1-integrin inhibitory mAb (a″). Normal and treated cells reveal acinar formation (reverted in a″), uniform and polarized (roundish) nuclei. The bright star-like structures in a and a″ are filamentous F-actin. The mock treated T4-2 cells on the other hand exhibited disorganized actin and pleiomorphic nuclei (a′). (b–b″) are confocal immunofluorescence microscopy images of E-cadherins and β-catenins. Because of lack of color the two stains (FITC for E-cadherins and Texas red for β-catenins) are not discernible in this figure. From the original color images, E-cadherins and β-catenins were colocalized and superimposed at the cell–cell junctions in S-1 (b) and T4-β1 reverted acini (b″). (Adapted from Weaver, V.M., et al. 1997, *J. Cell Biol.* 137(1): 231–245.)

(Bissell et al., 2003). The acinus is now an acceptable experimental system that provides complex physiological relevance as outlined in Chapter 3. If the acinus were to serve as a basis for an HTS assay for discovering compounds with "reversion activities," it would have to be implemented in a 3D format.

12.3   THE MATRIX METALLOPROTEINASE INHIBITORS CASE

12.3.1   EXTRACELLULAR MATRIX METALLOPROTEINASES (MMPs)

This family of zinc-dependent enzymes has already been discussed in Chapter 5 as an important component of the ECM. As a target, MMPs have been implicated in numerous diseases, such as neuroinflammation and neurodegenerative disorders (Rosenberg, 2009; Yong, 2005), cardiovascular disorders (Hamirani et al., 2009; Gueders et al., 2006), diabetic nephropathy (Thraikill et al., 2009), and cancer (for recent reviews, see Folgueras et al., 2004; Rydlova et al., 2008; Jezierska and Motyl, 2009). MMP expression is generally higher in and around malignant cancers in comparison to normal, benign, or premalignant tissues, which suggested the classical role of MMPs in cancer to be that of removal of the ECM to allow cancer cells to invade and metastasize in late invasive stages. However, identification of novel biological functions of MMPs has provided the rationale for evaluating MMP's role in cancer beyond the classical role. For example, tumor-associated MMP expression and activity are the result not only of tumor cell expression, but includes contributions from stromal cells (McKerrow et al., 2000). More recent studies have shown that MMP activity occurs at all stages of tumor development, from angiogenesis to metastasis and growth at secondary sites (Engeblad and Werb, 2002). In addition, there is evidence in support of a protective role for some MMPs in tumor progression (Martin and Matrisian, 2007). Taken together, the evidence in the literature supports the notion that the overall effect of MMPs on cancer is the sum of the MMP effects on individual processes (Corbitt et al., 2007).

12.3.2   MMP Inhibitors (MMPIs)

Endogenous inhibitors control MMP activities. Some of these inhibitors are general protease inhibitors, for example, α2-macroglobulin, which mainly blocks MMP activities nonspecifically in plasma and tissue fluids. Other more specific inhibitors are tissue inhibitors of metalloproteinases—so far four (TIMP-1, TIMP-2, TIMP-3, and TIMP-4) have been identified in vertebrates (Brew et al., 2000). The relationship between high MMP expression and tumor progression suggested the potential for synthetic MMP inhibitors (MMPIs) as a strategy for blocking enzyme proteolytic activities. As a family, all MMPs were considered druggable targets (Overall and Kleifeld, 2006; Overall 2002). Over the past three decades, many drug discovery labs have spent a great deal of money and time in search of small molecule MMPIs. Unfortunately, most clinical trials using MMPIs have yielded disappointing results (Coussens et al., 2002; Overall and Lopez-Otin, 2002; Pavlaki and Zucker, 2003).

12.3.3   Experimental System: Fibrosarcoma Cells in Collagen Gels

In different *in vitro* migration and invasion assays, MMPIs successfully impaired invasive tumor behavior (Mignatti et al., 1986; Kurschat et al., 1999; Ntayi et al., 2001), supporting the concept of proteolytic path generation as one of the means by which tumor cells migrate within tissues. But in other assays, significant residual migration of individual cells was observed (Deryugina et al., 1997; Hiraoka et al., 1998), in agreement with disappointing clinical trials of MMPIs.

Wolfe et al. (2003, p. 268) hypothesized that "... proteolytic compensation could be provided by enzymes not inhibited in the above studies; alternatively, cells could sustain motility via unknown protease-independent compensation strategies." To test these hypotheses, Wolfe et al. (2003) used 3D collagen lattices, proteolytically potent HT-1080 fibrosarcoma (stably transfected with membrane-anchored MT1-MMP), and either an inhibitory anti-β1-integrin mAb or a cocktail comprising six protease inhibitors (BB-2516, E-64, Pepsin A, Leupepstin, and Aprotinin). Cells were labeled with calcein-AM and observed with 3D time-lapse confocal microscopy. With computer-assisted cell-tracking, it was possible to calculate the cell population speed.

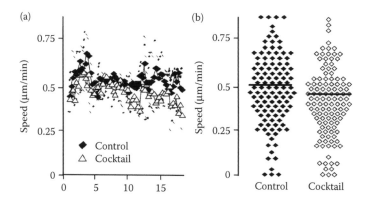

FIGURE 12.3 Protease inhibitor cocktail-independent cell migration: steady-state cell population speed (a) and individual cell mean speed (b) (120 cells, three replicates, and cells monitored for 20 h). (From Wolfe, K. et al. 2003. *J. Cell Biol.* 160: 270. With permission.)

MT1-MMP and β1-integrins were detected by immunofluorescence (rhodamine and FITC, respectively).

12.3.4 TREATMENT WITH PERICELLULAR PROTEOLYSIS INHIBITORS IN 3D CULTURES AND *IN VIVO* DID NOT PREVENT CELL MIGRATION OR METASTASIS

In control experiments the inhibition of collagenolysis by a protease inhibitor cocktail was nearly complete, but as shown in Figure 12.3, the migration efficiency of HT-1080 was barely reduced, suggesting the existence of compensation strategy to counterbalance the loss of pericellular proteolysis. Under impaired collagenolysis, cells migrated by changing shape (e.g., elongation) to pass through smaller matrix gaps. Figure 12.4 illustrates the assumed amoebic morphology that was accompanied by linear surface

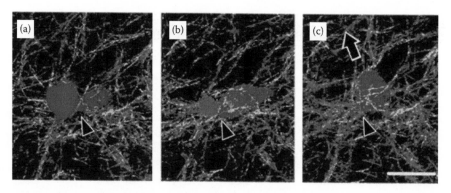

FIGURE 12.4 Amoebic cellular migration through a narrow gap in 3D collagen matrix achieved by morphological adaptation of HT-1080 cell (black arrowhead). The cell forms a constriction ring. The full arrow in (c) indicates cell direction after clearing the constriction labeled with the arrow heads in (a)–(c). (Adapted from Wolfe, K. et al. 2003. *J. Cell Biol.* 160(2): 267–277.)

distribution of β1-integrins at contacts to collagen (not shown). Similar findings were confirmed *in vivo* by injecting proteolytically competent HT-1080 cells into loose connective tissue of the mouse dermis and monitoring cell positions by multiphoton microscopy. Preincubation of cells with a protease inhibitor cocktail for 6 h resulted in stable amoebic movement for 10 h before reverting to protease-dependent migration.

Although the above findings make a good case for 3D cultures in early drug discovery, admittedly, the amoebic migration story is probably a partial explaination of the lack of success with the first and second generation MMPIs. It is now well known that MMPs have a wider functionality including homeostatic regulation of the extracellular environment and controlling innate immunity (Parks et al., 2004; Clark et al., 2008), and not simply to degrade ECM. The complex roles these enzymes play during physiological and pathological conditions are definitely the other partial explanation for the lack of success. The increased knowledge, especially at the molecular level, will lead to better druggable targets in the proteolytic system towards third generation MMPIs for cancer and other therapies (Overall and Kleifeld, 2006; Clark et al., 2008).

12.4 RESISTANCE TO THE CHEMOTHERAPEUTIC AGENTS CASE

12.4.1 EXPERIMENTAL SYSTEM: MULTICELLULAR TUMOR SPHEROID (MCTS)

Spheroids are a form of 3D culture, made possible by *in vitro* conditions that promote the formation of cell aggregates. These aggregates are predominantly produced with stem cells or tumor cells from malignant cell lines or fragments of human tumors (Yamada and Cukierman, 2007). The earliest uses of spheroids were for embryonic cell aggregation and *in vitro* malignancy assay (Moscona and Moscona, 1966). However, the first recognized MCTS in suspension culture was accidentally produced by Sutherland et al. (1971), whose original intention was to grow single cells in culture. Further investigations established the strong similarities in morphology and many functional characteristics between MCTS and *in vivo* solid tumors (e.g., Sutherland et al., 1977).

The analogy between MCTS and tumor regions is illustrated in Figure 12.5. For more than 30 years, numerous investigators have conducted comparative studies between *in vivo* tumors and MCTS. This volume of work has been summarized in a number of excellent reviews (e.g., Mueller-Klieser, 2000; Padrón et al., 2000; Satini et al., 2000; Kunz-Schughart et al., 2004). In more recent reviews, MCTS has been presented in the context of the overall 3D culture models (e.g., Kim, 2005; Yamada and Cukierman, 2007). An interesting review by Kunz-Schughart et al. (2004) evaluated MCTS for HTS and came up with four important features in support of developing MCTS for HTS of chemotherapeutic drug candidates as follows: (1) MCTS re-establish morphological, function, and mass transport features of corresponding tissue *in vivo*; (2) MCTS approximate many characteristics of avascular tumor nodules, micrometastases, or intravascular regions of large solid tumors with regard to both tumor growth kinetics and pathophysiological micromilieu; (3) well-defined, spherical symmetric geometry of MCTS allows direct comparison of structure to function, that is, the microenvironmental gradients that determine spheroid morphology and spatially correlated with changes in cellular physiology; and (4) spheroids are amenable to

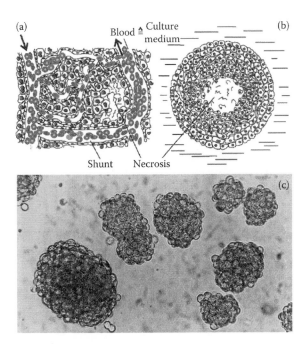

FIGURE 12.5 Comparison of tumor and MCTS microenvironments: (a) *in vivo* tumor microenvironments are characterized by irregular tissue architecture and abnormal microcirculations, which includes a loss of vascular hierarchy, large variability in vessel tortuousness, diameter and length, disturbances in the physiological endothelial function, pronounced shunt perfusion, and local perfusion; (b) MCTS exhibit a spherical geometry with a concentric arrangement of large proliferating cells in the periphery and smaller nonproliferating cells than in deeper regions. A central necrosis emerges at a certain spheroid size, which is largely variable depending on the cell line or type and on the culturing conditions; and (c) phase contrast micrograph illustrating human colon adenocarcinoma LoVo spheroids grown for 3 days by the gyratory rotation method—magnification is 1400×. [(a) and (b) Adapted from Mueller-Klieser, W. 2000. *Crit. Rev. Oncol. Hematol.* 36: 125. With permission; (c) Adapted from Satini, M.T., Rainaldi, G., and Indovina, P.L. 2000. *Crit. Rev. Oncol./Hematol.* 36: 75–87.]

coculture of different cell types, in particular tumor cell and normal cells, such as stromalfibroblast, endothelial cells, or cells of the hematopoietic/immune system.

12.4.2 MCTS More Accurately Approximate *In Vivo* Resistance to Chemotherapeutic Agents

MCTS have been used in numerous drug resistance studies. Examples include taxol (Frankel, 1997) and doxorubicin (Wartenberg et al., 1998). Classically, drug resistance was established *in vivo* in EMT6 tumors in mice. Kobayashi et al. (1993) conducted an elegant experiment that recapitulated the resistance in MCTS that was completely lost when the EMT-6 mammary cancer cells were cultured as a monolayer. The experiment was conducted with three alkylating agents: cyclophosphamide (CTX),

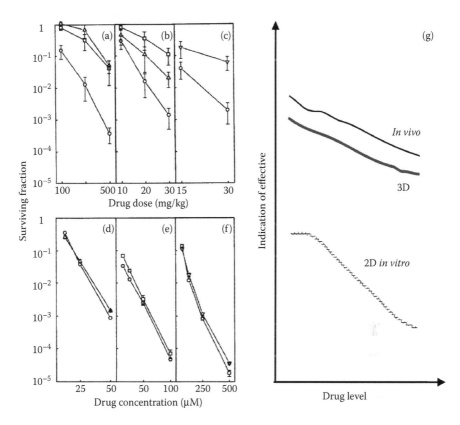

FIGURE 12.6 *In vivo* (a–c) and *in vitro* (d–f) colony-forming assay results using the parental tumor cell line EMT-6 (o) and its drug-resistant derivatives, EMT-6/CTX (Δ), EMT-6/CDDP (□), and EMT-6/Thio (▼). (g) A visual schematic representation of the results on the same drug level scale. *Abbreviations:* CTX = cyclophosphamide, CDDP = *cis*-diamminedichloroplatinum(II), and Thio = *N,N′,N″*-triethylenethiophosphoramide (thiopeta). (From Kobayashi, H. et al. 1993. *Proc. Natl. Acad. Sci. USA* 90: 3296. With permission.)

cis-diamminedichloroplatinum (II) (CDDP), and *N, N′, N″*-triethylenethiophosphoramide (thiotepa). Three alkaylating agent-resistant variants of the EMT-6 parent tumor (EMT-6/CTS, EMT-6/CDDP, and EMT-6/Thio) were generated by serial treatment of mice bearing the tumor. Figure 12.6a–c shows the sensitivity of each tumor line to the *in vivo* drug exposure assay based on the ability to form colonies *in vitro*. "Surviving fraction" was calculated by dividing the number of colony forming units from excised tumors of treated animals by colony forming units from control (untreated) animals. Two million cells each (EMT-6 parent tumor, EMT-6/CTS, EMT-6/CDDP, EMT-6/Thio) were inoculated intramuscularly into the hind legs or subcutaneously into the flank of 8–10-week-old female BALB/c mice. When the tumors volume reached approximately 100 mm³, the animals were treated with drugs by i.p. injection: CTX (Figure 12.6a), CDDP (Figure 12.6b), and Thio (Figure 12.6c). In the three cases, surviving fraction curves for cells from animals with the parental tumor were significantly lower in comparison to curves for cells from animals with the drug-resistant tumors. By calculating the resistance ratio as the surviving fraction of the resistant line divided by the surviving fraction of the parental line at each drug dose (i.e., dividing numbers from the top curve by numbers from the bottom curve in Figures 12.6a), resistant tumor line EMT-6/CTX was found 7.3, 52, and 140 times more resistant to 100, 300, and 500 mg of CTX/kg than the parental tumor line, respectively. A similar calculation from curves in Figure 12.6b suggested that the CDDP-resistant line was 2.7, 23, and 79 times more resistant to 10, 20, and 30 mg of CDDP/kg, respectively. Also, the thiotepa-resistant line was 4.5 and 31 times more resistant to thiotepa at 15 and 30 mg/kg, respectively. When the resistant lines were exposed to each of the drugs for 1 h as monolayer (2D) cultures (Figure 12.6d–f), resistance ratios were less than 2.1 times. However, when the resistant lines were cultured on 1% agarose-coated plates to form MTCS and exposed to drugs for 1 h, EMT-CTX tumor line was 8.1, 250, and 4900 times more resistant to 25, 50 and 100 µM CTX in comparison to the parental line, respectively; EMT-6/CDDP line was 18, 15, and 58 times more resistant to CDDP at 25, 50, and 100 µM, respectively; EMT/Thio line was 3.6, 12, and 65 times more resistant to thiopeta at 125, 250, and 500 µM, respectively. If the plasma concentrations of the drugs were known, it would have been possible to plot the *in vivo* and the *in vitro* results on the same plot and the expected results are schematically portrayed in Figure 10.6g, where the MTCS recapitulates the *in vivo* drug resistance (*in vivo* and 3D curves are not statistically different), when the *in vitro* 2D cultures do not (2D is significantly different from both the *in vivo* and the 3D curves). As outlined above, MCTS can be an excellent 3D model. Numerous 3D culture models have yielded results consistent with the presentation above. For example, mouse mammary tumor cells demonstrated a greater drug resistance to melphalan and 5-fluorouracil as multicellular aggregates in comparison to 2D cultures (Miller et al., 1985). When plated in monolayer cultures, MDA-MB-231 cells exhibited a much lower EC50 for cisplatinum in comparison to their spheroids counterpart. Additionally, MDA-MB-231 cisplatinum treatment induced expression of TGF-β1 nRNA expression as well as the protein upregulation similar to tumor cells *in vivo* that was not observed with 2D cultures (Ohmori et al., 1998). A review by Yamada and Cukierman (2007) summarized additional studies with similar outcomes. To be successfully implemented in HTS systems, several technical limitations will have to be addressed. For example, it takes relatively longer for MCTS to grow. Also, from the fact that spheroids are typically not adherent to the substrate, specialized microscopy systems may be required for readout, which may result in low throughput. Currently, MCTS should be ideal for secondary screen at relatively low throughput rates.

12.5 CONCLUDING REMARKS

The challenge in this chapter was to find cases where real drugs with known *in vivo* function replicated in 3D but not in 2D cultures and for which, if initially screened with 3D cultures early in discovery, savings in cost and time would have been realized. The cases presented meet the challenge. The question that is confronting proponents of 3D cultures in drug discovery is whether the three cases presented above are convincing enough for an HTS laboratory director to consider 3D cultures in their discovery program. Based on personal communications with several lab directors, more cases would be helpful, especially for the emerging targets. It has been impossible to obtain cases that fit the criteria outlined at the beginning of the chapter. Even if such cases exist in the pharmaceutical or biopharmaceutical company experience, the knowledge is more likely to be considered part of company trade secrets. An alternative approach is to experimentally generate the cases probably with public funding. In the following chapter the rationale and possible study, closely related to ongoing work in the author's laboratory, are outlined to encourage those interested to consider such experimentation for their favorite target(s).

REFERENCES

Alinari, L., Lapalombella, R., Andristsos, L., Baiocchi, R.A., Lin, T.S., and Byrd, J.C. 2007. Alemtuzumab (Campath-1H) in the treatment of chronic lymphocytic leukemia. *Oncogene* 26: 3644–3653.

Barok, M., et al. 2007. Trastuzumab causes antibody-dependent cellular cytotoxicity-mediated growth inhibition of supramacroscopic JIMT-1 breast cancer xenografts despite intrinsic drug resistance. *Mol. Cancer Ther.* 6: 2065–2072.

Baselga, J., et al. 2005. Phase II study of efficacy, safety, and pharmacokinetics of trastuzumab monotherapy administered on a 3-weekly schedule. *J. Clin. Oncol.* 23: 2162–2171.1.

Bissell, M.J., Rizki, A., and Mian, I.S. 2003. Tissue architecture: The ultimate regulator of breast epithelial function. *Curr. Opin. Cell Biol.* 15: 753–762.

Brew K., Dinakarprandian, D., and Nagase, H. 2000. Tissue inhibitors of metalloproteinases: Evolution, structure and function. *Biochem. Biophys. Acta* 1477: 267–268.

Briand, P., Petersen, O.W., and van Deurs, B. 1987. A new diploid nontumorigenic human breast epithelial cell line isolated and propagated in chemically defined medium. *In Vitro Cell Dev. Biol.* 23: 181–188.

Carter, P., et al. 1992. Humanization of an anti-p185HER2 antibody for human cancer therapy. *Proc. Natl. Acad. Sci. USA* 89: 4285–4289.

Chinn, P.C., Leonard, J.E., Rosenberg, J., Hanna, N., and Anderson, D.R. 1999. Preclinical evaluation of 90Y-labeled anti-CD20 monoclonal antibody for treatment of nonHodgkin's lymphoma. *Int. J. Oncol.* 15: 1017–1025.

Clark, I.M., Swingler, T.E., Sampieri, C.L., and Edwards, D.R. 2008. The regulation of matrix metalloproteinases and their inhibitors. *Int. J. Biochem. Cell Biol.* 40: 1362–1378.

Cohenuram, M. and Saif, M.W. 2007. Panitumumab the first fully human monoclonal antibody: From the bench to the clinic. *Anticancer Drugs* 18: 7–15.

Corbitt, C.A., Lin, J., and Lindsey, M.L. 2007. Mechanisms to inhibit matrix metalloproteinase activity: Where are we in the development of clinically relevant inhibitors? *Recent Patents Anti-Cancer Drug Discov.* 2: 135–142.

Coussens, L.M., Fingleton, B., and Matrisian, L.M. 2002. Matrix metalloproteinase inhibitors and cancer: Trials and tribulations. *Science* 295: 2387–2392.

da Silva, F.A, Corte-Real, S., and Goncalves, J. 2008. Recombinant antibodies as therapeutic agents—Pathways for modeling new biodrugs. *Biodrugs* 22(5): 301–314.

Davies, A.J. 2007. Radioimmunotherapy for B-cell lymphoma: Y90 ibritumomab tiuxetan and I (131) tositumomab. *Oncogene* 26: 3614–3628.

Dayam, R., Aiello, F., Deng, J., Garofalo, A., Chen, X., and Neamati, N. 2006. Discovery of small molecule integrins $\alpha_v\beta_3$ antagonists as novel anticancer agents. *J. Med. Chem.* 49: 4526–4534.

Deryugina, E.I., Luo, G.X., Reisfeld, R.A., Bourdon, M.A., and Strongin, A. 1997. Tumor cell invasion through Matrigel is regulated by activated matrix metalloproteinase-2. *Anticancer Res.* 17: 3201–3210.

Engeblad, M. and Werb, Z. 2002. New functions for the matrix metalloproteinases in cancer progression. *Nat. Rev. Cancer* 2: 161–174.

Folgueras, A.R., Pendás, A.M., Sánches, L.M., and Otín, C.L. 2004. Matrix metalloproteinases in cancer: From new functions to improved inhibition strategies. *Int. J. Dev. Biol.* 48: 411–424.

Frankel, A., Buchman, R., and Kerbel, R.S. 1997. Abrogation of taxol-induced G(2)-M arrest and apoptosis in human ovarian cancer cells grown as multicellular tumor spheroids. *Cancer Res.* 57(12): 2388–2393.

Furlan, M., Robles, R., Solenthaler, M., Wassmer, M., Sandoz, P., and Lammie, B. 1997. Deficient activity of von Millebrand factor-cleaving protease in chronic relapsing thrombotic thrombocytopenic purpura. *Blood* 89(9): 3097–3103.

Galizia, G., et al. 2007. Cetuximab, a chimeric human mouse anti-epidermal growth factor receptor monoclonal antibody, in the treatment of human colorectal cancer. *Oncogene* 26: 3654–3660.

Gerber, H.P., et al. 2007. Mice expressing a humanized form of VEGF-A may provide insights into the safety and efficacy of anti-VEGEF antibodies. *Proc. Natl. Acad. Sci. USA* 104: 3478–3483.

Glennie, M.J., French, R.R., Cragg, M.S., and Taylor, R.P. 2007. Mechanism of killing by anti-CD20 monoclonal antibodies. *Mol. Immunol.* 44: 3823–3837.

Gueders, M.M., Foidart, J.-M., Noel, A., and Catado, D.D. 2006. Matrix metalloproteinases (MMPs) and tissue inhibitors of MMPs in the respiratory tract: Potential implications in asthma and other lung diseases. *Eur. J. Pharmacol.* 533: 133–144.

Hale, G., Bright, S., Chumbley, G., Hoang, T., Metcalf, D., Munro, A.J., and Waldmann, H. 1983. Removal of T cells from bone marrow for transplantation: A monoclonal anti-lymphocyte antibody that fixes human complement. *Blood* 62: 873–882.

Hale, G., Clark, M.R., Marcus, R., Winter, G., Dyer, M.J.S., Phillips, J.M., Riechmann, J., and Waldmann, H. 1988. Remission induction in non-Hodgkin's lymphoma with reshaped monoclonal antibody CAMPATH-1H. *Lancet* 2: 1394–1399.

Hamann, P.R., Hinman, L.M., Hollander, I., Beyer, C.F., Lindh, D., Holcomb, R., Hallett, W., et al. 2002. Gemtuzumab ozogamicin conjugate for treatment of acute myeloid leukemia. *Bioconjug. Chem.* 13: 47–58.

Hamirani, Y.S., Pandey, S., Rivera, J.J., Ndumele, C. Budoff, M.J., Blumenthal, R.S., and Nasir, K. 2009. Markers of inflammation and coronary artery calcification: A systematic review. *Atherosclerosis* 201(1): 1–7.

Hiraoka, N., Allen, E., Apel, I.J., Gyetco, M.R., and Weiss, S.J. 1998. Matrix metalloproteinases regulate neovascularization by acting as pericellular fibrinolysins. *Cell* 95: 365–377.

Hu, J.L., Van den Steen, P.E., Sang, Q.X.A., and Opdenakker, G. 2007. Matrix metalloproteinase inhibitors as therapy for inflammatory and vascular diseases. *Nat. Rev. Drug Discov.* 6(6): 480–498.

Hudis, C.A. 2007. Trastuzumab-mechanism of action and use in clinical practice. *N. Eng. J. Med.* 357: 39–51.

Huxley-Jones, J., Foord, S.M., and Barnes, M.R. 2008. Drug discovery in the extracellular matrix. *Drug Discov. Today* 13(15/16): 685–694.

Izumi, Y., Xu, L., di Tomaso, E., Fukurama, D., and Jain, R.K. 2002. Tumor biology: Herceptin acts as an antiangiogenic cocktail. *Nature* 416: 279–280.

Jakobovits, A., Amado, R.G., Yang, X., Roskos, L., and Schwab, G. 2007. From XenoMouse technology to panitumumab, the first fully human antibody product from transgenic mice. *Nat. Biotechnol.* 25: 1134–1143.

Jazirehi, A.R., Huerta-Ypez, S., Cheng, G., and Bonavinda, B. 2005. Rituximab (chimeric anti-CD20 monoclonal antibody) inhibits the constitutive nuclear factor-{κ} B signaling pathway in non-Hodgkin's lymphoma B-cell lines: Role in sensitization to chemotherapeutic drug-induced apoptosis. *Cancer Res.* 65: 264–276.

Jazirehi, A.R., Vega, M.I., Chatterjee, D., Goodlick, L., and Bonavinda, B. 2004. Inhibition of Raf-MEK1/2-ERK1/2 signaling pathway, Bcl-xL down-regulation, and chemosensitization of non-Hodgkin's lymphoma B cell by Rituximab. *Cancer Res.* 64: 7117–7126.

Jefferis, R. 2009. Glycosylation as a strategy to improve antibody-based therapeutics. *Nat. Rev. Drug Discov.* 8(3): 226–234.

Jezierska, A. and Motyl, T. 2009. Matrix metalloproteinase-2 involvement in breast cancer progression: A mini-review. *Med. Sci. Monitor* 15(2): RA32–RA40.

Kaminski, M.S., et al. 2001. Pivotal study of iodine I 131 tositumomab for chemotherapy-refractory low-grade or transformed low-grade B-cell non-Hodgkin's lymphomas. *J. Clin. Oncol.* 19: 3918–3928.

Kawaguchi, Y., Kono, K., Mimura, K., Sugai, H., Akaike, H., and Fujii, H. 2007. Cetuximab induce antibody-dependent cellular cytotoxicity against EGFR-expressing esophageal squamous cell carcinoma. *Int. J. Cancer* 120: 781–787.

Kim, J.B. 2005. Three-dimensional tissue culture model in cancer biology. *Semin. Cancer Biol.* 15: 365–377.

Kim, K.J., et al. 1993. Inhibition of vascular endothelial growth factor-induced angiogenesis suppresses tumor growth *in vivo*. *Nature* 362: 841–844.

Kleinman, H.K. and Martin, G.R. 2005. Matrigel: Basement membrane matrix with biological activity. *Semin. Cancer Biol.* 15: 378–386.

Kobayashi, H., Man, S., Graham, C.H., Kapitan, S., Teicher, B.A., and Kerbel, R.S. 1993. Acquired multicellular-mediated resistance to alkylating agents in cancer. *Proc. Natl. Acad. Sci. USA* 90: 3294–3298.

Kohler, G. and Milstein, C. 1975. Continuous cultures of fused cells secreting antibody of predefined specificity. *Nature* 256: 495–497.

Kramer, I. and Lipp, H.P. 2007. Bevacizumab, a humanized anti-angiogenic monoclonal antibody for the treatment of colorectal cancer. *J. Clin. Pharm. Ther.* 32:1–14.

Kunq, P., Goldstein, G., Reinhen, E.L., and Schlossman, S.F. 1979. Monoclonal antibodies defining distinctive human T cell surface antigens. *Science* 206: 347–349.

Kunz-Schughart, L.A., Freyer, J.P., Hofstaedter, F., and Ebner, R. 2004. The use of 3D cultures for high-throughput screening: The multicellular spheroid model. *Soc. Biomol. Screen.* 9(4): 273–285.

Kurschat, P., Zigrino, P., Nischt, R., Breitkopf, K., Steurer, P., Hlein, C.E., Krieg, T., and Mauch, C. 1999. Tissue inhibitor of matrix metalloproteinase-2 regulates matrix metalloproteinase-2 activation by modulation of membrane type 1 matrix metalloproteinase activity in high and low invasive melanoma cell lines. *J. Biol. Chem.* 274: 21056–21062.

Lal, H., Verma, S.K., Foster, D.M., Golden, H.B., Reneau, J.C., Watson, L.E., Singh, H., and Dostal, D.E. 2009. Integrin and proximal signaling mechanisms in cardiovascular disease. *Front. Biosci.* 14: 2307–2334.

Landry, Y. and Gies, J.-P. 2008. Drugs and their molecular targets: An updated overview. *Fundam. Clin. Pharmacol.* 22: 1–18.

Liu, X.-Y., Pop, L.M., and Vitetta, E.S. 2008. Engineering therapeutic monoclonal antibodies. *Immunol. Rev.* 222: 9–27.

Loo, L., Robinson, M., and Adams, G.P. 2008. Antibody engineering principles and applications. *Cancer J.* 14(3): 149–153.

Martin, M.D. and Matrisian, L.M. 2007. The other side of MMPs: Protective roles in tumor progression. *Cancer Metastasis Rev.* 6(28): 16–18.

McKerrow, J.H., et al. 2000. A functional proteomics screen of proteases in colorectal carcinoma. *Mol. Med.* 5(6): 450–460.

Mendelsohn, J. 1997. Epidermal growth factor receptor inhibition by a monoclonal antibody as anticancer therapy. *Clin. Cancer Res.* 3: 2703–2707.

Mignatti, P., Robbins, E., and Rifkin, D.B. 1986. Tumor invasion through the human amniotic membrane: Requirement for a proteinase cascade. *Cell* 47:487–498.

Miller, B.E., Miler, F.R., and Heppner, G.H. 1985. Factors affecting growth and drug sensitivity of mouse mammary-tumor lines in collagen. *Cancer Res.* 45(9): 4200–4205.

Moscona, M.H. and Moscona, A.A. Inhibition of aggregation in vitro by puromycin. *Exp. Cell Res.* 41(3): 703–706.

Mosnier, J.F., Jarry, A., Bou-Hanna, C., Denis, M.G., Merlin, D., and Laboisse, C.L. 2006. ADAM15 upregulation and interaction with multiple binding partners in inflammatory bowel disease. *Lab. Invest.* 86(10): 1064–1073.

Mueller-Klieser, W. 2000. Tumor biology and experimental therapeutics. *Crit. Rev. Oncol./Hematol.* 36: 123–139.

Nadler, L.M., Ritz, J., Hardy, R., Pesando, J.M., Scholossman, S.F., and Stashenko, P. 1981. A unique cell surface antigen identifying lymphoid malignancies of B cell origin. *J. Clin. Invest.* 67: 134–140.

Nielsen, K.V. and Briand, P. 1989. Cytogenic analysis of *in vitro* karyotype evolution in a cell line established from nonmalignant human mammary epithelium. *Cancer Genet. Cytogenet.* 39: 103–118.

Ntayi, C., Lorimier, S., Berthier-Vergness, O., Hornebeck, W., and Bernard, P. 2001. Cumulative influence of matrix metalloproteinaise-1 and -2 in the migration of melanoma cells within three-dimensional type I collagen lattices. *Exp. Cell Res.* 270: 110–118.

Ohmori, T., Yang, J.L., Price, J.O., and Arteaga, C.L. 1998. Blockade of tumor cell transforming growth factor-beta s enhances cell cycle progression and sensitizes human breast carcinoma cells to cytotoxic chemotherapy. *Exp. Cell Res.* 245(2): 350–359.

Overall, C.M. 2002. Molecular determinants of metalloproteinase substrate specificity: Matrix metalloproteinase substrate binding domains, modules, and exocytes. *Mol. Biotechnol.* 22: 51–86.

Overall, C.M. and Kleifeld, O. 2006. Towards third generation matrix metalloproteinase inhibitors for cancer therapy. *Br. J. Cancer* 94: 941–946.

Overall, C.M. and Lopez-Otin, C. 2002. Strategies for MMP inhibition in cancer, innovations for the post-trailera. *Nat. Rev. Cancer* 2: 657–672.

Padrón, J.M., van der Wilt, C.L., Smid, K., Smitskamp-Wilms, E., Backus, H.J., Pizao, P.E., Giaccone, G., and Peters, G.J. 2000. The multilayered postconfluent cell culture as a model for drug screening. *Crit. Rev. Oncol./Hematol.* 36: 141–157.

Pagano, L., Fianchi, L., Caira, M., Rutella, S., and Leone, G. 2007. The role of gemtuzumab ozogamicin in the treatment of acute myeloid leukemia patients. *Oncogene* 26: 3679–3690.

Parks, W.C., Wilson, C.L., and López,-Boado, Y.S. 2004. Matrix metalloproteinases as modulators of inflammation and innate immunity. *Nat. Rev. Immunol.* 4:617–629.

Pavlaki, M. and Zucker, S. 2003. Matrix metalloproteinase inhibitors (MMPIs): The beginning of phase I or the termination of phase III clinical trials. *Cancer Metastasis Rev.* 22: 177–203.

Pescovitz, M.D. 2006. Rituximab, an anti-cd20 monoclonal antibody: History and mechanism of action. *Am. J. Transplant* 6: 859–866.

Presta, L.G., et al. 1997. Humanization of an antivascular endothelial growth factor monoclonal antibody for therapy of solid tumors and other disorders. *Cancer Res.* 57: 4593–4599.

Ravandi, F. and O'Brien, S. 2005. Alemtuzumab. *Expert Rev. Anticancer Ther.* 5:39–51.

Reff, M.E., et al. 1994. Depletion of B cells *in vivo* by chimeric mouse human monoclonal antibody to CD20. *Blood* 83: 435–445.

Ria, R., Vacca, A., Ribatti, D., Di Raimondo, F., Merchionne, F., and Dammacco, F. 2002. Alpha(v)beta(3) integrin engagement enhances cell invasiveness in human multiple myeloma. *Haematologica* 87(8): 836–845.

Rosenberg, G. 2009. Matrix metalloproteinases and their multiple roles in neurodegenerative disease. *Lancet Neurobiol.* 8: 205–216.

Rydlova, M., Holubec, L. Ludvikova, M., Kafert, D., Franekova, J., Povysil, C., and Ludvikova M. 2008. Biological activity and clinical implications of the matrix metalloproteinases. *Anticancer Res.* 28(2B): 1389–1397.

Satini, M.T., Rainaldi, G., and Indovina, P.L. 2000. Apoptosis, cell adhesion and the extracellular matrix in the three-dimensional growth multicellular tumor spheroids. *Crit. Rev. Oncol./Hematol.* 36: 75–87.

Stiehm, E.R., Keller, M.A., and Vyas, G.N. 2008. Preparation and use of therapeutic antibodies primarily of human origin. *Biologicals* 36: 363–374.

Stockwin, L.H. and Holmes, S. 2003. The role of therapeutic antibodies in drug discovery. *Drug bell, Discov. Des.* 31(2): 433–436.

Sutherland, R.M., McCredie, J.A., and Inch, W.R. 1971. Growth of multicellular spheroids in tissue culture as a model for nodular carcinoma. *J. Natl. Cancer Inst.* 46: 113–120.

Sutherland, R.M., McDonald, H.R., and Howell, R.L. 1977. Multicellular spheroids: A new model target for *in vitro* studies of immunity to solid tumor allografts. *J. Natl. Cancer Inst.* 58: 1849–1853.

Thraikill, K.M., Bunn, R.C., and Fowlkes, J.L. 2009. Matrix metalloproteinases: their potential role in the pathogenesis of diabetic nephropathy. *Endocrine* 35(1): 1–10.

Tjandra, J.J., Ramadi, L., and McKenzie, I.F.C. 1990. Development of human antimurine antibody (HAMA) response in patients. *Immunol. Cell Biol.* 68: 367–376.

Wang, F., Weaver, V.M., Petersen, O.W., Larabell, C.A., Dedhar, S., Briand, P., Lupu, R., and Bissell, M.J. 1998. Reciprocal interactions between β1-integrin and epidermal growth factor receptor in three-dimensional basement membrane breast cultures: A different perspective in epithelial biology. *Proc. Natl. Acad. Sci. USA* 95: 14821–14826.

Wartenberg, M., Hescheler, J., Acker, H., Diedershagen, H., and Sauer, H. 1998. Doxorubicin distribution in multicellular prostate cancer spheroids evaluated by confocal laser scanning microscopy and the "optical probe technique." *Cytometry* 31(2): 2353–2356.

Weaver, V.M., Petersen, O.W., Wang, F., Larabell, C.A., Briand, P., Damsky, C., and Bissell, M.J. 1997. Reversion of the malignant phenotype of human breast cells in three-dimensional cultures and *in vivo* by integrin blocking antibodies. *J. Cell Biol.* 137(1): 231–245.

Wolfe, K., Mazo, I., Leung, H., Engelke, K., von Andriana, U.H., Deryugina, E.I., Strongin, A.Y., Bröcker, E.B., and Friedl, P. 2003. Compensation mechanism in tumor cell migration: Mesenchymal-amoebic transition after blocking of pericellular proteolysis. *J. Cell Biol.* 160(2): 267–277.

Yamada, K.M. and Cukierman, E. 2007. Modeling tissue morphogenesis and cancer in 3D. *Cell* 130: 601–610.

Yong, V.W. 2005. Metalloproteinases: Mediators of pathology and regeneration in the CNS. *Nat. Rev. Neurosci.* 6: 931–944.

Zalutsky, M.R., Garg, P.K., and Narula, A.S. 1990. Labeling monoclonal antibodies with halogen nuclides. *Acta Radiol.* 374 (suppl.): 141–145.

13 Ideal Case Study Design

13.1 RATIONALE FOR THE CASE STUDY

The drug case studies provided in Chapter 12 make a case for at least considering 3D cell-based biosensing or bioassay in HTS. Despite these cases, a skeptical director of a drug discovery program from one of the major pharmaceutical companies once challenged this author after presenting these cases as a basis for the merits of adopting 3D cultures in drug discovery. "Show me a couple of drugs that were promising when tested with 2D cultures but failed after approval and 3D cultures testing after withdrawal reveals the drugs as nonpromising; and I will be totally convinced," the challenger said. A thorough search of the literature did not yield any results that satisfied the condition the challenger put forth. The presence of such data is considered critical to the narrative toward persuading pharmaceutical, biopharmaceutical, and biotechnological companies to deploy 3D biosensor technology in their drug discovery programs. Such a study is proposed herein. The execution and success of such a study will provide answers to the challenge and probably accelerate the adoption of 3D systems in drug discovery programs. The proposal focuses on hepatotoxicity with hepatocyte-like cells derived from human embryonic stem cells (hESCs) and drugs that were approved but later withdrawn. To be consistent with HTS applicability of the resultant knowledge, the study should be conducted in three commercially available 3D high-density (e.g., 96- and 1536-well palates) with a 2D high-density plate as control. The three subhypotheses driving the study are as follows: (1) stem cell-derived hepatocyte-like cells cultured in 3D microenvironments emulate the *in vivo* liver cells better than their 2D counterparts, (2) in comparison to 2D, 3D cell-based biosensors improve the hepatotoxicity predictability of experimental drugs, and (3) mitochondrial inner transmembrane potential ($\Delta\psi_m$) signal from 3D cells, in particular stem cell-derived hepatocyte-like cells, is easily distinguishable from background noise and can be used to yield a positive Z' factor in the proposed 3D plates.

This study is publicly offered here for two main reasons. First, to make it easier for others to either conduct this very study or other similarly designed studies with different cells, drugs, or scaffolds. Second, to encourage pharmaceutical, biopharmaceutical, and biotechnological companies that may already have unpublished results that can contribute to a database of cases that test the hypotheses above to share their results. The rest of the chapter is devoted to providing the necessary background, briefly, and the rationale in support of the key choices made.

13.2   WHY HEPATOTOXICITY?

13.2.1   Morphology of the Liver

The liver is the largest organ in the body, weighing approximately 1.5 kg in the average adult human, and performs many different functions including filtration and storage of blood; metabolism of carbohydrates, proteins, fats, hormones, and foreign chemicals; formation of bile (the yellowish brown or green fluid secreted by the liver and discharged into the duodenum, where it aids in emulsification of fats, increases peristalsis, and retards putrefactions—contains sodium glycocholate and sodium taurocholate, cholesterol, biliverdine and bilirubin, mucus, fat lecithin, and cells and cellular debris); storage of vitamins and iron; and formation of coagulation factors.

The liver lobule is the basic functional unit of the liver. It is a cylindrical structure several milliliters in length and 0.8–2 mL in diameter. The human liver contains 50,000–100,000 individual lobules. A schematic of the lobule is shown in Figure 13.1. The liver consists of six cell types. The hepatocytes, which perform the major metabolic and synthetic function of the liver, are of epithelial cell origin and account for 60% of the liver cells. Endothelial cells, which provide a barrier function regulating the entry of molecules to the hepatocytes, account for 20% of the liver cells. Kupffer cells, which are resident macrophages, account for about 15% of liver cells. Epithelial cells, which line the bile ducts, account for about 3–5% of liver cells. The endothelial lining of the sinusoids has extremely large pores (approximately 1.0 μm in diameter). Beneath this lining, lying between the endothelial cells and the hepatic cells are narrow tissue spaces called the spaces of Diss, which connect with the lymphatic vessels in the interlobular septa. Therefore, excess fluid in these spaces is removed through the lymphatics.

13.2.2   What Is Hepatotoxicity?

The human body identifies all drugs as foreign substances (i.e., xenobiotics) and makes them suitable for elimination by transforming them via various chemical processes (i.e., metabolism). Almost all tissues in the body have some ability to metabolize chemicals. However, the liver, and in particular its smooth endoplasmic reticulum, is the main "bioreactor" for transforming both endogenous chemicals (e.g., cholesterol, steroid hormones, fatty acids, proteins, and xenobiotics). The major outcome of metabolism in the liver is the formation of water-soluble metabolites that are removed from the body. Xenobiotic metabolism is achieved in two phases. In the first phase, the xenobiotic may be oxidized, reduced, hydrolyzed, or hydrated, or undergoes many other rare chemical reactions. In the second phase, the transformed xenobiotic is conjugated to highly polar molecules such as glucose, sulfate, cysteine, and glutathione, which takes place mainly in the cytosol, via transferase enzymes. Chemically active phase 1 products are rendered relatively inert and are transported directly from the hepatocytes into the biliary canaliculi (third-phase metabolism) to be excreted as bile, or are released back into the blood stream to be excreted into the urine via the kidneys. In the intestine, the metabolites may be deconjugated by gut bacterial flora and reabsorbed, leading to a repeat of the metabolic processes (enterohepatic recirculation).

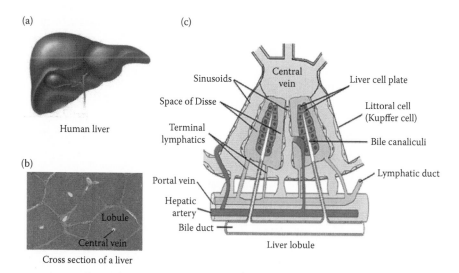

(a) Human liver

(b) Cross section of a liver

Lobule

Central vein

(c)

Sinusoids

Space of Disse

Terminal lymphatics

Portal vein

Hepatic artery

Bile duct

Central vein

Liver cell plate

Littoral cell (Kupffer cell)

Bile canaliculi

Lymphatic duct

Liver lobule

FIGURE 13.1 Schematic of a liver (a), cross section of the liver lobule (b) and the detailed hepatic cellular plates and blood vessels, the bile collecting system, and the lymph flow system composed of the spaces of Disse and the interlobular lymphatics (c). The lobule is constructed around a central vein that empties into the hepatic veins and then into the vena cava. The lobule is composed of many hepatic cellular plates (two of which are shown in this figure) that radiate from the central vein-like spokes in a wheel. Each hepatic plate is usually two cells thick, and between the adjacent cells lie small bile canaliculi that empty into bile ducts in the fibrous septa separating the adjacent liver lobules. Also in the septa are small portal venules that receive their blood mainly from the venous outflow of the gastrointestinal tract by way of portal vein. From these venules, blood flows into flat, branching hepatic sinusoids that lie between the hepatic plates and then into the central vein. Thus the hepatic cells are exposed continuously to portal venous blood. In addition to the portal venules, hepatic arterioles are present in the interlobular septa. These arterioles supply arterial blood to the septa tissues between the adjacent lobules, and many of the small arterioles also empty into the hepatic sinusoids, most frequently emptying into these about one-third of the distance away from the interlobular septa, as shown in this figure. (From Guyton, A.C. and Hall, J.E. 2000. *Textbook of Medical Physiology*, 10th edition. Philadelphia: W.B. Saunders Company. With permission.)

It should be pointed out that many innocuous chemicals can be transformed (activated) into toxic metabolites with the liver as the target organ of the toxicants.

The most important family of phase 1 metabolizing enzymes, located in the endoplasmic reticulum, are known as the cytochrome P450 (CYPs), a family of closely related 50 isoforms, six of which metabolize 90% of drugs (Lynch and Price, 2007). The key isoforms in human liver include CYP1A2, CYP2A6, CYP2C9, CYP2C19, CYP2D6, CYP2E1, and CYP3A4. CYP3A4 is believed to catalyze a large number of xenobiotics, with 50% of current human pharmaceuticals believed to be CYP3A4 substrates (Li et al., 1995). There is tremendous diversity of individual P450 gene products, and this heterogeneity allows for the liver to catalyze a vast array of chemical reactions. P450 genetic variation (polymorphism), the ability of drugs to induce or inhibit P450 activity and the ability of drugs to share the same P450 specificity and thus competitively block their transformation, singly or collectively, lead to drug-induced toxicity.

Hepatotoxicity is defined as injury to the liver by drugs or other foreign, noninfectious agents. There are over 900 drugs that have been implicated in causing liver injury after 2 years of marketing (Friedman et al., 2003). Of a total of 114 drug withdrawals from the US, Germany, France, and UK markets, due to clinical safety, during the period 1961–1992, 23 (20.2%) were due to hepatotoxicity, 15 were due to myelotoxicity, and 15 were due to neurotoxicity (Spriet-Pourra and Auriche, 1994). The top nine reasons for drug withdrawal in a separate study (Fung et al., 2001) for the period 1960–1999 were hepatic (26%), hematological (10.5%), cardiovascular (8.7%), dermatological (6.3%), carcinogenic issue (6.3%), renal (4.8%), drug interactions (4.1%), neurological (4.1%), and psychiatric (3.7%). Major hepatic events included all forms of hepatic injury, whereas proarrhythimias accounted for the majority of cardiovascular events. More recent drug withdrawals are presented in Table 13.1 for the period 1990–2007 and, as shown, drug-induced liver injury is the most common reason for withdrawal. It has been reported that drug-induced liver injury is responsible for 5% of all hospital admissions and 50% of all acute liver failures (Ostapowicz et al., 2002; Lee, 2003). Since the most common reason for withdrawal of drugs after preclinical/clinical evaluation is hepatotoxicity, all investigational drugs have to undergo hepatotoxicity evaluation, mandated by the FDA.

Adverse drug reactions are classified as type A (intrinsic or pharmacological) or type B (idiosyncratic) (Pirmohamed et al., 1998). Drugs or toxins that have a pharmacological-type hepatotoxicity are those that have predictable dose-response curves and well-characterized mechanisms of toxicity, such as directly damaging liver tissue or blocking the metabolic process. This type of injury occurs shortly after some threshold for toxicity is reached. Idiosyncratic adverse drug reactions (IADRs) occur without warning, when agents cause nonpredictable hepatotoxicity in susceptible individuals, which is not related to dose and has a variable latency period. This type of injury does not have a clear dose-response or temporal relationship, and most often does not have predictive models. IADRs have led to withdrawal of many drugs from the market post-FDA approval; Troglitazone and Trovafloxacin are two well-known examples. The underlying mechanism behind hepatic IADRs in humans is unknown. However, a widely accepted hypothesis is that mitochondrial dysfunction from oxidative stress leads to IADRs (Li, 2002; Shaw et al., 2007).

13.3   HEPATOTOXICITY AND hESC-DERIVED HEPATOCYTE-LIKE CELLS

13.3.1   Two Reasons Why IADRs Have Attracted Proposed Studies

Firstly, all investigational drugs have to undergo hepatotoxicity evaluation as mandated by FDA and therefore a viable 3D HTS cell-based assay platform will have industry-wide appeal irrespective of target or disease specialization. Secondly, hepatotoxic IADRs have provided the most extensive inventory of drugs withdrawn after FDA approval and as such provide many positive control candidates for establishing complex physiological relevance of 3D cultures in drug discovery as well as proof that these 3D cultures yield high value leads.

TABLE 13.1

Drug Withdrawals, Post Approval, from Various Markets (1990–2008) due to Safety Reasons

Drug	Year of Withdrawal; Company—if Known[a]	Use	Safety Problem
Dilevalol (Labetalol)	1990; Scheling-Plough	Antihypertension	Hepatotoxicity
Triazolan or Triazolam (Halcion)	1991; Upjohn	Sedative/sleep aid	Neuropsychiatric reactions
Terodiline (Bicor)	1991; Kabi Pharmacia	Urinary frequency/incontinence	QT interval prolongation and TdP[b]
Encainide (Enkaid)	1991; Bristol-Myers Squibb	Antiarrhythmic	Proarrhythmias
Fipexide (Vigilor)	1991; NK	Cognition activator for senile dementia	Hepatotoxicity
Temafloxacin (Omniflox)	1992; Abbott	Quinolone anti-infective	Hypoglycemia, hemolytic anemia, and renal failure
Benzarone (Fragivix)	1992; Sanofi-Winthrop	Treatment of peripheral venous disorders	Hepatotoxicity
Remoxipride	1993; Astra	D2 receptor blocker to treat schizophrenia	Aplastic anemia
Alpidem (Ananxyl)	1993; NK	Antianxiety	Hepatotoxicity
Flosequinan (Manoplax)	1993; Boots Company PLC	Vasodilator	Excess mortality possibly due to proarrhythmias
Bendazac	1993; Estechpharma Co. Ltd.	NSAID for joint pain and the prevention of cataracts	Hepatotoxicity
Soruvidine (Brovavir)	1993; Bristol-Myers Squibb	Antiviral used to treat herpes	Myelotoxicity following drug interaction
Chlormezanone (Trancopal)	1996; Sanofi-Winthrop (in the U.S.) (Daiichi Sankyo developed the drug)	Central acting muscle reactant	Hepatotoxicity and several skin reactions
Tolrestat (Alredase)	1996; American Home Products (later became Wyeth)	Aldose reductase inhibitor	Hepatotoxicity
Minaprine (Brantur, Cantor)	1996; Taishau Pharm. Co.	Psychotropic antidepressant	Convulsions
Pemoline (Cylert)	1997; Abbott	CNS stimulant	Hepatotoxicity
Dexfenfluramine (Redux)	1998; Wyeth	Diet aid	Cardiac valvulopathy and pulmonary hypertension

continued

TABLE 13.1 (continued)
Drug Withdrawals, Post Approval, from Various Markets (1990–2008) due to Safety Reasons

Drug	Year of Withdrawal; Company—if Known[a]	Use	Safety Problem
Fenfluramine (Pondimin)	1998; Wyeth	Diet aid/antiobesity	Cardiac valvulopathy and pulmonary hypertension
Terfenadine (Seldane)	1998; Hoechst M-R	Antihistamine for allergies	Drug interaction, QT interval prolongation, and TdP
Bromfenac (Duract)	1998; Wyeth	NSAID pain reliever	Hepatotoxicity following prolonged administration
Ebrotidine (Ebrocit)	1998; Ferrer	H2 blocker to reduce stomach acid	Hepatotoxicity
Sertindole (Serdolect)	1998; Abott	Atypical antipsychotic to treat schizophrenia	QT interval prolongation and potential for TdP
Mibefradil (Posicor)	1998; Roche	Hypertension	Statin-induced rhabdomyolysis following drug interaction and concerns on other potential drug interactions, including the risk of TdP
Tolcapone (Tasmar)	1998; Valeant Pharmaceuticals	COMT inhibitor to treat Parkinson's	Hepatotoxicity
Astermizole (Hismanal)	1999; Johnson & Johnson, Jansen	Antihistamine	Drug interaction, QT interval prolongation, and TdP
Trovafloxacin (Omniflox)	1999; Abbott	Broad-spectrum antibiotic	Hepatotoxicity
Grepafloxacin (Raxar)	1999; GlaxoWelcome	Broad-spectrum antibiotic	QT interval prolongation and TdPO
Troglitazone (Rezuline)	2000; Pfizer/ Warner-Lambert	Diabetes	Hepatotoxicity
Alosetron (Lotronex)	2000; GlaxoSmithKline	Irritable bowel syndrome	Ischemic colitis
Cisapride (Propulsid)	2000; Janssen	PUD/nighttime heartburn	Drug interactions, QT interval prolongation, and TdP
Droperidol (Inapsine)	2001; Akorn	Antiemetic to prevent postoperative nausea and vomiting	QT interval prolongation and TdP
Levacetylmethadol (abbr. LAAM) (Orlaam)	2001; Roxane	Long-acting opioid (to prevent opioid withdrawal)	Drug interactions, QT interval prolongation, and TdP

Drug	Year; Company	Indication	Adverse effect
Cerivastin (Baycol)	2001; Bayer	Cholesterolemia	Rhabdomyolysis following drug interactions
Dofetilide (Tikosyn)	2004; Pfizer	Improve heart rhythm by relaxing overactive heart	Drug interactions, QT interval prolongation, and TdP
Rofecoxib (Vioxx)	2004; Merck & Co.	NSAID for osteoarthritis and pain	Myocardial infarction and strokes
Valdecoxib (Bextra)	2005; G.D. Searle & Company	NSAID for arthritis and painful menstruation	Myocardial infarction and skin reactions
Thiordazine (Mellari)	2005; Norvartis	Antipsychotic schizophrenia and dementia	QT interval prolongation and TdP
Telithromycin (Ketek)	2006; Sanofi-Aventis	Respiratory infections—pneumonia, bronchitis, and sinusitis	Hepatotoxicity
Ximelagatran (Exanta)	2006; Astra-Zeneca	Anticoagulant for thrombotic disorders (to replace warfarin)	Hepatotoxicity
Aprotinin (Trasylol)	2007; Bayer AG	Control bleeding during heart surgery	Kidney failure, heart attack, and stroke
Pergolide (Permax)	2007; Valeant Pharmaceuticals	Parkinson's disease—used in combination with levodopa and carbidopa to manage tremors and slowness of movement	Mitral valve regurgitation
Tegaserod (Zelnorm)	2007; Norvartis	Constipation—predominantly irritable bowel syndrome	Heart attack
Rosiglitazone (Avandia)	2007; GlaxoSmithKline	Diabetes	Cardiovascular risk or heart attack

[a] NK = Not known.

[b] QT interval of the EGC represents the duration of ventricular action potential, determined by a net balance between inward depolarizing and outward repolarizing currents, especially during phase 3 of the action potential. The major determinant of phase 3 outward repolarizing current is the rapid component of delayed rectifier potassium current, mediated by a voltage-gated potassium channel (IKr). Excessive prolongation of QT interval, resulting from dysfunctional IKr, often leads to potentially fatal TdP (Torsade de pointes).

Hepatoblastoma cell line (HepG2) cultured in 2D format is the most recognized hepatocyte-like cells routinely used for drug screening. However, these cell types entirely lack or have very low levels of many of the drug-metabolizing enzymes (cytochrome P450s—CYPs) and transporters found in hepatocytes *in vivo* (Wilkening et al., 2003). It has been demonstrated at mRNA and protein expression levels that HepG2 in general expresses lower amounts of drug-metabolizing enzymes and many liver-specific genes (Ek et al., 2007) including CYPs (Table 13.2). Primary human hepatocytes are preferred owing to their high-level expression of both phases 1 and 2 drug-metabolizing enzymes (Raucy et al., 2002; Schuetz et al., 1993). However, primary hepatocytes are difficult to obtain and maintain. Primary hepatocytes usually have no ability to replicate sufficiently *in vitro* to meet the high demand for drug screening. Sometimes they even fail to maintain their differentiated properties *in vitro*. Additionally, these cells exhibit high variability depending on source. Although animals can be sacrificed to obtain animal hepatocytes, the process is time- and labor-intensive. Besides, animal hepatocytes lack certain important characteristics for human disease-related studies.

hESC-derived hepatocytes are emerging as an ideal cell choice. hESC-derived hepatocytes proliferate extensively *in vitro* in an undifferentiated state, with the potential to differentiate into a variety of cell lineages (i.e., ectodermal, mesodermal, and endodermal) (Hoffman and Carpenter, 2005). A three-stage differentiation protocol has been reported by Cai et al. (2007). This three-stage differentiation protocol gradually differentiates hESCs into matured hepatic cells in 18 days. This differentiation process resembles natural liver development; thus it can provide high efficiency of differentiation in serum-free media. More than 70% of the differentiated cells are albumin-positive cells, with expressions of several other important hepatocyte markers with mature hepatic functions. hESCs offer a potential unlimited source of functional human hepatocytes, since they can differentiate into hepatocyte-like cells displaying a characteristic hepatic morphology and expressing several hepatic markers (Ek et al., 2007) as shown in Table 13.2.

3D hepatocyte-like cell cultures have already been achieved in bioreactors (Morsiani et al., 2001; Miyashita et al., 2000) and sandwich cultures (see Chapter 11, Section 11.3.2). The purpose of the bioreactor cultured hepatocytes was to construct high-density 3D tissue structures and, therefore, to fabricate a functional bioartificial organ for transplantation. The high-density organization has been provided as an explanation for the high *in vivo* emulation characteristics exhibited. Better, if not, identical *in vivo* emulation is expected in the proposed plates in this study because beyond high density, the plates provide spatial (3D), chemical (e.g., limited tissue size imposed by pore size to avoid oxygen limitation in the absence of vasculaire), and suitable substrate material (polystyrene) (Green and Yamada, 2007). *In vivo* emulation can be confirmed with a low-density array (LDA) as outlined in the procedures below.

13.3.2 IADRs and Mitochondrial Inner Transmembrane Potential ($\Delta\psi_m$)

As pointed out before, a widely accepted hypothesis is that mitochondrial dysfunction from oxidative stress leads to hepatotoxic IADRs (Li, 2002; Shaw et al., 2007). Since mitochondrial oxidative stress can be detected through reduction in the inner

TABLE 13.2

Relative Gene Expression for Liver-Related Genes as Measured by LDA Assay[a]

Gene	SA167 Hepatocyte-like	SA002 Hepatocyte-like	HepG2	Primary Hepatocyte
Cytochrome P450s				
CYP1A1				
CYP1A2				
CYP1B1				
CYP2A6/2A7/2A13				
CYP2B6				
CYP2C8				
CYP2C9				
CYP2C19				
CYP2D6				
CYP2E1				
CYP3A4				
CYP3A5				
UDP-glucoronosyl-transferases				
UGT1A3				
UGT1A6				
UGT1A8				
UGT2B7				
UGT1A				
Transporters				
NTCP				
OATP-A				
OATP-C				
OCT1				
MDR1				
MDR3				
BSEP				
MRP2				
Transcription factors				
PXR				
CAR				
FXR				
RXR$_\alpha$				
RXR$_\beta$				
RXR$_\gamma$				
HNF1$_\alpha$				
HNF3$_\alpha$				
HNF3$_\beta$				
HNF4$_\alpha$				
HNF6				

continued

TABLE 13.2 (continued)
Relative Gene Expression for Liver-Related Genes as Measured by LDA Assay[a]

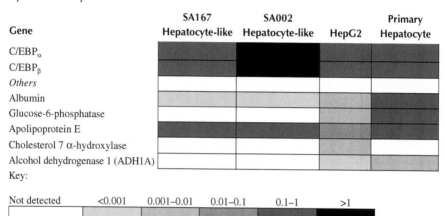

Gene	SA167 Hepatocyte-like	SA002 Hepatocyte-like	HepG2	Primary Hepatocyte
C/EBP$_\alpha$				
C/EBP$_\beta$				
Others				
Albumin				
Glucose-6-phosphatase				
Apolipoprotein E				
Cholesterol 7 α-hydroxylase				
Alcohol dehydrogenase 1 (ADH1A)				

Key:

Not detected	<0.001	0.001–0.01	0.01–0.1	0.1–1	>1

Source: Adapted from Ek, M. et al. 2007. *Biochem. Pharmacol.* 74: 496. With permission.

[a] All samples were run on LDA cards containing different genes associated with drug metabolism in the liver. Gene expressions were normalized against the expression of GAPDH in each sample. The expression levels in each sample are compared to the expression levels in human liver samples, which are set to 1.0 for all genes.

transmembrane potential ($\Delta\psi_m$), it is reasonable to expect that $\Delta\psi_m$ reduction will provide a sensitive signal (readout) detectable in current HTS systems. Review of the literature for a link between IADRs and $\Delta\psi_m$ reduction found supportive evidence in numerous studies. For example, Dykens et al. (2008) used Nefazodone (a triasolo-pyridine antidepressant that was withdrawn in 2004 due to idiosyncratic liver injury incidence of 28.9/1000 patients) and Buspirone (an antidepressant that is a 5-HT1A receptor partial agonist with no known hepatotoxicity) as a negative control and observed near $\Delta\psi_m$ abolishment in Nefazodone®-exposed sandwich-cultured human primary hepatocytes. Buspirone produced similar results as the 0.1% DMSO vehicle control—healthy mitochondria with normal $\Delta\psi_m$. The membrane potential-sensitive fluorescence probe, tetramethyl rhodamine methyl ester (TMRM), was used to detect $\Delta\psi_m$. In a second recent study, Lim et al. (2008) used Troglitazone (a first-generation thiazolidinedione insulin sensitizer that was withdrawn from the market due to unacceptable idiosyncratic liver injury) and 0.1% DMSO vehicle as the control. Troglitazone triggered mitochondrial permeabilization as detected by dissipated $\Delta\psi_m$, which was visualized by JC-1, a cationic probe that selectively accumulates within the mitochondrial matrix.

The feasibility to detect $\Delta\psi_m$ in HTS environments has been provided by Huang (2002), who described a high-throughput assay that measured $\Delta\psi_m$ in T47D (breast cancer cells) stained with TMRM. Carbonylcyanide *m*-chlorophenyl hydrazone (CCCP), a potent chemical uncoupler, was used at 10 µM as the positive control in column 24 and DMSO at 1% (v/v) was used as the negative control in columns 1–23 of a 2D 384-well plate. The experiment was replicated 14 times. Before TMRM fluorescence was read in an LJL analyst, plates were washed four times to remove 95% of the free TMRM. The Z' factor, an indicator of assay quality, was between 0.5 and 0.62, suggesting an excellent assay for HTS format. No report has been found on $\Delta\psi_m$ assay from 3D cultures. An innovative feature of this study case is the extension of $\Delta\psi_m$ signal to 3D HTS environments.

13.4 STUDY DESIGN AND METHODS

13.4.1 EXPERIMENTAL DESIGN AND RATIONALE

hESCs-derived hepatocyte-like cells are recommended. These cells can be cultured in all four commercially available 3D plates as well as the 2D control. In all cases, cells should be exposed to at least two positive controls, for example trovafloxacin (TVX), a fluoroquinolone antibiotic linked to idiosyncratic hepatotoxicity in humans, and CCCP, a potent chemical uncoupler. Cells should also be exposed to at least two negative controls, for example levofloxacin (LVX), a fluoroquinolone without liability for causing IADRs in humans, as well as 0.1% DMSO vehicle. Lipopolysaccharide (LPS) should be used to render the cells sensitive to TVX—LPS has been successfully used in rodents for the same purpose (Shaw et al., 2007). First, the investigator should examine the extent to which the 3D plates facilitate *in vivo* emulation by examining the expression of drug-metabolizing enzymes. Next, the investigator should examine the effect of the drugs by monitoring the increase in alanine aminotransferase (ATL) activity in the media as well as the mitochondrial membrane potential ($\Delta\psi_m$). The nine experimental runs outlined in Table 13.3 should be replicated at least three times.

TABLE 13.3
Experimental Design

Run Number	DMSO (Vehicle)	LVX (Negative Control)	LPS (Sensitizer)	CCCP (Positive Control)	TVX (Positive Control)
1 (three 3D and one 2D control plates)[a]	+	−	−	−	−
2	+	+	−	−	−
3	+	−	+	−	−
4	+	−	−	+	−
5	+	−	−	−	+
6	+	+	+	−	−
7	+	−	+	+	−
8	+	−	+	−	+
9	+	−	+	−	+

[a] The three plates include the following commercially available 3D 96-well plates: Algimetrix™ (Invitrogen); UltraWeb™ (Corning); and Extracel™ (Glycosan) (see Chapter 10).

13.4.2   CELL CULTURE AND DRUG EXPOSURE

hESC line obtained from WiCell Research Institute (Madison, WI) is recommended and should be maintained on irradiated mouse embryonic fibroblasts, in hESC medium: DMEM/F12 medium supplemented with 20% knockout serum replacement, 1 mM L-glutamine, 1% nonessential amino acids, 0.1 mM β-mercaptoethanol, and 4 ng/mL bFGF, under a standard humidified incubator with 5% CO_2. For hepatic differentiation, procedures described by Cai et al. (2007) should be followed. Briefly, hESCs should be cultured in 1640 medium (Hyclone, Logan, UT) supplemented with 0.5 mg/mL albumin fraction V (Sigma-Aldrich), and 100 ng/mL Activin A for 1 day. On the following 2 days, 0.1% and 1% insulin-transferrinselenium (ITS) (Sigma-Aldrich) should be added to this medium. After 3 days of Activin A treatment, the differentiated cells should be cultured in hepatocyte culture medium (HCM) (Cambrex, Baltimore, MD) containing 30 ng/mL FGF4 and 20 ng/mL BMP2 for 5 days. Then, the differentiated cells should be further matured in HCM containing 20 ng/mL HGF for 5 days and 10 ng/mL OSM plus 0.1 μM Dex from then on. The medium should be changed every day. Cells should be exposed to single doses of compounds at noncytotoxic and nonhepatotoxic concentrations typical of therapeutic plasma concentrations for 24 h.

13.4.3   EXPRESSION OF DRUG-METABOLIZING ENZYMES

The LDA (TaqMan, Applied Biosystems, Foster City, CA), real-time PCR, and Western blotting, following procedures described by Ek et al. (2007) and Cheng et al. (2008), are a cost-effective approach. Each set of the array contains housekeeping genes, of which glyceraldehydes-3-phosphate dehydrogenase (GAPDH) and hypoxanthine phosphoribosyltransferase 1 (HPRT1) should be used for normalization.

13.4.4   ALANINE AMINOTRANSFERASE (ATL) ACTIVITY ASSAY

The ATL assay was first scaled to 96-well plates by Bruin et al. (1995) for blood samples. ATL kits are now commercially available for blood samples (e.g., Abbott's 7D56-20 kit). Given the plasma similarity to media, the kit should be applicable in this situation without any modifications. The plates can be read on a standard 96-well absorption reader after transferring the reactants into a standard 2D 96-well plate.

13.4.5   MITOCHONDRIAL MEMBRANE POTENTIAL ($\Delta\psi_M$) MEASUREMENT

The cells should be loaded with a mitochondrial membrane potential dye (TMRM), and the overall well fluorescence should be monitored with a suitable instrument. The FlexStation™ (a bench-scale version of the industrial FLIPR Instrument, Molecular Devices Corporation, Sunnyvale, CA) works very well. To load the cells, procedures published by O'Brien et al. (2006) can be followed. The author has used TMRM in several studies in his laboratory (e.g., Wu et al., 2006a, 2006b; Mao and Kisaalita, 2004a, 2004b; Desai et al., 2006; Hernandez and Kisaalita, 1996). Procedures from these studies can be adopted.

13.5 ANALYSIS AND EXPECTED RESULTS

13.5.1 Quality Assessment of HTS Assays

In the design and validation of HTS assays, an assessment of the screening data, by measurements such as standard deviation (SD) or coefficient of variation (CV), is critical in determining whether an assay can identify hits with confidence. Zhang et al. (1999) introduced the Z' factor, a simple statistical dimensionless number that evaluates HTS assay quality with respect to identifying hits with a high degree of confidence.

$$Z' = 1 - \left[\frac{(3\sigma_{c+} + 3\sigma_{c-})}{(|\mu_{c+} - \mu_{c-}|)} \right], \tag{13.1}$$

where σ_{c+} and σ_{c-} denote signal SDs of positive and negative controls, respectively, and μ_{c+} and μ_{c-} denote signal means of the positive and negative controls, respectively.

Use of Z' factor is now an accepted industry standard. If the Z' factor is sufficiently large (>0) at the defined conditions, then the assay can be used in HTS. Many assay design papers have reported the quality of assays using the Z' factor. For example, Herrmann et al. (2008) have recently used the multicellular spheroid model (HCT116 colon carcinoma), considered to be of intermediate complexity between *in vivo* tumor and *in vitro* monolayer cultures, to screen for compounds that induce tumor cell apoptosis. Multicellular spheroids were generated in a 96-well plate, and apoptosis was determined using the M30-Apoptosense™ enzyme-linked immunosorbent assay method. A Z' factor of 0.5 was observed, attesting to the robustness of the assay and its suitability for HTS.

13.5.2 Expected Results

Results can be presented as mean ± standard deviation. Statistical comparisons with ANOVA can be conducted, where appropriate, after data normalization. All pairwise comparisons can be made using Dunn's method. The criterion for significance should be $p < 0.05$ for all studies. Z' factors should be calculated using the equation presented above. The experiments are designed to test three subhypotheses. The first subhypothesis is that cells cultured in 3D plates will better emulate primary human hepatocytes. This is based on the assumption that 3D cultures are physiologically more relevant. As such, the expression of cytochrome P450s, transporters, transcription factors, and other proteins such as albumin should be more aligned with human liver cell expression profiles. Consequently, all 3D plates are expected to better distinguish between drugs with and without the propensity to cause idiosyncratic liver injury when compared to 2D plate control (the second subhypothesis). The third subhypothesis is that $\Delta\psi_m$ signal from 3D plates will yield positive Z' factors, attesting to their robustness and accuracy of the 3D HTS assay platform.

13.5.3 POTENTIAL PITFALLS

The choice of drugs in this study case is critical. A list to choose from is provided in Table 13.1. A good number of these drugs may not have been withdrawn not because the resulting adverse drug reactions were not detectable in the preclinical and clinical studies, but because they were ignored. Well-known drugs in this category include Paxil, Vioxx, Zentia, Baycol, and Trasylol (Zarin and Tse, 2008). Without access to preclinical studies, it is difficult to evaluate whether potential problems might have been ignored. However, of all reasons for withdrawal, ISADs are the least likely to be detectable during preclinical studies, making them a more suitable withdrawal criterion for the selection of drugs to include in this study.

REFERENCES

Bruin des, R.N., Pouilles, F., Roche, M., and Claret, G. 1995. A microtiter plate assay for measurement of serum alanine aminotransferase in blood donors. *Transfection* 35(4): 331–334.

Cai, J., Zhao, Y., Liu, Y., Ye, F., Song, Z., Qin, H., Meng, S., Chen, Y., Zhou, R., Song, X., Gou, Y., Ding, M., and Deng, H. 2007. Direct differentiation of human embryonic stem cell into functional hepatic cells. *Hepatology* 45: 1229–1239.

Cheng, K., Lai, Y., and Kisaalita, W.S. 2008. 3D polymer scaffolds for high-throughput cell-based assay systems. *Biomaterials* 29: 2802–2812.

Desai, A., Kisaalita, W.S., Keith, C., and Wu, Z.-Z. 2006. Human neuroblastoma (SH-SY5Y) cell culture and differentiation in 3D collagen hydrogels for cell-based biosensing. *Biosens. Bioelectron.* 21: 1483–1492.

Dykens, J.A., Jamieson, J.D., Marroquin, L.D., Nadanaciva, S., Xu, J.J., Dunn, M.C., Smith, A.R., and Will, Y. 2008. *In vitro* assessment of mitochondrial dysfunction and cytotoxicity of nefazodone, trazodone, and buspirone. *Toxicol. Sci.* 103(2): 335–345.

Ek, M., Soderdahl, T., Kuppers-Munther, B., Edsbagge, J., Andersson, T.B., Bjorquist, P., Cotgreave, I., Jernstrom, B., Ingelman-Sundberg, M., and Johansson, I. 2007. Expression of drug metabolizing enzymes in hepatocyte-like cells. *Chem. Pharmacol.* 74: 496–503.

Friedman, S.E., Grendel, J.H., and McQuaid, K.R. 2003. *Current Diagnostics and Treatment in Gastroenterology*, pp. 664–679. New York: Lang Medical Books/McGraw-Hill.

Fung, M., Thornton, A., Mybeck, K., Wu, H.-H., Hornbuckle, K., and Muniz, E. 2001. Evaluation of the characteristics of safety withdrawals of prescription drugs from worldwide pharmaceutical markets—1960 to 1999. *Drug Inform. J.* 35: 293–317.

Green, J.A. and Yamada, K.M. 2007. Three-dimensional microenvironments modulate fibroblast signaling responses. *Adv. Drug Deliv. Rev.* 59(13): 1293–1298.

Guyton, A.C. and Hall, J.E. 2000. *Textbook of Medical Physiology*, 10th edition. Philadelphia: W.B. Saunders Company.

Hernandez, M. and Kisaalita, W.S. 1996. Comparative evaluation of the susceptibility of neuronal (N1E-115) and non-neuronal (HeLa) cells to Acetylsalicylic Acid (ASA) cytotoxicity by confocal microscopy. *Toxicol. In Vitro* 10: 447–453.

Herrmann, R., Fayad, W., Schwarz, S., Berndtsson, M., and Linder, S. 2008. Screening for compounds that induce apoptosis of cancer cells grown as multicellular spheroids. *J. Biomol. Screen.* 13(1): 1–8.

Hoffman, L.M. and Carpenter, M.K. 2005. Characterization and culture of human embryonic stem cells. *Nat. Biotechnol.* 23: 699–708.

Huang, S.-G. 2002. Development of a high throughput screening assay for mitochondrial membrane potential in living cells. *J. Biomol. Screen.* 7(4): 383–389.

Lee, W.M. 2003. Drug-induced hepatotoxicity. *N. Engl. J. Med.* 349: 474–485.

Li, A.P. 2002. A review of the common properties of drugs with idiosyncratic hepatotoxicity and the "multiple determinant hypothesis" for the manifestation of idiosyncratic drug toxicity. *Chem. Biol. Interact.* 142: 7–23.

Li, A.P., Rasmussen, A., and Kaminski, D.L. 1995. Substrates of human hepatic cytochrome P4503A4. *Toxicology* 104: 1–8.

Lim, P.L.K., Liu, J., Go, M.L., and Boelsterli, U.A. 2008. The mitochondrial superoxide/thioredoxin-2/Ask1 signaling pathway is critically involved in troglitazone-induced cell injury to human hepatocytes. *Toxicol. Sci.* 101(2): 341–349.

Lynch, T. and Price, A. 2007. The effect of cytochrome P450 metabolism on drug response, interactions, and adverse effects. *Am. Fam. Physician* 76(3): 391–396.

Mao, C. and Kisaalita, W.S. 2004a. Characterization of 3-D collagen gels for functional cell-based biosensing. *Biosens. Bioelectron.* 19: 1075–1088.

Mao, C. and Kisaalita, W.S. 2004b. Determination of resting membrane potential of individual neuroblastoma cells (IMR-32) using a potentiometric dye (TMRM) and confocal microscopy. *J. Fluoresc.* 14(6): 739–743.

Miyashita, T., Enosawa, S., Suzuki, S., Tamura, H., Amemiya, H., Matsumura, T., Omasa, T., Suga, K., Aoki, T., and Koyanagi, Y. 2000. Development of a bioartificial liver with glutamine synthetase-transduced recombinant human hepatoblastoma cell line, HepG2. *Transplant Proc.* 32(7): 2355–2358.

Morsiani, E., Brogli, M., Galavotti, D., Bellini, T., Ricci, D., Pazzi, P., and Puviani, A.C. 2001. Long-term expression of highly differentiated functions by isolated porcine hepatocytes perfused in a radial-flow bioreactor. *Artif. Organs* 25(9): 740–748.

O'Brien, P.J., Irwin, W., Diaz, D., Howard-Cofield, E., Krejsa, C.M., Slaughter, M.R., Gao, B. et al. 2006. High concordance of drug-induced human hepatotoxicity with *in vitro* cytotoxicity measured in novel cell-based model using high content screening. *Arch. Toxicol.* 80: 580–604.

Ostapowicz, G., Fontana, R.J., and Schiødt, F.V. 2002. Results of a prospective study of acute liver failure at 17 tertiary care centers in the United States, *Ann. Intern. Med.* 137(12): 947–954.

Pirmohamed, M., Breckenridge, A.M., Kitteringham, N.R., and Park, B.K. 1998. Adverse drug reactions. *BMJ* 316(7140): 1295–1298.

Raucy, J.L., Mueller, L., Duan, K., Allen, S.W., Strom, S., and Lasker, J.M. 2002. Expression and induction of CYP2C P450 enzymes in primary cultures of human hepatocytes. *J. Pharmacol. Exp. Ther.* 302: 475–482.

Schuetz, E.G., Schuetz, J.D., Strom, S.C., Thompson, M.T., Fisher, R.A., Molowa, D.T., Li, D., and Guzelian, P.S. 1993. Regulation of human liver cytochromes P-450 in family 3A in primary and continuous culture of human hepatocytes. *Hepatology* 18: 1254–1262.

Shaw, P.J., Hopfensperger, M.J., Ganey, P.E., and Roth, R.A. 2007. Lipopolysaccharide and trovafloxacin co exposure in mice causes idiosyncrasy-like liver injury dependent on tumor necrosis factor-alpha. *Toxicol. Sci.* 100(1): 259–266.

Spriet-Pourra, C. and Auriche, M. 1994. *SCRIP Report on Drug Withdrawal from Sale*, 2nd edition. Richmond, UK: PJB Publications Ltd.

Wilkening, W., Stahl, F., and Bader, A. 2003. Comparison of primary human hepatocytes and hepatoma cell line with regard to their biotransformation properties. *Drug Metab. Dispos.* 31(8): 1035–1042.

Wu, Z.-Z., Zhao, Y.-P., and Kisaalita, W.S. 2006a. A packed Cytodex microbead array for three-dimensional cell-based biosensing. *Biosens. Bioelectron.* 22: 685–693.

Wu, Z.-Z., Zhao, Y.-P., and Kisaalita, W.S. 2006b. Interfacing SH-SY5Y neuroblastoma cells with SU-8 microstructures. *Colloids Surf. B* 52: 14–21.

Zarin, D.A. and Tse, T. 2008. Moving toward transparency of clinical trials. *Science* 319(5868): 1340–1342.

Zhang, J.-H., Chung, T.D.Y., and Oldenburg, K.R. 1999. A simple statistical parameter for use in evaluation and validation of high throughput screening assays. *J. Biomol. Screen.* 4(2): 67–73.

Appendix A: Patents for 3D Scaffolds

3D Scaffolds-Related Patents Issued between 1996 and 2006

US Patent No.	Title	Patent Assignee	Reference
7160726	Compositions comprising conditioned cell culture media and uses thereof	Skin Medica, Inc. (Carlsbad, CA)	Mansbridge (2007)
7160719	Bioartificial liver system	Mayo Foundation (Rochester, MN)	Nyberg (2007)
7052720	Spheroid preparation	University of Wales College of Medicine (Cardiff, GB)	Jones (2006)
7122371	Modular cell culture bioreactor	The Florida State University Research Foundation, Inc. (Tallahassee, FL)	Ma (2006)
7022523	Carrier for cell culture	Fuji Photo Film Co., Ltd (Kanagawa, JP)	Tsuzuki (2006)
6861087	Preparation method of biodegradable porous polymer scaffolds having an improved cell compatibility for tissue engineering	Korea Institute of Science and Technology (Seoul, KR), Solco Biomedical Co., Ltd (Gyeonngi-do, KR)	Han (2005)
6911201	Method of producing undifferentiated hemopoietic stem cells using a stationary phase plug-flow bioreactor	Technion Research and Development Foundation Ltd. (Haifa, IL)	Merchav (2005)
6943008	Bioreactor for cell culture	Florida State University Research Foundation, Inc. (Tallahassee, FL)	Ma (2005a)
6875605	Modular cell culture bioreactor and associated methods	Florida State University Research Foundation, Inc. (Tallahassee, FL)	Ma (2005b)
6803037	Hyaluronic acid derivative-based cell culture and biodegradable 3D matrix	Fidia Advanced Biopolymers S.r.l. (Brindisi, IT)	Abatangelo (2004)

continued

3D Scaffolds-Related Patents Issued between 1996 and 2006   (continued)

US Patent No.	Title	Patent Assignee	Reference
6737270	Long-term 3D tissue culture system	University of Pittsburgh of the Commonwealth System of Higher Education (Pittsburgh, PA)	Michalopoulos (2004)
6777227	Bioreactor and cell culture surface with microgeometric surfaces	N/A	Ricci (2004)
6730315	Medium and matrix for long-term proliferation of cells	Encelle, Inc. (Greenville, NC)	Usala (2004)
6642050	3D cell culture material having sugar polymer containing cell recognition sugar chain	Amcite Research, Ltd (Kanagawa-ken, JP)	Goto (2003)
6479066	Device having a microporous membrane-lined deformable wall for implanting cell cultures	RST Implanted Cell Technology, LLC (Arden Hills, MN)	Harpstead (2002)
6465205	*In vitro* cell culture device including cartilage and methods of using the same	The Research Foundation of State University of New York (Amherst, NY)	Hicks (2002)
6312952	*In vitro* cell culture device including cartilage and methods of using the same	The Research Foundation of State University of New York (Amherst, NY)	Hicks (2001)
6315994	Medium and matrix for long-term proliferation of cells	N/A	Usala (2001a)
6231881	Medium and matrix for long-term proliferation of cells	N/A	Usala (2001b)
6103528	Reversible gelling culture media for *in vitro* cell culture in 3D matrices	Battelle Memorial Institute (Richland, WA)	An (2000)
6039972	Wound dressing containing mammalian cells anchored on hydrophobic synthetic polymer film	Smith & Nephew PLC (GB)	Barlow (2000)
6120875	Transparent micro perforated material and preparation process	Cyclopore SA (BE)	Haumon (2000)
6037171	Cell culture microchambers in a grid matrix sandwiched between a planar base and semipermeable membrane	Microcloning CCCD AB (Jonkoping, SE)	Larsson (2000)
6140039	3D filamentous tissue having tendon or ligament function	Advanced Tissue Sciences, Inc. (La Jolla, CA)	Naughton (2000a)

3D Scaffolds-Related Patents Issued between 1996 and 2006 (continued)

US Patent No.	Title	Patent Assignee	Reference
6022743	3D culture of pancreatic parenchymal cells cultured on living stromal tissue prepared *in vitro*	Advanced Tissue Sciences, Inc. (La Jolla, CA)	Naughton (2000b)
6008047	Cell culturing method and medium	Livercell LLC (East Sebago, ME)	Curcio (1999)
5863984	Biostable porous material comprising composite biopolymers	Universite Laval, Cite Universitaire (Quebec, CA)	Doillon (1999)
5964745	Implantable system for cell growth control	Med USA (San Antonio, TX)	Lyles (1999)
5863531	*In vitro* preparation of tubular tissue structures by stromal cell culture on a 3D framework	Advanced Tissue Sciences, Inc. (La Jolla, CA)	Naughton (1999a)
5858721	3D cell and tissue culture system	Advanced Tissue Sciences, Inc. (La Jolla, CA)	Naughton (1999b)
5780299	Method of altering blood sugar levels using nontransformed human pancreatic cells that have been expanded in culture	Human Cell Cultures, Inc. (East Sebago, ME)	Coon (1998)
5786216	Inner-supported, biocompatible cell capsules	Cytotherapeutics, Inc.	Dionne (1998)
5785964	3D genetically engineered cell and tissue culture system	Advanced Tissue Sciences, Inc. (La Jolla, CA)	Naughton (1998)
5843741	Method for altering the differentiation of anchorage-dependent cells on an electrically conducting polymer	Massachusetts Institute of Technology (Cambridge, MA)	Wong (1998)
5780281	Method of preparing a low-density porous fused-fiber matrix	Lockheed Martin Corporation (Bethesda, MD)	Yasukawa (1998)
5646035	Method for preparing an expanded culture and clonal strains of pancreatic, thyroid, or parathyroid cells	Human Cell Cultures, Inc. (Gaithersburg, MD)	Coon (1997)
5702945	Culture vessel for cell cultures on a carrier	Heraeus Instruments GmbH (Hanau, DE)	Nagels (1997)
5624839	Process culturing hepatocytes for forming spheroids	Seikagaku Corporation, Kogyo, Kabushiki, Kaisha	Yada (1997)

continued

3D Scaffolds-Related Patents Issued between 1996 and 2006   (continued)

US Patent No.	Title	Patent Assignee	Reference
5629186	Porous matrix and method of its production	Lockheed Martin Corporation (Bethesda, MD)	Yasukawa (1997)
5580781	3D tumor cell and tissue culture system	Advanced Tissue Sciences, Inc. (La Jolla, CA)	Naughton (1996a)
5578485	3D blood–brain barrier cell and tissue culture system	Advanced Tissue Sciences, Inc. (La Jolla, CA)	Naughton (1996b)
5541107	3D bone marrow cell and tissue culture system	Advanced Tissue Sciences, Inc. (La Jolla, CA)	Naughton (1996c)
5518915	3D mucosal cell and tissue culture system	Advanced Tissue Sciences, Inc. (La Jolla, CA)	Naughton (1996d)
5516681	3D pancreatic cell and tissue culture system	Advanced Tissue Sciences, Inc. (La Jolla, CA)	Naughton (1996e)
5516680	3D kidney cell and tissue culture system	Advanced Tissue Sciences, Inc. (La Jolla, CA)	Naughton (1996f)
5512475	3D skin cell and tissue culture system	Advanced Tissue Sciences, Inc. (La Jolla, CA)	Naughton (1996g)
5443950	3D cell and tissue culture system	Advanced Tissue Sciences, Inc. (La Jolla, CA)	Naughton (1995)
5478739	3D stromal cell and tissue culture system	Advanced Tissue Sciences, Inc. (La Jolla, CA)	Slivka (1995)
5308704	Cell adhesive material and method for producing the same	Sony Corporation (Tokyo, JP)	Suzuki (1994)
5244799	Preparation of a polymeric hydrogel containing micropores and macropores for use as a cell culture substrate	N/A	Anderson (1993)
5260211	Process for cultivating adhesive cells in a packed bed of solid cell matrix	Snow Brand Milk Products Co., Ltd. (Hokkaido, JP)	Matsuda (1993)
5266480	3D skin culture system	Advanced Tissue Sciences, Inc. (La Jolla, CA)	Naughton (1993)

3D Scaffolds-Related Patents Issued between 1996 and 2006 (continued)

US Patent No.	Title	Patent Assignee	Reference
5173421	Cell culture carriers	Mitsubishi Kasei Corporation (Tokyo, JP)	Kiniwa (1992)
5160490	3D cell and tissue culture apparatus	Marrow-Tech Incorporated (La Jolla, CA)	Naughton (1992)
5153133	Method for culturing mammalian cells in a horizontally rotated bioreactor	The United States of America as represented by the Administrator (Washington, DC)	Schwarz (1992)
5155035	Method for culturing mammalian cells in a perfused bioreactor	The United States of America as represented by the Administrator (Washington, DC)	Schwarz (1992)
5155034	3D cell to tissue assembly process	The United States of America as represented by the Administrator (Washington, DC)	Wolf (1992)
5006467	Cell culture microcarriers	Mitsubishi Kasei Corporation (Tokyo, JP)	Kusano (1991)
5032508	3D cell and tissue culture system	Marrow-Tech, Inc. (La Jolla, CA)	Naughton (1991)
4988623	Rotating bioreactor cell culture apparatus	The United States of America as represented by the Administrator (Washington, DC)	Schwarz (1991)
4963489	3D cell and tissue culture system	Marrow-Tech, Inc. (La Jolla, CA)	Naughton (1990)
4789634	Carrier for the cultivation of human and/or animal cells in a fermenter	Dr. Muller-Lierheim KG Biologische Laboratorien (Planegg, DE)	Muller-Lierheim (1988)
4546083	Method and device for cell culture growth	Stolle Research and Development Corporation (Cincinnati, OH)	Meyers (1985)
4201845	Cell culture reactor	Monsanto Company (St. Louis, MO)	Feder (1980)
4206015	Method of simulation of lymphatic drainage utilizing a dual circuit, woven artificial capillary bundle	United States of America (Washington, DC)	Knazek (1980a)

continued

3D Scaffolds-Related Patents Issued between 1996 and 2006   (continued)

US Patent No.	Title	Patent Assignee	Reference
4184922	Dual circuit, woven artificial capillary bundle for cell culture	The Government of the United States (Washington, DC)	Knazek (1980b)
4087327	Mammalian cell culture process	Monsanto Company (St. Louis, MO)	Feder (1978)

Source: Adapted from Lai, Y., et al. 2008. *Recent Pat. Biomed. Eng.*, 1(2): 104. With permission.

REFERENCES

Abatangelo, G. 2004. Hyaluronic acid derivative based cell culture and biodegradable three-dimensional matrix. US6803037.

An, Y.H. 2000. Reversible gelling culture media for *in vitro* cell culture in three-dimensional matrices. US6103528.

Anderson, D.M. 1993. Preparation of a polymeric hydrogel containing micropores and macropores for use as a cell culture substrate. US5244799.

Barlow, Y.M. 2000. Wound dressing containing mammalian cells anchored on hydrophobic synthetic polymer film. US6039972.

Coon, H.G. 1997. Method for preparing an expanded culture and clonal strains of pancreatic, thyroid, or parathyroid cells. US5646035.

Coon, H.G. 1998. Method of altering blood sugar levels using nontransformed human pancreatic cells that have been expanded in culture. US5780299.

Curcio, F. 1999. Cell culturing method and medium. US6008047.

Dionne, K.E. 1998. Inner-supported, biocompatible cell capsules. US5786216.

Doillon, C.J. 1999. Biostable porous material comprising composite biopolymers. US5863984.

Feder, J. 1978. Mammalian cell culture process. US4087327.

Feder, J. 1980. Cell culture reactor. US4201845.

Goto, M. 2003. Three-dimensional cell culture material having sugar polymer containing cell recognition sugar chain. US6642050.

Han, D.K. 2005. Preparation method of biodegradable porous polymer scaffolds having an improved cell compatibility for tissue engineering. US6861087.

Harpstead, S.D. 2002. Device having a microporous membrane lined deformable wall for implanting cell cultures. US6479066.

Haumon, C. 2000. Transparent microperforated material and preparation process. US6120875.

Hicks, J. and Wesley L. 2001. *In vitro* cell culture device including cartilage and methods of using the same. US6312952.

Hicks, J.W.L. 2002. *In vitro* cell culture device including cartilage and methods of using the same. US6465205.

Jones, J.R.D.L. 2006. Spheroid preparation. US7052720.

Kiniwa, H. 1992. Cell culture carriers. US5173421.

Knazek, R.A. 1980a. Method of simulation of lymphatic drainage utilizing a dual circuit, woven artificial capillary bundle. US4206015.

Knazek, R.A. 1980b. Dual circuit, woven artificial capillary bundle for cell culture. US4184922.

Kusano, H. 1991. Cell culture microcarriers. US5006467.

Larsson, B. 2000. Cell culture microchambers in a grid matrix sandwiched between a planar base and semipermeable membrane. US6037171.

Lyles, M.B. 1999. Implantable system for cell growth control. US5964745.

Ma, T. 2005a. Bioreactor for cell culture. US6943008.

Ma, T. 2005b. Modular cell culture bioreactor and associated methods. US6875605.

Ma, T. 2006. Modular cell culture bioreactor. US7122371.

Matsuda, Y. 1993. Process for cultivating adhesive cells in a packed bed of solid cell matrix. US5260211.

Merchav, S. 2005. Method of producing undifferentiated hemopoietic stem cells using a stationary phase plug-flow bioreactor. US6911201.

Meyers, W.E. 1985. Method and device for cell culture growth. US4546083.

Michalopoulos, G. 2004. Long-term three-dimensional tissue culture system. US6737270.

Muller-Lierheim, W.G.K. 1988. Carrier for the cultivation of human and/or animal cells in a fermenter. US4789634.

Nagels, H.-O. 1997. Culture vessel for cell cultures on a carrier. US5702945.

Naughton, G.K. 1990. Three-dimensional cell and tissue culture system. US4963489.

Naughton, G.K. 1991. Three-dimensional cell and tissue culture system. US5032508.

Naughton, G.K. 1992. Three-dimensional cell and tissue culture apparatus. US5160490.

Naughton, G.K. 1993. Three-dimensional skin culture system. US5266480.

Naughton, G.K. 1995. Three-dimensional cell and tissue culture system. US5443950.

Naughton, G.K. 1996a. Three-dimensional blood–brain barrier cell and tissue culture system. US5578485.

Naughton, G.K. 1996b. Three-dimensional bone marrow cell and tissue culture system. US5541107.

Naughton, G.K. 1996c. Three-dimensional kidney cell and tissue culture system. US5516680.

Naughton, G.K. 1996d. Three-dimensional mucosal cell and tissue culture system. US5518915.

Naughton, G.K. 1996e. Three-dimensional pancreatic cell and tissue culture system. US5516681.

Naughton, G.K. 1996f. Three-dimensional skin cell and tissue culture system. US5512475.

Naughton, G.K. 1996g. Three-dimensional tumor cell and tissue culture system. US5580781.

Naughton, G.K. 1998. Three-dimensional genetically engineered cell and tissue culture system. US5785964.

Naughton, G.K. 1999a. *In vitro* preparation of tubular tissue structures by stromal cell culture on a three-dimensional framework. US5863531.

Naughton, G.K. 1999b. Three-dimensional cell and tissue culture system. US5858721.

Naughton, G.K. 2000a. Three-dimensional filamentous tissue having tendon or ligament function. US6140039.

Naughton, G.K. 2000b. Three-dimensional culture of pancreatic parenchymal cells cultured living stromal tissue prepared *in vitro*. US6022743.

Nyberg 2007. Bioartificial liver system. US7160719.

Ricci, J.L. 2004. Bioreactor and cell culture surface with microgeometric surfaces. US6777227.

Schwarz, P.R. 1991. Rotating bioreactor cell culture apparatus. US4988623.

Schwarz, P.R. 1992. Method for culturing mammalian cells in a horizontally rotated bioreactor. US5153133.

Schwarz, P.R. 1992. Method for culturing mammalian cells in a perfused bioreactor. US5155035.

Slivka, S.R. 1995. Three-dimensional stromal cell and tissue culture system US5478739.

Suzuki, Y. 1994. Cell adhesive material and method for producing same. 5308704.

Tsuzuki, H. 2006. Carrier for cell culture. US7022523.

Usala, A.-L. 2001a. Medium and matrix for long-term proliferation of cells. US6315994.

Usala, A.-L. 2001b. Medium and matrix for long-term proliferation of cells. US6231881.

Usala, A.-L. 2004. Medium and matrix for long-term proliferation of cells. US6730315.

Wolf, D.A.1992. Three-dimensional cell to tissue assembly process. US5155034.

Wong, J.Y. 1998. Method for altering the differentiation of anchorage dependent cells on an electrically conducting polymer. US5843741.

Yada 1997. Process for culturing hepatocytes for formation of spheroids. US5624839.

Yasukawa, R.D. 1997. Porous matrix and method of its production. US5629186.

Yasukawa, R.D. 1998. Method of preparing a low-density porous fused-fiber matrix. US5780281.

Appendix B: Current Drug Targets

The following nine tables provide the nature of drug targets, a list of approved drugs substances, and their approved therapeutical use in the following categories: enzymes (Table B1), substrates, metabolites and proteins (Table B2), receptors (Table B3), ion channels (Table B4), transport proteins (Table B5), DNA/RNA and the ribosome (Table B6), targets of monoclonal antibodies (Table B7), various physical chemical mechanisms (Table B8), and drugs with unknown mechanisms (Table B9).

TABLE B1

Enzymes

Type	Drug Activity	Drug Example	Disease
		Oxidoreductases	
Aldehyde dehydrogenase	Inhibitor	Disulfiram	Chronic alcoholism
Monoamine oxidases (MAOs)	MAO-A inhibitor	Tranylcypromine, moclobemide	Depression
	MAO-B inhibitor	Tranylcypromine	Depression
Cyclo-oxygenases (COXs)	COX1 inhibitor	Acetylsalicylic acid, profens, acetaminophen, and dipyrone (as arachidonylamides)	Pain, inflammation, blood clotting, fever, heart attacks, strokes
	COX2 inhibitor	Acetylsalicylic acid, profens, acetaminophen and dipyrone (as arachidonylamides)	Pain, inflammation, blood clotting, fever, heart attacks, strokes
Vitamin K epoxide reductase	Inhibitor	Warfarin, phenprocoumon	Blood clotting
Aromatase	Inhibitor	Exemestane	Hormonally responsive breast cancer in postmenopausal women
Lanosterol demethylase (fungal)	Inhibitor	Azole antifungals	Serious fungus infections
Lipoxygenases	Inhibitor	Mesalazine	Chrohn's disease, ulcerative colitis
	5-Lipoxygenase inhibitor	Zileuton	Asthma (difficulty breathing, chest tightness, wheezing, and coughing) not an attack
Thyroidal peroxidase	Inhibitor	Thiouracil	Hyperthyroidism
Iodothyronine-5' deiodinase	Inhibitor	Propylthiouracil	Hyperthyroidism, Graves' disease
Inosine monophosphate dehydrogenase	Inhibitor	Mycophenolate mofetil	Immunosuppressant in organ transplantation
HMG-CoA reductase	Inhibitor	Statins	Cardiovascular disease
5α-Testosterone reductase	Inhibitor	Finasteride, dutasteride	Benign prostatic hyperplasia (BPH) in low doses, prostate cancer in higher doses
Dihydrofolate reductase (bacterial)	Inhibitor	Trimethoprim	Urinary tract infections

Dihydrofolate reductase (human)	Inhibitor	Methotrexate, pemetrexed	Cancer, autoimmune diseases (methotrexate), pleural mesothelioma and nonsmall lung cancer (pemetrexed)
Dihydrofolate reductase (parasitic)	Inhibitor	Proguanil	Malaria
Dihydroorotate reductase	Inhibitor	Leflunomide	Rheumatoid arthritis and psoriatic arthritis
Enoyl reductase (mycobacterial)	Inhibitor	Isoniazid	Tuberculosis
Squalene epoxidase (fungal)	Inhibitor	Terbinafin	Jock itch, athlete's foot, fungal infections
$\Delta 14$ reductase (fungal)	Inhibitor	Amorolfine	Fungus infections
4-Hydroxyphenylpyruvate dioxygenase	Inhibitor	Allopurinol	Chronic gout, excess uric acid
Ribonucleoside diphosphate reductase	Inhibitor	Nitisinone	Tyrosinemia
Transferases			
Protein kinase C	Inhibitor	Miltefosine	Protozoan infection, tumors
Bacterial peptidyl transferase	Inhibitor	Chloramphenicol	Bacteria, microbes (fungi, viruses)
Catecholamine-O-methyltransferase	Inhibitor	Entacapone	Parkinson's disease
RNA polymerase (bacterial)	Inhibitor	Ansamycins	Gram-positive and gram-negative bacteria
Reverse transcriptases (viral)	Competitive inhibitors	Zidovudine	HIV
	Allosteric inhibitors	Efavirenz	HIV type 1
DNA polymerases	Inhibitor	Acyclovir, suramin	Herpes simplex virus, viral infections (acyclovir), sleeping sickness, onchocerciasis, other diseases caused by trypanosomes and worms (suramin)
GABA transaminase	Inhibitor	Valproic acid, vigabatrin	Epileptic seizures, epilepsy, bipolar disorder, clinical depression (valproic acid), epilepsy, substance dependence, panic disorder, seizures (vigabatrin)

continued

TABLE B1   (continued)

Enzymes

Type	Drug Activity	Drug Example	Disease
Tyrosine kinases	PDGFR/ABL/KIT inhibitor	Imatinib	Types of leukemia, other cancers of the blood cells
	EGFR inhibitor	Erlotinib	Nonsmall cell lung cancer, pancreatic cancer, several other types of cancer
	VEGFR2/PDGFRβ/KIT/FLT3	Sunitinib	Renal cell carcinoma, gastrointestinal stromal tumor
	VEGFR2/PDGFRβ/RAF	Sorafenib	Advanced renal cell carcinoma
Glycinamide ribonucleotide formyl transferase	Inhibitor	Pemetrexed	Pleural mesothelioma and nonsmall lung cancer (pemetrexed)
Phosphoenolpyruvate transferase (murA, bacterial)	Inhibitor	Fosfomycin	Urinary tract infections
Human cytosolic branched-chain aminotransferase (hBCATc)	Inhibitor	Gabapentin	Epilepsy, neuropathic pain
Hydrolases (proteases)			
Aspartyl proteases (viral)	HIV protease inhibitor	Saquinavir, indinavir	HIV
Hydrolases (serine proteases)			
Unspecific	Unspecific inhibitors	Aprotinin	Bleeding during surgery
Bacterial serine protease	Direct inhibitor	β-lactams	Bacteria
Bacterial serine protease	Indirect inhibitor	Glycopeptides	Bacteria
Bacterial lactamases	Direct inhibitor	Sulbactam	Bacteria
Human antithrombin	Activator	Heparins	Coagulation
Human plasminogen	Activator	Streptokinase	Coagulation
Human coagulation factor	Activator	Factor IX complex, Factor VIII	Hemophilia B

Target	Action	Drug	Indication
Human factor Xa	Inhibitor	Fondaparinux	Blood clots
Hydrolases (metalloproteases)			
Human ACE	Inhibitor	Captopril	Hypertension, some types of congestive heart failure
Human HRD	Inhibitor	Cilastatin	Bacteria
Human carboxypeptidase A (Zn)	Inhibitor	Penicillamine	Wilson's disease, rheumatoid arthritis
Human enkephalinase	Inhibitor	Racecadotril	Diarrhea
Hydrolases (other)			
26S Proteasome	Inhibitor	Bortezomib	Myeloma, mantle cell lymphoma
Esterases	AchE inhibitor	Physostigmine	Types of glaucoma
	AchE reactivators	Obidoxime	Nerve gas poisoning
	PDE inhibitor	Caffeine	Drowsiness
	PDE3 inhibitor	Amrinone, milrinone	Heart failure
	PDE4 inhibitor	Papaverine	Visceral spasm, vasospasm, erectile dysfunction
	PDE5 inhibitor	Sildenafil	Erectile dysfunction, pulmonary arterial hypertension
	HDAC inhibitor	Valproic acid	Epileptic seizures, epilepsy, bipolar disorder, clinical depression
	HDAC3/HDAC7 inhibitor	Carbamazepine	Epilepsy, bipolar disorder
Glycosidases (viral)	α-Glycosidase inhibitor	Zanamivir, oseltamivir	Prophylaxis of Influenzavirus A and Influenzavirus B
Glycosidases (human)	α-Glycosidase inhibitor	Acarbose	Diabetes mellitus, prediabetes
Lipases	Gastrointestinal lipases inhibitor	Orlistat	Obesity
Phosphatases	Calcineurin inhibitor	Cyclosporin	Organ rejection
	Inositol polyphosphate phosphatase inhibitor	Lithium ions	Bipolar disorder, depression, mania

continued

TABLE B1 (continued)
Enzymes

Type	Drug Activity	Drug Example	Disease
GTPases	Rac1 inhibitor	6-Thio-GTP (azathioprine metabolite)	Rheumatoid arthritis, pemphigus, inflammatory bowel disease (Crohn's disease, and ulcerative colitis)
Phosphorylases	Bacterial C55-lipid phosphate dephosphorylase inhibitor	Bacitracin	Bacterial infections, pneumonias
Lyases			
DOPA decarboxylase	Inhibitor	Carbidopa	Parkinson's disease
Carbonic anhydrase	Inhibitor	Acetazolamide	Glaucoma, epileptic seizures, benign intracranial hypertension, altitude sickness, cystinuria, dural ectasia
Histidine decarboxylase	Inhibitor	Tritoqualine	Hypersensitivity reactions and pruritus
Ornithine decarboxylase	Inhibitor	Eflornithine	Sleeping sickness
Soluble guanylyl cyclase	Activator	Nitric acid esters, molsidomine	Angina pectoris
Isomerases			
Alanine racemase	Inhibitor	D-Cycloserine	Chronic neuropathic pain, tuberculosis
DNA gyrases (bacterial)	Inhibitor	Fluoroquinolones	Bacteria
Topoisomerases	Topoisomerase I inhibitor	Irinotecan	Colon cancer
	Topoisomerase II inhibitor	Etoposide	Ewing's sarcoma, lung cancer, testicular cancer, lymphoma, nonlymphocytic leukemia, glioblastoma multiforme

Δ8,7 isomerase	Inhibitor	Amorolfin	Fungal infections
Ligases (also known as synthases)			
Dihydropteroate synthase	Inhibitor	Sulfonamides	Bacterial infections and some fungal infections
Thymidylate synthase (fungal and human)	Inhibitor	Fluorouracil	Cancer
Thymidylate synthase (human)	Inhibitor	Methotrexate, pemetrexed	Cancer and autoimmune diseases (methotrexate), pleural mesothelioma and nonsmall lung cancer (pemetrexed)
Phosphofructokinase	Inhibitor	Antimony compounds	Tropical parasites
mTOR	Inhibitor	Rapamycin	Rejection in organ transplantation (especially kidney)
Heme polymerase (*Plasmodium*)	Inhibitor	Quinoline antimalarials	Malaria
1,3-β-D-glucansynthase (fungi)	Inhibitor	Caspofungin	Aspergillosis, candidemia, Candida infections, candidiasis
Glucosylceramise synthase	Inhibitor	Miglustat	Gaucher's disease

Source: Modified from Imming, P., Sinning, C., and Meyer, A. 2006. *Drug Discov.* 5: 821–834.

TABLE B2
Substrates, Metabolites, and Proteins

Substrate	Drug Substance	Disease
Asparagine	Asparaginase	Acute lymphoblastic leukemia
Urate	Rasburicase (a urate oxidase)	Tumor lysis syndrome
VAMP-synaptobrevin, SNAP25, syntaxin	Light chain of the botulinum neurotoxin (Zn-endopeptidase)	Painful muscle spasms

Source: Modified from Imming, P., Sinning, C., and Meyer, A. 2006. *Drug Discov.* 5: 821–834.

TABLE B3
Receptors

Type	Activity of Drug	Drug Examples	Disease
		Direct ligand-gated ion channel receptors	
GABA$_A$ receptors	Barbiturate binding site agonists	Barbiturate	Seizures
	Benzodiazepine binding site agonists	Benzodiazepines	Anxiety, insomnia, agitation, seizures, muscle spasms, alcohol withdrawal
	Benzodiazepine binding site antagonists	Flumazenil	Benzodiazepine overdose
Acetylcholine receptors	Nicotinic receptor agonists	Pyrantel (of Angiostrongylus), levamisole	Roundworm, hookworm, pinworm, other worm infections
	Nicotinic receptor stabilizing antagonists	Alcuronium	Muscle spasm, pain, hyperreflexia
	Nicotinic receptor depolarizing antagonists	Suxamethonium	Muscle spasm, pain, hyperreflexia
	Nicotinic receptor allosteric modulators	Galantamine	Alzheimer's disease
Glutamate receptors (ionotropic)	NMDA subtype antagonists	Memantine	Alzheimer's disease
	NMDA subtype expression modulators	Acamprosate	Alcohol dependence
	NMDA subtype phencyclidine binding site antagonists	Ketamine	Bronchospasm, pain, low blood pressure
		G-protein-coupled receptors	
Acetylcholine receptors	Muscarinic receptor agonists	Pilocarpine	Glaucoma, dry mouth, diagnosing cystic fibrosis
	Muscarinic receptor antagonists	Tropane derivatives	HIV, genetically related retroviral infections, inflammatory diseases
	Muscarinic receptor M$_3$ antagonists	Darifenacin	Overactive bladder

Adenosine receptors	Agonists	Adenosine	Supraventricular tachycardia
	Adenosine A_1 receptor agonists	Lignans from valerian	Insomnia, sleeping disorders
	Adenosine A_1 receptor antagonists	Caffeine, theophylline	Drowsiness, respiratory diseases (COPD or asthma) (theophylline)
	Adenosine A_{2A} receptor antagonists	Caffeine, theophylline	Drowsiness, respiratory diseases (COPD or asthma) (theophylline)
Adrenoceptors	Agonists	Adrenaline, noradrenaline, ephedrine	Hypotension (noradrenaline), breathing problems, nasal congestion, low blood pressure problems, myasthenia gravis, narcolepsy, menstrual problems, urine-control problems (ephedrine), acute anaphylaxis (adrenaline)
	α_1- and α_2-receptors agonists	Xylometazoline	Congestion (allergies), hay fever, sinus irritation, common cold
	α_1-receptor antagonists	Ergotamine	Migraines, acute attacks, postpartum hemorrhage
	α_2-receptor, central agonists	Methyldopa (as methylnoradrenaline)	High blood pressure
	β-adrenoceptor antagonists	Isoprenaline	Asthma, chronic bronchitis, emphysema, torsades de pointes
	β_1-receptor antagonists	Propranolol, atenolol	Tremors, angina, hypertension, heart rhythm disorders, heart or circulatory conditions, heart attacks, migraine headaches (propranolol), high blood pressure, angina, heart attacks (atenolol)
	β_2-receptor agonists	Salbutamol	Acute asthma, COPD, hyperkalemia, cystic fibrosis
	β_2-receptor antagonists	Propranolol	Tremors, angina, hypertension, heart rhythm disorders, heart or circulatory conditions, heart attacks, migraine headaches
Angiotensin receptors	AT_1—receptors antagonists	Sartans	Hypertension, diabetic nephropathy, congestive heart failure
Calcium-sensing receptor	Agonists	Strontium ions	
	Allosteric activators	Cinacalcet	Hyperparathyroidism, chronic renal failure

continued

TABLE B3 (continued)

Receptors

Type	Activity of Drug	Drug Examples	Disease
Cannabinoid receptors	CB_1- and CB_2-receptors agonists	Dronabinol	Loss of appetite in people with AIDS, severe nausea, vomiting (cancer chemotherapy)
Cysteinyl-leukotriene receptors	Antagonists	Montelukast	Difficulty breathing, chest tightness, wheezing and coughing (asthma), bronchospasm, seasonal and perennial allergic rhinitis
Dopamine receptors	Dopamine receptor subtype direct agonists	Dopamine, levodopa	Parkinson's disease, dystonia (dopamine), Parkinson's disease (levodopa)
	D_2, D_3, and D_4 agonists	Apomorphine	Parkinson's disease, erectile dysfunction
	D_2, D_3, and D_4 antagonists	Chlorpromazine, fluphenazine, haloperidol, metoclopramide, ziprasidone	Hiccups, nausea (chlorpromazine), schizophrenia, acute manic phases of bipolar disorder (fluphenazine), schizophrenia, acute psychotic states, delirium (haloperidol), nausea, vomiting, facilitating gastric emptying (metoclopramide), schizophrenia, mania, mixed states (ziprasidone)
Endothelin receptors (ET_A, ET_B)	Antagonists	Bosentan	Pulmonary artery hypertension
$GABA_B$ receptors	Agonists	Baclofen	Spasticity (spinal cord injury, spastic diplegia, multiple sclerosis, amyotrophic lateral sclerosis, trigeminal neuralgia)
Glucagon receptors	Agonists	Glucagon	Insulin coma, hypoglycemia
Glucagon-like peptide-1 receptor	Agonists	Exenatide	Type 2 diabetes, insulin resistance
Histamine receptors	H_1-receptors	Diphenhydramine	Red, irritated, itchy, watery eyes, sneezing, runny nose (hay fever, allergies, common cold), couch, motion sickness, insomnia, abnormal movements (early-stage parkinsonian syndrome)

	H$_2$-receptors	Cimetidine	Heartburn, peptic ulcers, dyspepsia
Opioid receptors	μ-opioid agonists	Morphine, buprenorphine	Chronic pain
	μ-, κ-, and δ-opioid antagonists	Naltrexone	Alcohol dependence, opioid dependence, sexual dysfunction, Crohn's disease, self-injury
	κ-opioid antagonists	Buprenorphine	Chronic pain
Neurokinin receptors	NK$_1$ receptor antagonists	Aprepitant	Nausea, vomiting
Prostanoid receptors	Agonists	Misoprostol, sulprostone, iloprost	Ulcers (misoprostol), pulmonary arterial hypertension, scleroderma, Raynaud's phenomenon, ischemia (iloprost)
Prostamide receptors	Agonists	Bimatoprost	Glaucoma, ocular hypertension
Purinergic receptors	P$_2$Y$_{12}$ antagonists	Clopidogrel	Coronary artery disease, peripheral vascular disease, cerebrovascular disease
Serotonin receptors	Subtype-specific (partial) agonists	Ergometrine, ergotamine	Migraines, acute attacks, postpartum hemorrhage (ergotamine), postpartum hemorrhage (ergometrine)
	5-HT$_{1A}$ partial agonists	Buspirone	Generalized anxiety disorder
	5-HT$_{1B/1D}$ agonists	Triptans	Migraine, cluster headaches
	5-HT$_{2A}$ antagonists	Quetiapine, ziprasidone	Schizophrenia, acute manic episodes (bipolar I disorder)
	5-HT$_3$ antagonists	Granisetron	Nausea and vomiting following chemotherapy, cyclic vomiting syndrome
	5-HT$_4$ partial agonists	Tegaserod	Abdominal discomfort, bloating, and constipation (irritable bowel syndrome), chronic idiopathic constipation
Vasopressin receptors	Agonists	Vasopressin	Diabetes insipidus, stomach after surgery, or during abdominal x-rays
	V$_1$ agonists	Terlipressin	Hypotension, norepinephrine-resistant septic shock, hepatorenal syndrome, bleeding esophageal varices
	V$_2$ agonists	Desmopressin	Diabetes insipidus, bedwetting
	OT agonists	Oxytocin	Postpartum hemorrhaging, weak labor
	OT antagonists	Atosiban	Stops premature labor

continued

TABLE B3 (continued)
Receptors

Type	Activity of Drug	Drug Examples	Disease
		Cytokine receptors	
Class I cytokine receptors	Growth hormone receptor antagonists	Pegvisomant	Acromegaly
	Erythropoietin receptor agonists	Erythropoietin	Anemia from chronic kidney disease, from treatment of cancer (chemotherapy and radiation), heart failure
	Granulocyte colony stimulating factor agonists	Filgrastim	Neutropenia, infection (cancer), leukapheresis, bone marrow transplants
	Granulocyte–macrophage colony stimulating factor agonists	Molgramostim	Low levels of white blood cells (WBCs), bone marrow transplants
	Interleukin-1 receptor antagonists	Anakinra	Rheumatoid arthritis (pain and swelling)
	Interleukin-2 receptor agonists	Aldesleukin	Metastatic renal cell cancer
TNFα receptors	Mimetics (soluble)	Etanercept	Certain autoimmune disorders (rheumatoid arthritis, psoriatic arthritis, juvenile idiopathic arthritis, ankylosing spondylitis, chronic plaque psoriases)
		Integrin receptors	
Glycoprotein IIb/IIIa receptor	Antagonists	Tirofiban	Unstable angina, non-Q-wave myocardial infarction
Receptors associated with a tyrosine kinase			
Insulin receptor	Direct agonists	Insulin	High levels of glucose, type-1 diabetes
Insulin receptors	Sensitizers	Biguanides	Diabetes mellitus, prediabetes, malaria
		Nuclear receptors (steroid hormone receptors)	
Mineralocorticoid receptor	Agonists	Aldosterone	Low blood pressure
	Antagonists	Spironolactone	Too much aldosterone, edema (congestive heart failure), cirrhosis of the liver, nephrotic syndrome, hypokalemia

Receptor	Type	Drug	Indication
Glucocorticoid receptor	Agonists	Glucocorticoids	Inflammation, pain
Oestrogen receptor	Agonists	Oestrogens	Menopause symptoms, postmenopausal osteoporosis, ovarian failure, breast cancer, advanced cancer of the prostate, osteoporosis, abnormal bleeding of uterus, vaginal irritation, female castration, Turner's syndrome
	(Partial) antagonists	Clomifene	Female infertility (anovulation)
	Antagonists	Fulvestrant	Hormone receptor-positive metastatic breast cancer
	Modulators	Tamoxifen, raloxifene	Breast cancer (tamoxifen), osteoporosis (raloxifene)
Androgen receptor	Agonists	Testosterone	AIDS wasting syndrome, low muscle mass, increased body fat in HIV patients
	Antagonists	Cyproterone acetate	Acne, hitsutism, androgen-dependent hair loss, overactive oil glands
Vitamin D receptor	Agonists	Retinoids	Inflammatory skin disorders, skin cancers, disorders of increased cell turnover, acne
ACTH receptor agonists	Agonists	Tetracosactide (cosyntropin)	Diagnosis of Addison's disease and cortisol disorders
Nuclear receptors (other)			
Retinoic acid receptors	RARα agonists	Isotretinoin	Severe recalcitrant nodular acne, certain skin cancers, Harlquin-type ichthyosis
	RARβ agonists	Adapalene, isotretinoin	Acne, keratosis pilaris (adapalene), Severe recalcitrant nodular acne, certain skin cancers, Harlquin-type ichthyosis (isotretinoin)
	RARγ agonists	Adapalene, isotretinoin	Acne, keratosis pilaris (adapalene), severe recalcitrant nodular acne, certain skin cancers, Harlquin-type ichthyosis (isotretinoin)
Peroxisome proliferator-activated receptor (PPAR)	PPARα agonists	Fibrates	Metabolic disorders (mainly hypercholesterolemia)
	PPARγ agonists	Glitazones	Diabetes mellitus type-2
Thyroid hormone receptors	Agonists	L-Thyroxine	Hypothyroidism

Source: Modified from Imming, P., Sinning, C., and Meyer, A. 2006. *Drug Discov.* 5: 821–834.

TABLE B4
Ion Channels

Type	Activity of Drug	Drug Examples	Disease
			Voltage-gated Ca²⁺ channels
General	Inhibitor	Oxcarbazepine	Seizures
In *Schistosoma* sp.	Inhibitor	Praziquantel	Human schistosomiasis, paragonimiasis, echinococcosis, cysticercosis, intestinal tapeworms, liver flukes (except fascioliases)
L-type channels	Inhibitor	Dihydropyridines, diltiazem, lercanidipine, pregabalin, verapamil	Increased systemic vascular resistance and arterial pressure, hypertension (dihydropyridines), hypertension, angina pectoris, some types of arrhythmia (diltiazem), neuropathic pain, generalized anxiety disorder, chronic pain in fibromyalgia and spinal cord injury (pregabalin), hypertension, angina pectoris, cardiac arrhythmia, cluster headaches (verapamil)
T-type channels	Inhibitor	Succinimides	Seizures
			K⁺ channels
Epithelial K⁺ channels	Opener	Diazoxide, minoxidil	Acute hypertension, increased insulin in disease states (insulinoma) (diazoxide), androgenic alopecia, other baldness treatments (minoxidil)
	Inhibitor	Nateglinide, sulfonylureas	Type-2 diabetes (nateglinide), diabetes mellitus type-2 (sulfonylureas)
Voltage-gated K⁺ channels	Inhibitor	Amiodarone	Life-threatening heart rhythm disorders of the ventricles, ventricular tachycardia, or ventricular fibrillation
			Na⁺ channels
Epithelial Na⁺ channels (ENaC)	Inhibitor	Amiloride, bupivacaine, lidocaine, procainamide, quinidine	Hypertension, congestive heart failure (amiloride), pain (spinal block) (bupivacaine), itching, burning and pain from skin inflammations, pain (dental), minor surgery (lidocaine), cardiac arrhythmias (procainamide), abnormal heart rhythms, malaria (quinidine)

Voltage-gated Na$^+$ channels	Inhibitor	Carbamazepine, flecainide, lamotrigine, phenytoin, propafenone, topiramate, valproic acid	ADD, ADHD, schizophrenia, phantom limb syndrome, paroxysmal extreme pain disorder, trigeminal neuralgia (carbamazepine), cardiac arrhythmias (paroxysmal atrial fibrillation, paroxysmal supraventricular tachycardia, ventricular tachycardia) (flecainide), epilepsy (seizures), mood disorders, unipolar depression, neuropathy, and bipolar disorder (lamotrigine), seizures, trigeminal neuralgia, certain cardiac arrhythmias (phenytoin), astrial and ventricular arrhythmias (propafenone), epilepsy, obesity, bipolar disorder, migraines, Lennox–Gastaut syndrome (topiramate), epileptic seizures, epilepsy, bipolar disorder, clinical depression (valproic acid)
Ryanodine-inositol 1,4,5-triphosphate receptor Ca^{2+} channel (RIR-CaC) family			
Ryanodine receptors	Inhibitor	Dantrolene	Malignant hyperthermia, neuroleptic malignant syndrome, muscle spasticity, ecstasy intoxication, serotonin syndrome, 2,4-dinitrophenol poisoning
Transient receptor potential Ca^{2+} channel (TRP-CC) family			
TRPV1 receptors	Inhibitor	Acetaminophen (as arachidonylamide)	Moderate pain from headaches, muscle aches, menstrual periods, colds and sore throats, toothaches, backaches, reactions from vaccinations (shots), fever, osteoarthritis
Cl-channels			
Cl$^-$ channel	Inhibitor (mast cells)	Cromolyn sodium	Asthma, allergic rhinitis, allergic conjunctivitis, mastocytosis, dermatographic urticaria, ulcerative colitis
	Opener (parasites)	Ivermectin	Onchocerciasis, other worm infestations, mites, scabies

Source: Modified from Imming, P., Sinning, C., and Meyer, A. 2006. *Drug Discov.* 5: 821–834.

TABLE B5
Transport Proteins (Uniporters, Symporters, and Antiporters)

Type	Activity of Drug	Drug Examples	Disease
Cation-chloride cotransporter (CCC) family	Thiazide-sensitive NaCl symporter, human inhibitor	Thiazide diuretics	Hypertension
	Bumetanide-sensitive NaCl/KCl symporters, human inhibitor	Furosemide	Swelling, fluid retention, high blood pressure
Na$^+$/H$^+$ antiporters	Inhibitor	Amiloride, triamterene	Hypokalemia (amiloride), hypertension, and edema (triamterene)
Proton pumps	Ca^{2+}-dependent ATPase (PfATP6; Plasmodia) inhibitor	Artemisinin and derivatives	Parasitic infections, malaria, particular types of cancer (leukemia, colon, etc.)
	K$^+$/H$^+$-ATPase inhibitor	Omeprazole	Dyspepsia, peptic ulcer disease, gastroesophageal reflux disease, Zollinger–Ellison syndrome
Na$^+$/H$^+$ ATPase	Inhibitor	Cardiac glycosides	Congestive heart failure, cardiac arrhythmia
Eukaryotic (putative) sterol transporter (EST) family	Niemann–Pick C1-like 1 (NPC1L1) protein inhibitor	Ezetimibe	High amount of cholesterol and other fatty substances

Neurotransmitter/Na+ symporter (NSS) family			
	Serotonin/Na+ symporter inhibitor	Cocaine, tricyclic antidepressants, paroxetine	Pain (cocaine), clinical depression, neuropathic pain, nocturnal enuresis, ADHD, headache, anxiety, insomnia, smoking cessation, bulimia nervosa, irritable bowel syndrome, narcolepsy, pathological crying or laughing, hiccups, interstitial cystitis, ciguatera poisoning, schizophrenia (tricyclic antidepressants), depression, panic disorder, social anxiety disorder, obsessive compulsive disorder, posttraumatic stress disorder, PMDD (paroxetine)
	Noradrenaline/Na+ symporter inhibitor	Bupropion, venlafaxine	Depression, seasonal affective disorder, smoking (bupropion), clinical depression, anxiety disorders, social anxiety disorder, panic disorder, diabetic neuropathy (venlafaxine)
	Dopamine/Na+ symporter inhibitor	Tricyclic antidepressants, cocaine, amphetamines	Pain (cocaine), clinical depression, neuropathic pain, nocturnal enuresis, ADHD, headache, anxiety, insomnia, smoking cessation, bulimia nervosa, irritable bowel syndrome, narcolepsy, pathological crying or laughing, hiccups, interstitial cystitis, ciguatera poisoning, schizophrenia (tricyclic antidepressants), ADD, ADHD, traumatic brain injury, symptoms of narcolepsy, chronic fatigue syndrome (amphetamines)
	Vesicular monoamine transporter inhibitor	Reserpine	High blood pressure

Source: Modified from Imming, P., Sinning, C., and Meyer, A. 2006. *Drug Discov.* 5: 821–834.

TABLE B6
DNA/RNA and the Ribosome

Target	Activity of Drug	Example Drugs	Disease
Nucleic acids			
DNA and RNA	Alkylation	Chlorambucil, cyclophosphamide, dacarbazine	Chronic lymphocytic leukemia, some types of non-Hodgkin lymphoma, Waldenström macroglobulinemia, polycythemia vera, trophoblastic neoplasms, and ovarian carcinoma, inflammatory conditions (nephrotic syndrome) (chlorambucil), lymphomas, multiple myeloma, leukemias, mycosis fungoides, neuroblastoma, ovarian carcinoma, retinoblastoma, breast cancer (cyclophosphamide), melanoma, Hodgkin's disease, soft-tissue sarcoma (leiomyosarcoma, fibrosarcoma, and rhabdomyosarcoma; neuroblastomas; and malignant glucagonoma) (dacarbazine)
	Complexation	Cisplatin	Metastatic testicular tumors, metastatic ovarian tumors, advanced bladder carcinoma
	Intercalation	Doxorubicin	Breast cancer, ovarian cancer, transitional cell bladder cancer, bronchogenic lung cancer, thyroid cancer, gastric cancer, soft tissue and osteogenic sarcomas, neuroblastoma, Wilms' tumor, malignant lymphoma (Hodgkin's and non-Hodgkin's), acute myeloblastic leukemia, acute lymphoblastic leukemia, Kaposi's sarcoma related to acquired immunodeficiency syndrome (AIDS)
	Oxidative degradation	Bleomycin	Lymphomas, squamous cell carcinomas, testicular carcinomas, malignant pleural effusions

RNA	Strand breaks	Nitroimidazoles	Anaerobic bacterial and parasitic infections
	Interaction with 16S-rRNA	Aminoglycoside anti-infectives	Bacterial infections
	Interaction with 23S-rRNA	Macrolide anti-infectives	Bacterial infections
	23S-rRNA/tRNA/2-polypeptide complex	Oxazolidinone anti-infectives	Bacterial infections
Spindle	Inhibition of development	Vinca alkaloids	Cancer (leukemia)
	Inhibition of desaggregation	Taxanes	Cancer
Inhibition of mitosis	—		
Ribosome			
20S subunit (bacterial)	Inhibitors	Tetracyclines	Bacterial infections (pneumonia), other respiratory tract infections; acne; infections of skin, genital and urinary systems, and infection that causes stomach ulcers, Lyme disease, anthrax
50S subunit (bacterial)	Inhibitors	Lincosamides, quinupristin-dalfopristin	Staphylococci and streptococci, bacteroides fragilis, other anaerobes (lincosamides), staphylococci (quinupristin-dalfopristin)

Source: Modified from Imming, P., Sinning, C., and Meyer, A. 2006. *Drug Discov.* 5: 821–834.

TABLE B7

Targets of Monoclonal Antibodies

Target	Agent	Disease
Vascular endothelial growth factor	Bevacizumab	Cancer (metastatic colon cancer and nonsmall cell lung cancer, breast cancer)
Lymphocyte function-associated antigen 1	Efalizumab	Psoriasis
Epidermal growth factor receptor	Cetuximab	Metastatic colorectal cancer, head and neck cancer
Human epidermal growth factor receptor 2	Trastuzumab	Breast cancer
Immunoglobin (IgE)	Omalizumab	Allergy-related asthma
CD-3	Muromonab-CD3	Acute rejection (organ transplants)
CD-20	Rituximab, ibritumomab tiuxetan, I-tositumomab	B-cell non-Hodgkin's lymphoma, B-cell leukemia, some autoimmune disorders (rituximab), B-cell non-Hodgkin's lymphoma (ibritumomab tiuxetan), B-cell non-Hodgkin's lymphoma (I-tositumomab)
CD-33	Gemtuzumab	Cancer (acute myeloid leukaemia)
CD-52	Alemtuzumab	Chronic lymphocytic leukemia (CLL) and T-cell lymphoma
F protein of RSV subtypes A and B	Palivizumab	Respiratory syncytial virus infections
CD-25	Basiliximab, daclizumab	Rejection in organ transplantation (basiliximab and daclizumab)
Tumor necrosis factor-α	Adalimumab, infliximab	Rheumatoid arthritis, psoriatic arthritis, ankylosing spondylitis, Crohn's disease, moderate-to-severe chronic psoriasis and juvenile idiopathic arthritis (adalimumab), psoriasis, Crohn's disease, ankylosing spondylitis, psoriatic arthritis, rheumatoid arthritis, sarcoidosis and ulcerative colitis (infliximab)
Glycoprotein IIb/IIIa receptor	Abciximab	Heart attack
α4-Integrin subunit	Natalizumab	Multiple sclerosis and Crohn's disease

Source: Modified from Imming, P., Sinning, C., and Meyer, A. 2006. *Drug Discov.* 5: 821–834.

TABLE B8
Physicochemical Mechanisms

Mechanism	Agent	Disease
Ion exchange	Fluoride	Tooth decay, cavities
Acid binding	Magnesium hydroxide, aluminum hydroxide	Constipation (magnesium hydroxide), heartburn, sour stomach, and peptic ulcer pain (aluminum hydroxide)
Adsorptive	Charcoal, colestyramine	Hypercholaterolemia, diarrhea, ileal resection, Crohn's disease, vagotomy, diabetic vagal neuropathy, itchiness due to liver problems (colestyramine), stomach pain, itchiness (kidney dialysis), overdoses (charcoal)
Adstringent	Bismuth compounds	Diarrhea, other gastrointestinal diseases, eye infections, malodor (gas and feces)
Surface active	Simeticone, chlorhexidine, chloroxylene	Flatulence and abdominal discomfort (dyspepsia, gastroesophageal reflux disease) (simeticone), gingivitis (chlorhexidine)
Surface active on cell membranes	Coal tar	Psoriasis, dryness, redness, flaking, scaling, itching (temporary relief)
Surface active from fungi	Nystatin, amphotericin B	Fungal infections (skin, mouth, vagina, intestinal tract) (nystatin), fungal infections and in visceral leishmaniasis. aspergillosis, *Naegleria fowleri* primary amoebic meningoencephalitis, cryptococcus infections (amphotericin B)
Mucosal irritation	Anthrones, anthraquinones	Laxative (anthraquinones)
Osmotically active	Lactulose, dextran 70, polygeline, glucose, electrolyte solutions, mannitol	Chronic constipation (lactulose), hypovolemic shock (dextran 70), low blood volume (polygeline), low glucose levels (glucose), diarrhea-related dehydration (electrolyte solutions), certain kidney conditions, increased swelling of the brain, increased pressure in the eye (mannitol)

continued

TABLE B8 (continued)
Physicochemical Mechanisms

Mechanism	Agent	Disease
Water binding	Urea, ethanol	Certain conditions of the fingernails and toenails (e.g., hyperkeratotic conditions), certain other skin conditions (e.g., corns; calluses; rough, dry skin) (urea), toxic poisoning, and bacteria (ethanol)
UV absorbent	4-Aminobenzoic acid derivatives	Fibrotic skin disorders
Reflective	Zinc oxide, titanium dioxide	Diaper rash, minor burns, chapped skin, minor skin irritations, burning, irritation, other rectal discomfort caused by hemorrhoids, painful bowel movements (zinc oxide), external irritations (solar rays) (titanium dioxide)
Oxidative	Tannins, polyphenoles, dithranol, polyvidone iodine, silver nitrate, hypochlorite, permanganate, benzoylperoxide, nitroimidazoles, nitrofuranes, temoporfin (mainly via singlet oxygen, cytostatic drug), verteporfin (mainly via singlet oxygen, ophthalmic drug)	Virus, bacterial, parasitic infections, HIV replication (tannins), psoriasis (dithranol), skin and mucous membrane infections (polyvidone iodine), eye infections and blindness in infants (silver nitrate), infection (hypochlorite), ulcers, dermatitis, pompholyx, fungal infections (permanganate), acne (benzoylperoxide), bacterial and parasitic infections (nitroimidazoles), head and neck cancers (temoporfin), abnormal blood formation in eye (verteporfin)
Reduce disulfide bridges	D-Penicillamine, N-acetyl-cysteine	Joint inflammation (arthritis), lung inflammation, skin thickening (D-Penicillamine), liver failure, bronchitis, kidney damage (N-acetyl-cysteine)
Complexing agents	AL^{3+}, arsenic compounds	CFS, leukemia (arsenic compounds), hyperhidrosis (AL^{3+})
Salt formation	Sevelamer	High blood levels of phosphorus (kidney disease)
Modification of tertiary structure	Enfuvirtide (from HIV glycoprotein 41)	HIV infection

Source: Modified from Imming, P., Sinning, C., and Meyer, A. 2006. *Drug Discov.* 5: 821–834.

TABLE B9
Drugs with Unknown Mechanisms of Action

Possible Target	Suspect Activity	Drug Example	Disease
2-Amino-4-hydroxy-6-hydroxymethyldihydropteridine pyrophosphokinase[a]	Inhibitor	4-Aminosalicylic acid	Inflammatory bowel diseases, multidrug-resistant tuberculosis, Crohn's disease
Farnesyl pyrophosphate synthetase[b]	Inhibitor	Alendronate	Osteoporosis, Paget's disease
Guanylate cyclase[c]	Inhibitor	Ambroxol	All forms of tracheobronchitis, emphysema with bronchitis pneumoconiosis, chronic inflammatory pulmonary conditions, bronchiectasis, bronchitis with bronchospasm asthma
Aberrant PML–retinoic acid receptor α fusion protein[d]	Inhibitor	Arsenic trioxide	Leukemia
β platelet-derived growth factor receptor[e]	Agonist	Becaplermin	Ulcers of the foot, ankle, or leg in patients with diabetes
Retinoic acid receptor RXR-β[f]	Activator	Bexarotene	Cutaneous manifestations of cutaneous T-cell lymphoma
mRNA[g]	Inhibitor	Chloral hydrate	Insomnia, pain, alcohol withdrawal
Potassium transporter[h]	Inhibitor	Clofazimine	Leprosy
DNA[i]	Inhibitor	Dactinomycin (RNA synthesis inhibitor)	Wilms' tumor, rhabdomyosarcoma, Ewing's sarcoma, trophoblastic neoplasms, testicular carcinoma
Dihydrofolic acid[j]	Inhibitor	Dapsone (folic acid synthesis inhibitor)	Leprosy and skin infections
Arachidonate 5-lipoxygenase[k]	Inhibitor	Diethyl carbamazine	Bancroft's filariasis, eosinophilic lung, loiasis, river blindness
Alcohol dehydrogenase[l]	Inhibitor	Diethyl ether	Pain
Trophozoites of E. histolytica[m]	Inhibitor	Diloxanide	Amebiasis

continued

TABLE B9 (continued)
Drugs with Unknown Mechanisms of Action

Possible Target	Suspect Activity	Drug Example	Disease
NMDA glutamate receptors[n]	NMDA antagonist	Dinitric oxide	Pain
Arabinosyltransferase B[o]	Inhibitor	Ethambutol	Tuberculosis
DNA[p]	Inhibitor	Gential violet	Fungal infections of the mouth, vagina, or intertriginous areas
$\beta_2\gamma_{2L}$ GABA$_A$[q]	Antagonists	Ginkgolides	Dementia, acute exacerbations of multiple sclerosis
Tubulin β chain[r]	Inhibitor	Griseofulvin	Jock itch, athlete's foot, and ringworm; and fungal infections of the scalp, fingernails, and toenails
Vacuolar ATP synthase catalytic subunit A[s]	Synthesizer	Halofantrine	Malaria
Glutamate receptor 1[t]	Inducer	Halothane	Pain
DNA (methylation)[u]	Inhibitor	Hydrazinophthalazine	Human cervical cancer
Apicoplast[v]	Inhibitor	Lumefantrine (antimalarial; prevents heme polymerization)	Malaria
Sodium channel protein type 1 subunit α[w]	Inhibitor	Levetiracetam	Seizures
Tubulin α chain[x]	Modulator, Inhibitor	Mebendazole	Roundworm, hookworm, pinworm, whipworm, and other worm infections
Low-density lipoprotein receptor[y]	Synthesizer (singlet oxygen)	Methyl-(5-amino-4-oxopentanoate)	Actinic keratosis
Oxidative phosphorylation[z]	Inhibitor	Niclosamide	Broad or fish tapeworm, dwarf tapeworm, and beef tapeworm infections
DNA[aa]	Inhibitor	Pentamidine	*Pneumocystis carinii* pneumonia

continued

DNA topoisomerase 2-α[bb]	Modulator	Podophyllotoxin	Cancer, genital warts
DNA[cc]	Inhibitor	Procarbazine	Some types of cancer, Hodgkin lymphoma and brain tumors
P. ovale[dd]	Inhibitor	Selenium sulfide	Dandruff, seborrheic dermatitis, tinea versicolor

Source: Modified from Imming, P., Sinning, C., and Meyer, A. 2006. *Drug Discov.* 5: 821–834.

[a] http://www.drugbank.ca/cgi-bin/getCard.cgi?CARD=APRD00749

[b] http://www.drugbank.ca/cgi-bin/getCard.cgi?CARD=DB00630

[c] http://www.sciencedirect.com/science?_ob=ArticleURL&_udi=B6T1J-41G1T84-7&_user=10&_rdoc=1&_fmt=&_orig=search&_sort=d&view=c&_acct=C000050221&_version=1&_urlVersion=0&_userid=10&md5=70aadf23c4713b8fc8df4a2e010eaf24

[d] http://cancerres.aacrjournals.org/cgi/content/full/62/14/3893

[e] http://www.drugbank.ca/cgi-bin/getCard.cgi?CARD=DB00102, http://www.drugbank.ca/cgi-bin/getCard.cgi?CARD=DB00307

[f] http://www.drugbank.ca/search/search?query=Bexarotene

[g] http://www.springerlink.com/content/j14810873852407 4/, http://www.jbc.org/cgi/reprint/251/9/2637.pdf

[h] http://www.drugbank.ca/cgi-bin/getCard.cgi?CARD=DB00845

[i] http://www.drugbank.ca/cgi-bin/getCard.cgi?CARD=DB00970

[j] http://www.drugbank.ca/cgi-bin/getCard.cgi?CARD=DB00250

[k] http://www.drugbank.ca/cgi-bin/getCard.cgi?CARD=DB00711

[l] 10.1111/j.1530-0277.1987.tb01282.x

[m] http://www.drugs.com/mmx/furamide.html, http://books.google.com/books?id=UTXUwdX9fuoC&pg=PA57&lpg=PA57&dq=Diloxanide+mechanism+of+action&source=web&ots=MP5vxFblju&sig=E0hKvibletAFSLCKtVnVOrZq1AQ&hl=en&sa=X&oi=book_result&resnum=2&ct=result

[n] http://www.nature.com/nm/journal/v4/n4/abs/nm0498-460.html

[o] http://www.drugbank.ca/cgi-bin/getCard.cgi?CARD=DB00330

[p] http://www.drugbank.ca/cgi-bin/getCard.cgi?CARD=DB00406

[q] http://www.drugbank.ca/search/search?query=Ginkgolides

[r] http://www.drugbank.ca/cgi-bin/getCard.cgi?CARD=DB00400

[s] http://www.drugbank.ca/cgi-bin/getCard.cgi?CARD=DB01218

TABLE B9 (continued)
Drugs with Unknown Mechanisms of Action

t   http://www.drugbank.ca/cgi-bin/getCard.cgi?CARD=DB01159

u   http://www.ncbi.nlm.nih.gov/pubmed/18521605?ordinalpos=3&itool=EntrezSystem2.PEntrez.Pubmed.Pubmed_ResultsPanel.Pubmed_RVDocSum

v   http://www.malariaandhealth.com/professional/poster/scientificposter.htm, http://www.ncbi.nlm.nih.gov/pubmed/12244914, http://www.druglib.com/trial/38/NCT00620438.
html

w   http://www.drugbank.ca/cgi-bin/getCard.cgi?CARD=DB01202

x   http://www.drugbank.ca/cgi-bin/getCard.cgi?CARD=DB00643

y   http://www.drugbank.ca/cgi-bin/getCard.cgi?CARD=APRD01105

z   http://www.drugs.com/mmx/niclosamide.html

aa   http://www.mongabay.com/health/medications/Pentamidine.html

bb   http://www.drugbank.ca/cgi-bin/getCard.cgi?CARD=DB01179

cc   http://www.drugbank.ca/cgi-bin/getCard.cgi?CARD=DB01168

dd   http://www.drugbank.ca/cgi-bin/getCard.cgi?CARD=DB00971

Appendix C: Popular Cell Lines in Drug Discovery

In the 2007 survey of HTS laboratories conducted by HighTech Business Decisions (http://www.hightechdecisions.com), 59 well-established cell lines and 53 proprietary cell types were identified. The top 10 cell lines were: HEK 293 (75%), CHO (71%), HeLa (29%), U2OS (27%), HepG2 (24%), COS-7/CV-1 (20%), Yeast (20%), SH5Y (14%), CaCo (10%), and THP1/THP2 (10%). In the same survey, 27% of the respondents indicated use of stem cells at some point in drug discovery, mainly for discovering small molecules to differentiate stem cells and target improvement before clinical studies. For more details on stem cell use, see Appendix D. The purpose of this appendix is to provide the history and background for the top six cell lines used by at least 20% of the survey respondents.

C1   HEK 293

C1.1   BACKGROUND

The name HEK 293 came from the cell source, human embryo kidney, and "293" came from the originator's (Frank Graham) 293rd experiment. This cell line was developed by the transformation of normal human embryonic kidney cells with adenovirus 5 DNA. The cells were cultured from a healthy, aborted fetus by Alex Van der Eb in Leiden, Holland, in the early 1970s and were transformed in 1973 by Frank Graham. The kidneys were removed from the aborted fetus, minced with scissors, trypsinized, and then recovered. They were then cultured in 10% bovine serum in Glasgow modified Eagle's medium. The adenovirus was taken from grown virions and fragmented by shearing through a 22-gauge needle. A calcium phosphate technique was used to transfect the adenovirus 5 DNA fragments with salmon sperm. It took 33 days until two colonies showed a transformation. The resulting line now called HEK 293 developed from the one surviving colony (the other colony was lost in attempts to isolate it) (van der Eb, 2001; Graham et al., 1977). Getting human cells to transform in response to the adenovirus was difficult, as indicated by the fact that only two colonies succeeded out of approximately 160 cultures (eight experiments were run with approximately 20 cultures each). It has been speculated that the colony that did develop may have come from a rare neuronal cell present in the HEK culture (van der Eb, 2001). A recent study has shown that the cell lines display some properties associated with a neuronal lineage. This was initially revealed by the strong immunoreactivity HEK cells demonstrate for the four major neurofilament subunits normally expressed in neurons (Shaw et al., 2002). In addition to these neurofilament proteins, a total of 61 mRNAs, which are normally expressed only in neurons, can be detected in HEK293.

C1.2 Morphology and Ploidy

HEK 293 cells are epithelioid in character, larger than most adenotransformed cells, and exhibit a large variation in size (Graham et al., 1977). Heterokaryons are common in the cell line with tetraploid observed toward the beginning of culturing (passage 8). However, at later phases the cells were on average diploid (Graham et al., 1977).

C2 CHO

C2.1 Background

CHO cell line was derived from the ovaries of adult Chinese Hamsters (*Cricetulus griseus*) by Joe Hin Tjio and Theodore T. Puck in 1958, at the Department of Biophysics, the University of Colorado Medical Center in Denver (Tjio and Puck, 1958). Chinese hamsters were first used in scientific experiments in 1919 at the Peking Medical Hospital to identify pneumonia before treatment options were available. They were particularly used because they have a low chromosome number of $2n = 22$, but have since been replaced in experiments by rats and mice which are easier to keep and breed. These hamsters actually look more like rats with a long tail in proportion to their body, and a black stripe down their spine, which gives them a slimmer appearance. Adult Chinese hamsters are 7.5–9 cm in length and weigh 50–75 g. The females can be extremely aggressive, particularly with other females and are therefore, best kept away from same sex interactions and watched closely when around males. Chinese hamsters are mainly nocturnal. They have large cheek pouches for carrying food. Their original habitats are the dry climates of China and Mongolia (Frankel, 1996).

C2.2 Morphology and Ploidy

CHO cells are epithelial cells but are elongated and packed tightly together in parallel to form colonies, resembling fibroblast cells (Tjio and Puck, 1958). CHO cells are derived from a $2n$ (22 chromosomes) line, but express a large variation in chromosome number, generally from 21 chromosomes to 44 chromosomes. A large portion of CHO cells have 23 chromosomes. The drastic differences in size of the chromosomes of CHO cells (as compared to chromosome size differences in human cells) makes them wonderful models to test genetic variables.

C3 HeLa

C3.1 Background

HeLa cells were derived from the adult human cervical cancer of Henrietta Lacks. They were first developed in 1951 and were the first human cancer cells to grow indefinitely *in vitro* (Sharrer, 2006). Henrietta Pleasant was born in 1920 in Virginia. She married David Lacks in 1935 to become Mrs. Henrietta Pleasant Lacks and moved to Baltimore with him, their three sons, and two daughters in 1943. In January 1951, a cone biopsy confirmed that she had advanced cervical cancer, at the time the number one cause of death from cancer in women. A portion of this biopsy was sent, unknown to Mrs. Lacks, to the director of the Tissue Culture Research Laboratory at Johns Hopkins Hospital, George Gey. By using a chicken plasma nutrient medium, Henrietta Lacks' cells became the first cells to grow *in vitro*. These "HeLa" cells, named after Henrietta Lacks, were sent to the National Foundation for Infantile Paralysis to develop a polio vaccine (Scherer et al., 1953), which was later accomplished by Jonas Salk. HeLa cells went on to many medical researchers who used the aggressively growing cells to study a wide variety of diseases, including cancer. On October 4, 1951, Henrietta Lacks' cervical cancer resulted in her death. She was buried back in Virginia, but her HeLa cells survived and presently account for several hundred times the mass of Henrietta Lacks herself (Sharrer, 2006).

In 1974, Walter Nelson-Rees published an article on the likelihood of contamination of HeLa in 40 previously believed to be different cell culture lines, such as HEK. His article sparked controversy as many scientists did not want to believe that their labs were contaminated and that their hard work using a particular cell line was not valid. In 2004, however, he received a lifetime achievement award for his efforts from the Society for *In Vitro* Biology (Nelson-Rees et al., 1974).

C3.2 MORPHOLOGY AND PLOIDY

HeLa cells are epithelial cells which multiply very quickly and can multiply many times in cell culture, because they have an active telomerase, which prevents the shortening of telomeres that causes cell death. HeLa cells have several chromosomal mutations resulting in heteroploid (hypertriploid) cell type, with multiple copies of some chromosomes, and 20 clonally abnormal chromosomes that are HeLa markers (Macville et al., 1999).

C4 HepG2

C4.1 BACKGROUND

HepG2 cells were originally taken from the liver of a 15-year-old Caucasian male with hepatocellular carcinoma without hepatitis B virus in 1979 by David P. Aden and Barbara B. Knowles of the Wistar Institute of Anatomy and Biology (Aden et al., 1979). There is some dispute over whether it is a hepatocellular carcinoma or a hepatoblastoma. Hepatoblastomas are generally in children younger than 36–40 months, while hepatocellular carcinomas are found in adults. HepG2 cells were originally derived from human liver biopsies placed on feeder cultures of irradiated, embryonic, fibroblast mouse cell layers (STO cell line). HepG2 cells were passaged on the feeder layers for several months and then passaged onto cultures without the feeder layer (Javitt, 1990).

Barbara B. Knowles worked at the Wistar Institute from 1967 to 1993, where she helped to develop the cell line HepG2. In 1978, she also helped to discover the antibody to stage-specific embryonic antigen 1 (SSEA-1). This antibody has an affinity for embryonic stem cells and cancer cells in mice and neutrophil cells—white blood cells that are present in infection sites—in humans. SSEA-1 was incorporated into the drug NeutroSpec™ in 2004 to detect appendicitis. Currently, she is a Presidential

Professor at the University of Maine and is studying the transition from oocyte to embryo. Stanley A. Plotkin (M.D.) developed the Rubella vaccine, RA27/3, which is currently used in the United States at the Wistar Institute. In addition to the rubella vaccine, he also worked on developing polio, rabies, varicella, and cytomegalovirus vaccines. He was the Medical and Scientific Director of Aventis Pasteur and the 2002 Sabin Gold Medal Winner for his work with vaccines. He also worked at the Epidemic Intelligence Service of the Center for Disease Control and wrote a popular book titled *Vaccines*.

C4.2 MORPHOLOGY AND PLOIDY

These cells exhibit an epithelial morphology and have an average chromosome number of 55, with three copies of chromosomes 1, 6, 9, 15, and 17 and three or four copies of chromosome 2 (Javitt, 1990).

C5 U2OS

C5.1 BACKGROUND

U2OS cells were derived from the tibia of a 15-year-old female with osteosarcoma. The cell line was first established in 1964 and called "2T" by Ponten and Saksela (1967).

Jan Ponten in 2004 was with the Sahlgrenska University Hospital, Goteb örg, Sweden (studying the effects of ventricular fibrillation) (Sandstedt et al., 2004). Jan Ponten also worked extensively with cervical cancer determining the similarities and differences of precancer and cancer colonies at the University of Uppsala in Sweden (Guo et al., 2000) and published a book titled *Precancer: Biology, Importance, and Possible Prevention* (Ponten, 1998).

C5.2 MORPHOLOGY AND PLOIDY

U2OS cells are epithelial adherent cells. They contain the insulin-like growth factors IGF-I and IGF-II and the wild-type p53 protein (Ralie et al., 1994). The cells are hypertriploid with multiple chromosomal mutations.

C6 COS-7/CV-1

C6.1 BACKGROUND

CV-1 cells were cultured from an adult male African green monkey (*Cercopithecus aethiops*) by F. C. Jansen in 1964 at the Wistar Institute of Anatomy and Biology in Philadelphia, Pennsylvania (Jensen et al., 1964). COS-7 cells were then derived in 1981 by Yakov Gluzman from the CV-1 cell line (Gluzman, 1981). The CV-1 cell line was developed with the intention of propagating Rous sarcoma virus (RSV). An ape line was chosen for this task since RSV is known to create tumors in primates (Jensen et al., 1964). Some cells from the CV-1 line were later transfected with the origin-defective mutant of simian virus 40 (SV-40), an ape origin virus

5243 nucleotides long, which may be linked to mesothelioma. These cells should be treated as a biohazard.

African green monkeys were originally classified under the scientific name *Cercopithecus aethiops*, but the name has recently changed with a distinction made between six species all previously classified under the name *Cercopithecus aethiops*. Now, the African green monkey has the scientific name *Chlorocebus sabaeus*. Males in this species have an average height of 490 mm and weight of 5.5 kg, while females are about 426 mm tall and 4.1 kg. The name "green monkey" comes from the olive green hue in their golden brown fur. *Chlorocebus sabaeus* have pale hands and feet and a golden tip on their tail. They are omnivores eating a wide variety of foods, much like the diet of chimpanzees and baboons. In the wild, African green monkeys live in large groups of up to 80 members, spending most of the day on the ground and sleeping on trees. Their original range spread only in sub-Saharan Africa from Senegal and Ethiopia down to South Africa; however, some monkeys were taken as pets to the Caribbean and now reside on some Caribbean islands and even in Florida. The cell lines derived from these monkeys are particularly useful because they can support SIV, the ancestor of HIV (Cawthon-Lang, 2006).

C6.2 Morphology and Ploidy

COS-7/CV-1 cells are fibroblast-like cells and are adherent to both glass and plastic. COS-7 and CV-1 cells are pseudodiploid with a modal chromosome number of 60 (48% of cells) (Amiss et al., 2003).

REFERENCES

Aden, D.P., Fogel, A., Plotkin, S., Damjanov, I., and Knowles, B.B. 1979. Controlled synthesis of HBsAg in a differentiated human liver carcinoma derived cell line. *Nature* 282: 615–616.

Amiss, T.J., McCarty, D.M., Skulimowski, A., and Samulski, R.J. 2003. Identification and characterization of an adeno-associated virus integration site in CV-1 cells from the African monkey. *J. Virol.* 77(3): 1904–1915.

Cawthon-Lang, K.A. 2006. Primate Factsheets: Vervet (*Chlorocebus*) Taxonomy, Morphology, & Ecology (http://pin.primate.wisc.edu/factsheets/entry/vervet).

Frankel, B.J. 1996. Diabetes in the Chinese hamster. In *Lessons from Animal Diabetes*, E. Shafrir (ed.), 6th edition. Boston: Birkhäuser.

Gluzman, Y. 1981. SV40-transformed simian cells support the replication of early SV40 mutants. *Cell* 23: 175–182.

Graham, F.L., Smiley, J., Russel, W.C., and Nairn, R. 1977. Characteristics of human cell line transformed by DNA from human adenovirus type 5. *J. Gen. Virol.* 36: 59–74.

Guo, Z., Ponten, F., Wilander, E., and Ponten, J. 2000. Clonality of precursors of cervical cancer and their genetically links to invasive cancer. *Mod. Pathol.* 13: 606–613.

Javitt, N.B. 1990. Hep G2 cells as a resource for metabolic studies: Lipoprotein, cholesterol, and bile acids. *Fed. Am. Soc. Exp. Biol.* 4: 161–168.

Jensen, F.C., Girardi, A.J., Gilden, R.V., and Koprowski, H. 1964. Infection of human and simian tissue cultures with Rous Sarcoma Virus. *Proc. Nat. Acad. Sci. U.S.A.* 52: 53–59.

Macville, M., Schröck, E., Padilla-Nash, H., Keck, C., Ghadimi, B.M., Zimonjic, D., Popescu, N., and Ried, T. 1999. Comprehensive and definitive molecular cytogenetic characterization of HeLa cells by spectral karyotyping. *Cancer Res.* 59: 141–150.

Nelson-Rees, W.A., Flandermeyer, R.R., and Hawthorne, P.K. 1974. Banded marker chromosomes as indicators of intraspecies cellular contamination. *Science* 184: 1093–1096.

Ponten, J. 1998. *Precancer: Biology, Importance, & Possible Prevention*. New York: Cold Spring Harbor Laboratory Press.

Ponten, J. and Saksela, E. 1967. Two established *in vitro* cell lines from human mesenchymal tumors. *Int. J. Cancer* 2: 434–447.

Raile, K., Hoflich, A., Kessler, U., Yang, Y., Pfuender, M., Blum, W.F., Kolb, H., Schwarz, H.P., and Kiess, W. 1994. Human osteosarcoma (U-2 OS) cells express both insulin-like growth factor-I (IGF-1) receptors and insulin-like growth factor-II/mannose-6-phosphate (IGF-II/M6P) receptors and synthesize IGF-II: Autocrine growth stimulation by IGF-II via the IGF-I receptor. *J. Cell. Physiol.* 159(3): 531–541.

Sandstedt, B., Ponten, J., Olsson, S.B., and Edvardsson, N. 2004. Genuine effects of ventricular fibrillation upon myocardial blood flow, metabolism, and catecholamines in patients with aortic stenosis. *Scand. Cardiovasc. J.* 38(2): 113–120.

Scherer, W., Syverton, J.T., and Gey, G.O. 1953. Studies on the propagation *in vitro* of poliomyelitis viruses: IV, Viral multiplication in a stable strain of human malignant epithelial cells (strain HeLa) derived from an epidermoid carcinoma of the cervix. *J. Exp. Med.* 97: 695–710.

Sharrer, T. 2006. "HeLa" Herself. *The Scientist* 20(7): 22.

Shaw, G., Morse, S., Ararat, M., and Graham, F.L. 2002. Preferential transformation of human neuronal cells by human adenoviruses and the origin of HEK 293 cells. *FASEB J.* 16: 869–871.

Tjio, J.H. and Puck, T.T. 1958. Genetics of somatic mammalian cells. II. Chromosomal constitution of cells in tissue culture. *J. Exp. Med.* 108: 256–268.

van der Eb, A. 2001. Speech given at the United States of America Food and Drug Administration Vaccines and Related Biological Products Advisory Committee Meeting, Gaithersburg, Maryland, 77–100 (http://www.fda.gov/ohrms/dockets/ac/01/transcripts/3750t1_01.pdf).

Appendix D: Stem Cells in Drug Discovery

Stem cells represent populations that are found in all multicellular organs and they have the capacity to form a variety of different cell types (Keller, 2005). Pluripotent human embryonic stem cell (hESC) lines were successfully derived from the inner cell mass of human blastocysts in the late 1990s (Thomson et al., 1998). So far, the predominant focus of stem cell applications has been on regenerative medicine and cell therapies (Cezar, 2007; Améen et al., 2008). It is anticipated that embryonic stem cells (ESC) will become a source of all differentiated cells used in drug discovery and nonclinical development once reliable protocols for directed differentiation have been established (Pouton and Haynes, 2005). This is based on several factors. First, ESC can be grown in unlimited quantities and can be derived from transgenic animals engineered to express disease-causing proteins, which establishes a direct link between the target and *in vivo* function. Second, ESC can be directed to specific phenotype thus providing opportunities to identify tissue-selective compounds. Third, the ability to organize ESC into multiple cell types that more accurately emulate the native tissue of interest returns drug discovery back to the highly successful pharmacological methods of the past, in which organ and tissue-based systems were used, but with the advantage that in this case, these cells can be used in HTS formats (Eglen et al., 2008). The purpose in this appendix was to provide samples of strategies toward differentiating ESC to neurons, cardiomyocytes, and hepatocytes. The rationale for these three cell/tissue types has already been provided in Chapter 4. The reader should, however, be aware of the interesting developments not covered in this brief appendix involving endocrine, pancreatic insulin-secreting cells derived from human ESC (Pierma, 2006). Also, in a pioneering paper, Takahashi et al. (2007), described pluripotent stem cells obtained from adult human fibroblasts by retroviral insertion of Oct4, Sox2, Klf4, and c-Myc into human dermal fibroblasts. The morphology, proliferation, surface antigens, gene expression, and differentiation characteristics of the induced cells were similar to those of human ESC. However, DNA microarray analysis detected differences between the two cells.

The state-of-the-art of *in vitro* differentiation of embryonic and adult stem cells, mesenchymal stem/progenitor cells, and liver progenitor cells into hepatocytes has recently been reviewed by Snykers et al. (2009). Most approaches have been based on reconstructing the chemical and spatial/temporal microenvironmental factors such as soluble medium factors, cell-matrix chemistry, and cell–cell interactions through coatings and/or 3D scaffolding. A comprehensive catalog of strategies and molecular and hepatic functionality endpoints has been presented by Snykers et al. (2009). Habib et al. (2008) have reviewed the methods used to achieve cardiomyocytes from human ESC with specific emphasis on therapy and regenerative medicine. The most

TABLE D1

Growth Factors Affecting Cardiac Differentiation of Human Embryonic Stem Cells

Pathway	Growth Factor or Small Molecules	Role	Concentration (Additional Factors Used)	Media	Timing of Application	Effect	References
TGF-β superfamily (BMP-2, BMP-4, TGF-β1-activin)	TGF-β	N/A	2 ng/mL	SR	D5–15	↑ Cardiac actin expression	Schuldiner et al. (2000)
			50 ng/ml + activin A (50 ng/mL)	N2/B27	D1–4	↑ Cardiac marker expression (IHC, RT-PCR)	Yao et al. (2006)
			0.3 ng/mL	N2/B27	D3–4	↑ Posterior PS marker (Mesp1, Hoxb)	Nostro et al. (2008)
	BMP-4	*Early and short-term induce posterior PS cells formation	25 ng/mL	RPMI + 1%	D1	↑ Mesodermal markers	Zhang et al. (2008)
			0.5 ng/mL	ITS	D3 (0.5 ng/mL)	↑ Cardiac troponin T-positive cells	Yang et al. (2008)
		*Cardiac progenitors commitment	10 ng/mL + activin A bFGF (D4–6)	StemPro 34	D4–6 (10 ng/mL)		
			10 ng/mL + activin A (D1, 100 ng/mL)	RPMI-B27	D2–4	↑ Cardiomyocyte yield	Laflamme et al. (2007)
	BMP-2		25 ng/mL	Fetal bovine serum	D3/6–D16	↑ Cardiac marker expression (IF, RT-PCR)	Pal and Khanna (2007)

Pathway/Factor	Mechanism	Concentration	Medium	Day	Effect	Reference
Activin A	Induction of mesoendodermal differentiation	50 ng/mL + BMP-4 (50 ng/mL)	N2/B27		↑ Cardiac marker expression (IHC, TR-PCR)	Yao et al. (2006)
		3 ng/mL + BMP-4 bFGF (D4–6) 100 ng/mL + BMP-4 (D2–4, 10 ng/mL)	StemPro 34 RPMI-B27	D4–6 D2–4	↑ Cardiac troponin T-positive cells ↑ Cardiomyocyte yield	Yang et al. (2008) Laflamme et al. (2007)
END-2 conditioned medium	Blockage of trophoblast differentiation	20 ng/mL	SR	D5–15	↑ Mesodermal markers	Schuldiner et al. (2000)
	—	—	END-2 conditioned medium	D1 +	↑ Cardiomyocyte yield ↑ Cardiac mesoderm markers ↑ Cardiac marker expression (IF, RT-PCR, qRT-PCR)	Mummery et al. (2003), Passier et al. (2005)
Canonical Wnt Pathway Dkk1	Canonical wnt signaling inhibition	150 ng/mL + activin A, BMP-4, bFGF, VEGF	Stem Pro 34	D6–14	↑ Cardiac troponin T positive cells	Yang et al. (2008)
P38 MAP Kinase PGI2		0.2–2 μM	Insulin-free serum-free media	D1–12	↑ % of beating EBs ↑ α MHC expression	Xu et al. (2008)

continued

TABLE D1 (continued)
Growth Factors Affecting Cardiac Differentiation of Human Embryonic Stem Cells

Pathway	Growth Factor or Small Molecules	Role	Concentration (Additional Factors Used)	Media	Timing of Application	Effect	References
	SB203580	P38 Map Kinase inhibitor	5 μM	Insulin-free serum-free media	D1–12	↑ Cardiomyocytes	Xu et al. (2008)
		Enhance cardiac mesoderm formation (induction of neuroectoderm at high concentrations)		Insulin-free serum-free media, END2 CM	D1–12	↑ Cardiac mesoderm markers ↑ Cardiac marker expression (IF, RT-PCR, qRT-PCR)	Graichen et al. (2008)
	5-Aza-2'-deoxycytidine	Regulation of gene expression, demethylation	10 μM 0.1 μM	FBS 20% Serum	D6–8 D1–3	↑ Cardiac marker expression (qRT-PCR) ↑ % Beating areas	Xu et al. (2002) Yoon et al. (2006)
P13K/Akt	Insulin		ITS 1% or insulin 10 ng/mL	Serum-free or END-2 conditioned media	D1–12	↑ % Beating EBs	Xu et al. (2008)

Source: From Habib, M., Caspi, O., and Gepstein, L. 2008. *J. Mol. Cell. Cardiol.* 35: 467. With permission.

TABLE D2

Protocols for Human Embryonic Stem Cell Differentiation to Specific Neural Precursors

Growth Conditions	Growth Factors	Progenitor Cells	Markers	Final Differentiation	Markers	References
EB	EGF, bFGF, RA	OPC	OLIG1, A2B5, SOX10, NG2	Oligodendrocyte	GalC, RIP, O4	Keirstead et al. (2005)
EB	EGF, bFGF, PDGF, RA	OPC	FDGFR, A2B5, NG2Oligodendrocyte	Oligodendrocyte	O4, O1, MBP, PLP	Kang et al. (2007)
Suspension culture	RA, EGF, bFGF, Noggin, vitamin C, mouse laminin	OPC	PDGFR, NG2, OLIG1/2, SOX10	Oligodendrocyte	O4, O1, MBP, PLP	Izrael et al. (2007)
Coculture with MS5 stromal feeders	BDNF, GDNF, AA, RA, SHH, Noggin	Motoneuron progenitor	BF1, HOXB4, NKX6-1/6-2, OLIG1/2	Motoneuron	NKx6.1, OLIG2, NGN2, ISL1, ChAT, VAChT, HB9, LHX3, HOX	Lee et al. (2007)
EB	bFGF, RA, SHH, BDNF, GDNF, IGF-1	Motoneuron progenitor	OLIG1/2, NKX6-1/6-2, NGN2	Motoneuron	NKX6.1, OLIG2, NGN2, ISL1, ChAT, VAChT, HB9, synapsin	Li et al. (2005)
Coculture with MS5 stromal cells	SHH, FGF8, BDNF, AA, TGFβ3	DA precursor	PAX2, PAX5, LMX, EN1	DA neuron	MAP2, TH, AADC, VMAT, NURR1, PTX3	Perrier et al. (2004)
EB	FGF2 or FGF8, SHH, BDNF, GDNF, cAMP, AA	DA precursor	EN1, OTX2, WNT1, PAX2, GBX2	DA neuron	TH, GABA, EN1, AADC	Yan et al. (2005)
Coculture with telomerase-immortalized fetal midbrain	FGF2, FGF8, SHH, BDNF, GDNF, FBS	DA precursor	EN1, PAX2, OTX2	DA neuron	TH, TUJ-1	Roy et al. (2006)
Coculture PA6	SHH, FGF8, BDNF, GDNF, AA, IGF-1	DA precursor	PAX2, EN1, NURR1, LMX1B	DA neuron	TH, EN1, AACD	Park et al. (2005)
Two-stage method using cyclopamine	Cyclopamine, human astrocyte medium	–	–	Astrocyte	GFAP, S100β, GLAST, BDNF, GDNF	Lee et al. (2006)
Coculture with PA6 or MS5 stroma	Noggin, NGF	Neural precursor	NCAM, TUJ-1, SNAIL, dHAND, SOX9	Peripheral sensory neuron	Peripherin, BRN3, TH, TRK-1	Brokhman (2008)
EB	Noggin, Dickkopf-1, IGF-1	Retinal progenitor	RX, PAX6, LHX2, SIX3	–	–	Lamba et al. (2006)

Source: From Erceg, S., Ronaghi, M., and Stojković, M. 2009. *Stem Cells 27:* 80. With permission.

common method used to initiate differentiation is culturing in the absence of the self-renewal signal provided by the mouse embryonic fibroblast feeder layer or leukemia inhibitory factor, followed by cultivation in suspension to form 3D differentiating cell clusters (embryoid bodies, EBs). By allowing EBs to attach to a matrix for further differentiation, identifiable rhythmic contacting appears within one to four days (Xu et al., 2002). As shown in Table D1, growth factor families have been shown to enhance cardiomyogenesis. Bone morphogenetic proteins, members of the transforming growth factor β (TGF-β) superfamily have been found to be required in all species studied so far for cardiomyogenesis (Olson, 2001). Human ESC have been differentiated to oligodendrocytes (nonneuronal cells located in white matter and have a vital role in the support and maintenance of the central nervous system by insulating the axons from the nerve cells), spinal motor neurons, doperminergic neurons, astrocytes (nonneuronal cells that offer supportive function during development, such as secreting different neurotrophic factors such as BDNF) peripheral neurons, and retinal progenitor cells (Erceg et al., 2009). Progenitors are proliferative cells with limited capacity for self-renewal and are often unipotent (Seaberg and van der Kooy, 2003). The strategies for *in vitro* neural differentiation of human ESC have predominantly involved adding growth factors, growth factor antagonists, and morphogens (signaling molecules that acts directly on cells to produce cellular responses depending on morphogen concentration, e.g., TGF-β) (Carpenter et al., 2001; Trounson, 2006) as shown in Table D2.

So far, practical use of ESC in screening studies has focused on discovery of small molecules, from compound libraries, capable of altering the human ESC fate (Ding et al., 2003; Sartipy et al., 2008). Given the advantages human ESC offer over the traditional cell lines as well as primary cells and the research efforts shifting from solely focusing on regenerative medicine and cell therapy applications to the development and deployment of cell-based biosensors, human ESC should find more practical use in drug discovery and nonclinical development (Steel et al., 2009; Nirmalanandhan and Sittampalam, 2009).

REFERENCES

Améen, C., Strehl, R., Bjorquist, P., Lindahl, A., Hyllner, J., and Sartipy, P. 2008. Human embryonic stem cells: Current technologies and emerging industrial applications. *Crit. Rev. Oncol. Hematol.* 65(1): 54–80.

Brokhman, I., Gamarnik-Ziegler, L., Pomp, O., Aharonowiz, M., Reubinoff, B.E., and Goldstein, R.S. 2008. Peripheral sensory neurons differentiate from neural precursors derived from human embryonic stem cells. *Differentiation* 76(2): 145–155.

Carpenter, M.K., Inokuma, M.S., Denham, J., Mujtaba, T., Chiu, C.P., and Rao, M.S. 2001. Enrichment of neurons and neural precursors from human embryonic stem cells. *Exp. Neurol.* 172(2): 383–397.

Cezar, G.G. 2007. Can human embryonic stem cells contribute to the discovery of safer and more effective drugs? *Curr. Opin. Chem. Biol.* 11(4): 405–409.

Ding, S., Wu, T.Y., Brinker, A., Peters, E.C., Hur, W., Gray, N.S., and Schultz, P.G. 2003. Synthetic small molecules that control stem cell fate. *Proc. Natl. Acad. Sci. USA* 100(13): 7632–7637.

Erceg, S., Ronaghi, M., and Stojković, M. 2009. Human embryonic stem cell differentiation toward regional specific neural precursors. *Stem Cells* 27(1): 78–87.

Eglen, R.M., Gilchrist, A., and Reisine, T. 2008. An overview of drug screening using primary and embryonic stem cells. *Combinatorial Chemistry & High Throughput* 11(7): 566–572.

Graichen, R., Xu, X., Braam, S.R., Balakrishnan, T., Norfiza, S., Sieh, S., Soo, S.Y., et al. 2008. Enhanced cardiomyogenesis of human embryonic stem cells by a small molecular inhibitor of p38 MAPK. *Differentiation* 76(4): 357–370.

Habib, M., Caspi, O., and Gepstein, L. 2008. Human embryonic stem cells for cardiomyogenesis. *J. Mol. Cell Cardiol.* 45(4): 462–474.

Izrael, M., Zhang, P., Kaufman, R., Shinder, V., Ella, R., Amit, M., Itskovitz-Eldor, J., Chebath, J., and Revel, M. 2007. Human oligodendrocytes derived from embryonic stem cells: Effect of noggin on phenotypic differentiation *in vitro* and on myelination *in vivo*. *Mol. Cell Neurosci.* 34(3): 310–323.

Kang, S.M., Cho, M.S., Seo, H., Yoon, C.J., Oh, S.K., Choi, Y.M., and Kim, D.W. 2007. Efficient induction of oligodendrocytes from human embryonic stem cells. *Stem Cells* 25(2): 419–424.

Keirstead, H.S., Nistor, G., Bernal, G., Totoiu, M., Cloutier, F., Sharp, K., and Steward, O. 2005. Human embryonic stem cell-derived oligodendrocyte progenitor cell transplants remyelinate and restore locomotion after spinal cord injury. *J. Neurosci.* 25(19): 4694–4705.

Laflamme, M.A., Chen, K.Y., Naumova, A.V., Muskheli, V., Fugate, J.A., Dupras, S.K., Reinecke, H., et al. 2007. Cardiomyocytes derived from human embryonic stem cells in prosurvival factors enhance function of infarcted rat hearts. *Nat. Biotechnol.* 25(9): 1015–1024.

Lamba, D.A., Karl, M.O., Ware, C.B., and Reh, T.A. 2006. Efficient generation of retinal progenitor cells from human embryonic stem cells. *Proc. Natl. Acad. Sci. USA* 103(34): 12769–12774.

Lee, D.S., Yu, K., Rho, J.Y., Lee, E., Han, J.S., Koo, D.B., Cho, Y.S., Kim, J., Lee, K.K., and Han, Y.M. 2006. Cyclopamine treatment of human embryonic stem cells followed by culture in human astrocyte medium promotes differentiation into nestin- and GFAP-expressing astrocytic lineage. *Life Sci.* 80(2): 154–159.

Lee, H., Shamy, G.A., Elkabetz, Y., Schofield, C.M., Harrsion, N.L., Panagiotakos, G., Socci, N.D., Tabar, V., and Studer, L. 2007. Directed differentiation and transplantation of human embryonic stem cell-derived motoneurons. *Stem Cells* 25(8): 1931–1939.

Li, X.J., Du, Z.W., Zarnowska, E.D., Pankratz, M., Hansen, L.O., Pearce, R.A., and Zhang, S.C. 2005. Specification of motoneurons from human embryonic stem cells. *Nat. Biotechnol.* 23(2): 215–221.

Mummery, C., Ward-van Oostwaard, D., Doevendans, P., Spijker, R., van den Brink, S., Hassink, R., van der Heyden, M., et al. 2003. Differentiation of human embryonic stem cells to cardiomyocytes: Role of coculture with visceral endoderm-like cells. *Circulation* 107(21): 2733–2740.

Nirmalanandhan, V.S. and Sittampalam, G.S. 2009. Stem cells in drug discovery, tissue engineering, and regenerative medicine: Emerging opportunities and challenges. *J. Biomol. Screening* 14(7): 755–768.

Nostro, M.C., Cheng, X., Keller, G.M., and Gadue, P. 2008. Wnt, activin, and BMP signaling regulate distinct stages in the developmental pathway from embryonic stem cells to blood. *Cell Stem Cell* 2(1): 60–71.

Olson, E.N. 2001. Development. The path to the heart and the road not taken. *Science* 291(5512): 2327–2328.

Pal, R. and Khanna, A. 2007. Similar pattern in cardiac differentiation of human embryonic stem cell lines, BG01V, and ReliCellhES1, under low serum concentration supplemented with bone morphogenetic protein-2. *Differentiation* 75(2): 112–122.

Park, C.H., Minn, Y.K., Lee, J.Y., Choi, D.H., Chang, M.Y., Shim, J.W., Ko, J.Y., et al. *In vitro* and *in vivo* analyses of human embryonic stem cell-derived dopamine neurons. *J. Neurochem.* 92(5): 1265–1276.

Passier, R., Oostwaard, D.W., Snapper, J., Kloots, J., Hassink, R.J., Kuijk, E., Roelen, B., de la Riviere, A.B., and Mummery, C. 2005. Increased cardiomyocyte differentiation from human embryonic stem cells in serum-free cultures. *Stem Cells* 23(6): 772–780.

Perrier, A.L., Tabar, V., Barberi, T., Rubio, M.E., Bruses, J., Topf, N., Harrison, N.L., and Studer, L. 2004. Derivation of midbrain dopamine neurons from human embryonic stem cells. *Proc. Natl. Acad. Sci. USA* 101(34): 12543–12548.

Pouton, C.W. and Haynes, J.M. 2005. Pharmaceutical applications of embryonic stem cells. *Adv. Drug Deliv. Rev.* 57(13): 1918–1934.

Roy, N.S., Cleren, C., Singh, S.K., Yang, L., Beal, M.F., and Goldman, S.A. 2006. Functional engraftment of human ES cell-derived dopaminergic neurons enriched by coculture with telomerase-immortalized midbrain astrocytes. *Nat. Med.* 12(11): 1259–1268.

Sartipy, P., Strehl, R., Bjorquist, P., and Hyllner, J. 2008. Low molecular weight compounds for *in vitro* fate determination of human embryonic stem cells. *Pharmacol. Res.* 58(2): 152–157.

Schuldiner, M., Yanuka, O., Itskovitz-Eldor, J., Melton, D.A., and Benvenisty, N. 2000. Effects of eight growth factors on the differentiation of cells derived from human embryonic stem cells. *Proc. Natl. Acad. Sci. USA* 97(21): 11307–11312.

Seaberg, R.M. and van der Kooy, D. 2003. Stem and progenitor cells: The premature desertion of rigorous definitions. *Trends Neurosci.* 26(3): 125–131.

Snykers, S., De Kock, J., Rogiers, V., and Vanhaecke, T. 2009. *In vitro* differentiation of embryonic and adult stem cells into hepatocytes: State of the art. *Stem Cells* 27(3): 577–605.

Steel, D., Hyllner, J., and Sartipy, P. 2009. Cardiomyocytes derived from human embryonic stem cells—characteristics and utility for drug discovery. *Curr. Opin. Drug Discov. Dev.* 12(1): 133–140.

Takahashi, K., Tanabe, K., Ohnuki, M., Narita, M., Ichisaka, T., Tomoda, K., and Yamanaka, S. 2007. Induction of pluripotent stem cells from adult human fibroblasts by defined factors. *Cell* 131(5): 861–872.

Thomson, J.A., Itskovitz-Eldor, J., Shapiro, S.S., Waknitz, M.A., Swiergiel, J.J., Marshall, V.S., and Jones, J.M. 1998. Embryonic stem cell lines derived from human blastocysts. *Science* 282(5391): 1145–1147.

Trounson, A. 2006. The production and directed differentiation of human embryonic stem cells. *Endocr. Rev.* 27(2): 208–219.

Xu, C., Police, S., Rao, N., and Carpenter, M.K. 2002. Characterization and enrichment of cardiomyocytes derived from human embryonic stem cells. *Circ. Res.* 91(6): 501–508.

Xu, X.Q., Graichen, R., Soo, S.Y., Balakrishnan, T., Rahmat, S.N., Sieh, S., et al. 2008. Chemically defined medium supporting cardiomyocyte differentiation of human embryonic stem cells. *Differentiation* 76(9): 958–970.

Yan, Y., Yang, D., Zarnowska, E.D., Du, Z., Werbel, B., Valliere, C., Pearce, R.A., Thomson, J.A., and Zhang, S.C. 2005. Directed differentiation of dopaminergic neuronal subtypes from human embryonic stem cells. *Stem Cells* 23(6): 781–790.

Yang, L., Soonpaa, M.H., Adler, E.D., Roepke, T.K., Kattman, S.J., Kennedy, M., Henckaerts, E., et al. 2008. Human cardiovascular progenitor cells develop from a KDR+ embryonic-stem-cell-derived population. *Nature* 453(7194): 524–528.

Yao, S., Chen, S., Clark, J., Hao, E., Beattie, G.M., Hayek, A., and Ding, S. 2006. Long-term self-renewal and directed differentiation of human embryonic stem cells in chemically defined conditions. *Proc. Natl. Acad. Sci. USA* 103(18): 6907–6912.

Yoon, B.S., Yoo, S.J., Lee, J.E., You, S., Lee, H.T., and Yoon, H.S. 2006. Enhanced differentiation of human embryonic stem cells into cardiomyocytes by combining hanging drop culture and 5-azacytidine treatment. *Differentiation* 74(4): 149–159.

Zhang, P., Li, J., Tan, Z., Wang, C., Liu, T., Chen, L., Yong, J., Jiang, W., Sun, X., Du, L., Ding, M., and Deng, H. 2008. Short-term BMP-4 treatment initiates mesoderm induction in human embryonic stem cells. *Blood* 111(4): 1933–1941.

Index

Company Index